Green Chemistry and Sustainable Practices

Green Chemistry and Sustainable Practices

Dr. Gabor Kovacs

ARCLER
PRESS

www.arclerpress.com

Green Chemistry and Sustainable Practices

Dr. Gabor Kovacs

Arcler Press
4164 Lakeshore Road
Burlington, ON L7L 1A4
Canada
www.arclerpress.com
Email: orders@arclereducation.com

ISBN: 978-1-77956-955-4 (Hardcover)

ABOUT THE AUTHOR

Dr. Gábor Kovács is a lecturer at the Sapientia Hungarian University of Transylvania, Faculty of Technical and Human Sciences, Department of Horticulture in Târgu Mureş, Romania. He has been in this role since 2021.

Kovács obtained his B.Sc. and M.Sc. in Chemistry/Applied Electrochemistry from Babeş-Bolyai University in Cluj-Napoca, Romania, in 2007 and 2008, respectively. He later earned his PhD in Chemistry from Babeş-Bolyai University in 2012.

Before his current lecturing position, Kovács had several research roles. He was an assistant Researcher at the Babeş-Bolyai University Faculty of Physics / Institute of Interdisciplinary Bio-Nano Sciences in Cluj-Napoca. Kovács also held postdoctoral research positions, including at the Department of Applied and Environmental Chemistry at the University of Szeged in Hungary, and at the Institute of Environmental Science and Technology of the same university.

Kovács has taught a variety of chemistry-related courses at the B.Sc. and M.Sc. levels, including General/Inorganic Chemistry, Electrochemistry, Corrosion and Anticorrosive Protection, Organic Chemistry, Biochemistry, Phytosanitary Chemistry, and Toxicology.

Kovács has led several research projects funded by the Hungarian Academy of Sciences and the National Research, Development and Innovation Office of Hungary. These projects have focused on areas such as the synthesis, characterization, and applicability of photocatalytic nanostructures, carbon nanotube-based composites, and investigations dealing with the ecotoxicological aspects of these materials.

TABLE OF CONTENTS

LIST OF FIGURES

PREFACE

Green chemistry is a philosophy of science and an operating toolbox for reducing harm as it brings people the molecules, materials, and energy services they rely on. It recasts the question chemists ask from "Can we make it?" to "Can we make it efficiently, safely, and with low impact on people and the planet? By combining prevention, efficiency, and circularity in the initial stages of discovery and process design, green chemistry prevents waste from being created, minimizes hazards rather than treating them once they are created, and optimizes resource productivity across the whole life cycle."

The approach ties synthesis to meaningful outcomes: less greenhouse gas release, fewer toxics releases, less solvent consumption, better energy performance, and less regulatory intensity. In industry, the rewards become shorter scale-up times, less environmental footprint, and increased profitability signaled by the utilization of metrics such as atom economy, E-factor, process mass intensity, and life-cycle impact. The field also opens up the design space of chemistry. Safer solvents and solventless technology open new reaction windows. Catalysis replaces stoichiometric reagents, lowering emissions and increasing selectivity. Renewable feedstocks introduce functional handles that minimize synthesis. Process intensification, continuous flow, and real-time analysis make manufacturing leaner and more consistent. Digital tools—screening, mechanism modeling, and data-driven optimization now accelerate greener route design. Underpinning these scientific advances are corporate sustainability goals, regulation, and stakeholder expectation. From fine chemicals and pharmaceuticals to polymers, agrochemicals, and materials for energy storage, green chemistry guides decision with performance as a function of responsibility and innovation as a function of accountability.

Green Chemistry and Sustainable Practices is created as a brief, practice-oriented guidebook for students, scientists, and practitioners wanting a clear-cut pathway from principles to practice. The book takes ideals to bench-top decisions, pilot plant decisions, and production plant decisions. It combines theory with design heuristics, decision architecture, and solved examples of how to compare routes, choose solvents, choose catalysts, and optimize for safer, cleaner production. Each chapter interweaves underlying principles with measurement and improvement tools, so readers can examine trade-offs openly and react with assurance in academic research labs, start-ups, or established operations.

Chapter 1 is where the scope and value of green chemistry are introduced after the way prevention, efficiency, and hazard decrease become design requirements and introducing leading metrics—atom economy, E-factor, process mass intensity, carbon intensity, and qualitative hazard bands—that place research and development as well as manufacturing choice in context.

Chapter 2 considers principles of sustainability in chemistry, explaining the Twelve Principles in terms of synthonic guidance on synthesis, formulation, and process design and connecting them with circular economy paradigms of reuse, remanufacture, and end-of-life design. Chapter 3 explains atom economy and process efficiency, showing how route choice, step economy, protection group avoidance, telescoping, and flow processing optimize yields, reduce auxiliaries, and optimize throughput by using explicit calculations and decision trees.

Chapter 4 covers safer solvents and reaction media, providing solvent selection indexes, substitution strategies, and alternatives such as water, bio-based solvents, supercritical fluids, ionic liquids, deep eutectic systems, and solventless routes, along with guidance on drying, recovery, and emissions control. Chapter 5 focuses on catalysis in green chemistry—homogeneous, heterogeneous, biocatalysis, organocatalysis, and photocatalysis—explaining how catalysts increase selectivity, decrease energy input, and simplify separations, with the emphasis on catalyst choice, recyclability, and control of deactivation. Chapter 6 addresses renewable feedstocks and materials, surveying bio-based platform molecules, CO_2 and waste-derived carbon, and recycled polymers, and dealing with reliability of supply, purity, variability, and how to incorporate renewables without sacrifice of performance.

Chapter 7 defines green chemical synthesis and design, constructing route prospecting strategies, reagent and solvent replacement, protecting-group minimization, alternative activation (microwave, mechanochemistry, electrochemistry), and building-block approaches that minimize steps and limit hazard classes. Chapter 8 is about waste minimization and pollution prevention, from in-process control and pinch points to real-time analysis, closed-loop solvent systems, byproduct valorization, and hierarchy-based solutions that prioritize inherently cleaner alternatives above end-of-pipe controls.

Chapter 9 addresses toxicology, safety, and environmental impact, getting readers ready to determine hazard endpoints, interpret safety data, apply predictive toxicology and in silico alerts, and create safer molecules, with process safety aspects such as thermal hazards, inherent safety, and emergency preparedness included.

Chapter 10 snapshots new directions and trends, such as data-driven route design, AI-enabled retrosynthesis, electrification of synthesis, cleaner battery and semiconductor chemistry, low-global-warming-potential refrigerants, plastic circularity, and policy incentives that favor prevention and performance.

This book has a number of different audiences. Undergraduate and graduate students will possess a systematic framework for the discipline and a reliable guide to laboratory decisions. Process chemists and young researchers will possess practical models for route selection, solvent selection, catalytic selection, and scale-up that will satisfy quality, safety, and environmental specifications. Instructors can use the book to connect lecture topics and lab exercises with metric-based assignments and decisionaided assignments. R&D managers and business managers can use a ready guide for introducing sustainability into product and process flows without losing velocity.

Regulatory professionals and EHS personnel will be able to bridge hazard assessment and compliance with design-stage choices that prevent problems from arising. Finally, business leaders and sustainability officers will grasp how green chemistry facilitates business strategy through effectiveness, risk reduction, brand value, and measurable environmental benefits.

The goal is straightforward: bring better chemistry to design, measure, and apply. With principles linked to decisions, and decisions linked to performance, *Green Chemistry and Sustainable Practices* aims to help readers convert good intentions into reliable results—molecules made with less risk, less waste, and more durable economic returns, poised for a world that requires scientific greatness and environmental care.

CHAPTER 1

INTRODUCTION TO GREEN CHEMISTRY

Contents

INTRODUCTION

The term green chemistry has now become a crucial aspect of environmental science and chemical education. Green chemistry itself revolves around the synthesis of chemical products and processes that consume and release fewer toxic materials. It is a futuristic approach aimed at reducing the environmental impact of traditional chemical processes, being a part of an overall sustainability agenda. With the integration of principles like waste minimization, energy efficiency, and renewable feedstocks, green chemistry not only promotes innovation but also responds to urgent global issues such as climate change and resource depletion. Notably, the incorporation of environmental literacy into curricula in educational institutions is crucial; it empowers future generations with the science and capabilities to work effectively with these concerns (Sholpan et al., 2024). On an industrial scale, green chemistry signifies a technological revolution in industries such as the automotive and pharma industries. Environmental innovations, which are driven by regulations, are transforming the manner in which industries produce chemicals and develop products. Through the adoption of green chemistry, companies can cut down on waste output, have increased energy efficiency, and be able to meet environmental standards while simultaneously stimulating economic growth. The automotive sector, for example, is increasingly counting on such regulations to develop green technologies that reduce emissions and maximize the use of resources (Maïder et al., 2025). With more businesses realizing the economic and environmental benefits of green chemistry, its application will expand further, to the realization of a greener industrial future. The addition of green chemistry to curricula further enhances its future impact. By putting more emphasis on environmental literacy in chemistry courses, educators can potentially empower a generation of informed citizens who are problem solvers amidst environmental challenges. This connection between green chemistry and ecological literacy not only prepares students to understand the complexity of their world, but also allows them to innovate and campaign for better practices in chemistry and other disciplines. As recognized in current studies, the incorporation of green chemistry principles into school curriculum frameworks aligns with the rising demand for learning systems that prepare students to make significant contributions to urgent global issues (Sholpan et al., 2024). In the years to come, collaboration between academia and industry will have a central role in ensuring this teaching methodology so that green chemistry remains a central component of scientific inquiry and practice.

Explanation of Green Chemistry Principles

The principles of green chemistry are a collection of philosophies designed to minimize the health and environmental impact of chemical products and processes. Green chemistry ends up innovating to the point where it reduces waste, conserves energy, and eliminates the use of hazardous materials. Through promotion of more efficient and sustainable synthetic pathways, green chemistry not only seeks to safeguard human health but also enhance the sustainability of industrial processes. Such a forward-thinking approach is supplemented with utilities such as in silico models for toxicity prediction that convey data on the environmental and toxicological fingerprinting of chemicals so that safer product design is enabled (Banerjee et al., 2024). Therefore, awareness and compliance with these principles can largely enhance scientific progress and conservation of the environment, guiding scientists and industries to develop more sustainable practices. Its twelve guiding principles are at the core of green chemistry, serving as benchmarks for environmentally friendly chemical processes. Among these are selections such as designing safer chemicals, maximizing atom economy, and employing renewable feedstocks. Each principle focuses on minimizing the harmful effects of the manufacture and use of chemicals, thus encouraging a holistic approach towards sustainability. For instance, energy efficiency comes under the principle with provisions to conserve energy in chemical reactions and hence cut down greenhouse gas emissions and combat the problem of climate change.

Figure 1.1. Green chemistry aims to innovate in such a way that reduces waste, conserves energy, and lessens the use of hazardous substances.

Source: https://www.azolifesciences.com/images/Article_Images/ImageForArticle_822_16521879951603306.jpg.

Embracing these principles is not only vital for traditional industries but also for emerging industries encompassing those with advanced technologies and green business models aligned with circular economy goals (Khan et al., 2021). These root principles thus become the guide for both chemists and captains of industry in making intelligent decisions that are beneficial to the world. Green chemistry's impact is not confined to the laboratory and spreads into broader circles of society through regulatory policy and educational initiatives. With growing concern for environmental affairs, education programs incorporating the principles of green chemistry are more necessary than ever before. These educate future chemists to make sustainability a part of their research and applications, and future generations stand ready to tackle environmental issues effectively. Additionally, integration of green chemistry principles is in compliance with global initiatives for sustainable development. Under such circumstances, use of advanced computational technology facilitates prediction of chemical toxicity, thereby safety profiles for newly synthesized compounds and materials (Banerjee et al., 2024). Thus, green chemistry encourages not only chemical technology innovation but also encourages systematic changes in policy and education that ensure sustainable processes on a global scale (Figure 1.1).

Historical Context and Development

The evolution of green chemistry over history is closely linked with the development of chemical science and the growth of ecological awareness. Beginning in the early 1990s, the concept emerged as a response to increasing apprehension about the ecological impact of traditional chemical methods. It sought to minimize harmful chemicals and maximize the production of cleaner, more sustainable chemical processes. This shift was concurrent with the broader societal movement towards environmental stewardship and the encouragement it brought for scientists to create methodologies that value not only efficiency but also ecosystem and human health. As the discipline has evolved, chemical process complexity has been increasingly in the limelight, necessitating innovations in areas such as catalysis and waste minimization. This historical development provided a background that ultimately influenced modern green chemistry standards and paved the way for future green practice. Moreover, the progression of green chemistry has been widely influenced by emerging technology and public policy. Central to this progress has been the discovery of metrics employed to determine the greenness of chemical processes, such as the E factor and atom economy, which quantify waste generation and resource consumption. As sustainability ideals gained popularity, regulatory agencies began incorporating green chemistry principles in chemical manufacturing processes. This regulatory focus has created a climate in which greener options are encouraged for industries through the promotion of catalysis and waste management innovation. Improvements in biocatalysis and alternative solvents show how emerging technologies promote the precepts of green chemistry. Further, the incorporation of life cycle analysis in chemical production reflects the historical shift towards an overall comprehension of sustainability, with environmental impact considered in every stage in the life of a product. The incorporation of white biotechnology within green chemistry's history represents a major breakthrough in the pursuit of sustainable operation. Compared to the traditional synthetic procedures, which generate high waste levels and utilize raw materials that are not renewable, white biotechnology uses biocatalytic reactions for facilitating chemical conversion with enzymes or whole microbial cells. This fulfills the requirements of green chemistry since it inherently avoids toxic byproducts while maximizing resource efficiency. The production of drugs such as atorvastatin and sitagliptin is a good indication of the effectiveness of biocatalysis in generating eco-friendly, sustainable products. In addition, the use of the renewable biomass in the production of chemicals and biofuels marks the effort in achieving sustainable sources of material through the paradigm of green chemistry. Through such developments in history, the inclusion of white biotechnology not only boosts the area of green chemistry but also demonstrates its ability to reduce existing environmental problems, affording the importance of sustainability in chemical reactions (Sheldon, 2015).

Importance in Modern Chemistry

The significance of green chemistry in today's chemical practice cannot be overstated. As ecological concerns grow with the global demand for environmentally friendly solutions, the principles of green chemistry work to eliminate toxic substances and waste in chemical processes. Green chemistry promotes the creation of chemical products and chemical processes that minimize environmental impact, facilitating the shift towards sustainable resource utilization. Through its emphasis on the elimination or limitation of toxic reagents and solvents, green chemistry not only lowers risks to chemical manufacture but also enhances the safety profile of processes. When industries begin to adopt these concepts, they do not just comply with the law; they also benefit from improved public perception and liability decrease, ultimately to result in safer communities and a healthful world. Electrochemical synthesis has been a groundbreaking technology in the realm of modern chemistry, appropriately mirroring green chemistry principles. Electricity has been used in organic compound syntheses with recent advancements choosing reaction processes that do not involve the use of stoichiometric reagents, thereby avoiding waste (Wiebe et al., 2018). The processes also operate under mild conditions, enhancing the reaction efficiency and safety. The regenerative nature of electrochemical reactions enables recycling of reagents, which means they possess a naturally lower environmental footprint. Of particular interest, when powered by renewable energy sources, electrochemical processes promote sustainability on multiple fronts well demonstrating how emerging trends can find practical applications in chemistry and also reduce greenhouse gas emissions and overall ecological footprint. On top of this, computational packages such as Quantum ESPRESSO also play an important role in green chemistry development. With this collection of software programs, scientists can carry out quantum simulations to allow better information on chemical reactions and material properties at the atomic level (Giannozzi et al., 2017). Through its capacity to model various properties and support new strategies, Quantum ESPRESSO facilitates scientists to design cleaner, better-performing chemical reactions right from the beginning. Through such advanced computational tools, scientists can optimize existing reactions and determine new compounds that are in compliance with green chemistry metrics. This intermarriage of theoretical knowledge and practical application highlights the complementary relationship between green chemistry and advanced technology, ultimately leading modern chemical research towards a more environmentally conscious path (Figure 1.2).

Figure 1.2. As environmental concerns grow alongside the global demand for sustainable solutions, the principles of green chemistry aim to minimize hazardous substances and reduce waste throughout chemical processes.

Source: https://www.frontiersin.org/files/Articles/1045345/frym-11-1045345-HTML-r1/image_m/main.jpg.

1.1. PRINCIPLES OF GREEN CHEMISTRY

The principles of green chemistry represent a paradigm change in the chemical sciences and promote a system that minimizes environmental damage while maximizing the efficiency of chemical processes. Foremost among them is the philosophy of sustainable development, which urges chemists to give preference to those processes that are not only economically viable but also sustainable. This synergy with green processes is crucial because companies are becoming increasingly pressured by regulatory bodies and consumers to adopt green approaches. Through the integration of green chemistry principles in research and development, chemists are able to create safer, more environmentally friendly chemicals that are responsive to market demands without sacrificing environmental integrity. Such a policy does not just encourage innovation but by its nature alone addresses the most pressing issues concerning environmental degradation and resource depletion, as framed by continuous endeavors delineating future-forward industry standards. Moreover, embracing a circular economy strongly advances the essence of green chemistry by encouraging a system wherein materials are constantly reused, remanufactured, and recycled. This paradigm encourages a shift away from the traditional linear models of production that are centered on disposal toward green-focused practice based on longevity and maximizing resources. As the concept of circular economy has been in growing demand globally—most explicitly among policy circles and company strategy—it is a necessary backdrop for green chemistry. This compatibility can be realized as stakeholders overcome the challenges of material management, production efficiency, and waste minimization. As illustrated in recent studies, the comprehension of the interplay among these two principles allows for the implementation of more integrated strategies beyond simple recycling programs, promoting an organizational revolution in how industries conceptualize sustainability and their impact on the environment (Haynes et al., 2015). Furthermore, the green chemistry principles also address the social dimensions of sustainable development, confirming that chemical processes not only seek to tackle issues of environmental health but also must have a positive influence on human well-being. This combined perspective broadens the definition of success in chemical manufacture beyond economic success. Such principles advocate for the merits of interdisciplinary cooperation in bringing together ideas from ecology, economics, and social science to inform decisions in chemistry. An integrated design may lead to innovative solutions taking into account societal impact in addition to environmental impact. For instance, the utilization of the principles of green chemistry to develop safer pesticides is able to avoid agricultural runoff and provide for public health and so enhance ecosystem performance as well as human quality of life (Dewulf et al., 2010). Ultimately, embracing these far-reaching guidelines will drive significant advances toward the creation of sustainable chemistries that benefit society and the planet.

1.1.1. Overview of the 12 Principles

12 green chemistry principles are the fundamental principles that aim to maximize the sustainability of chemical processes. The principles promote the creation of chemical products and processes that prevent or minimize the generation of unwanted substances, thereby reducing the impact on the

environment. Interestingly, the first principle targets the avoidance of waste instead of treatment, challenging chemists to solve issues of waste from the design phase. Effective application of this principle can eventually lead to breakthrough solutions that end up saving resources and expense. Breakthroughs that such can manifest can be evidenced in processes developed to optimize the atom economy, a case in point for how effective planning can be used to achieve more efficiency in the use of resources.

Thus, the translation of these principles into the overall practice not only contributes to environmental health but also economic sustainability with the focus on the dual benefit of taking up green chemistry practices. Apart from avoiding waste, principles of green chemistry also focus on energy efficiency in chemical processes. Principles five and six address energy requirements and renewable feedstocks, respectively. The careful choice of energy avoids the overall environmental load, encouraging practices that utilize lower pressures and temperatures, which can lower energy requirements significantly. The use of renewable feedstocks instead of depleting fossil resources also reflects a significant shift toward sustainable chemical production. This strategy not only addresses bigger-scale climate change goals but also allows industries to transition towards circular economies. As companies change their ways, the use of renewable resources illustrates a widespread interest in environmental balance and awareness. Combined with energy efficiency and renewable feedstock priority, green chemistry ensures that it not only addresses chemical hazards but also meets short-term global demands for sustainability. Upon closing the debate on the 12 principles, researchers place a strong emphasis on studying toxicology in green chemistry, as outlined in principles eight and nine. Such principles mandate safer design of chemicals and consideration of chemical hazard information in order to minimize risks to human health and the environment.

This demands the integration of rigorous risk assessment into the product design stage, which can avert or minimize upcoming hazards before they materialize. For instance, application of computational facilities such as KiNetX can potentially provide insight into the kinetics of chemical reactions while simultaneously coping with uncertainties in predictions accompanying electronic structure calculations (Proppe et al., 2018). The overall objective is to design chemicals that are not only efficacious when used but also possess a lower toxicological profile. Equilibrating these principles within the chemical industry assists in fostering safety and responsibility culture—ultimately consolidating the core ethos of green chemistry as a method of sustainable development (Dewulf et al., 2010) (Figure 1.3).

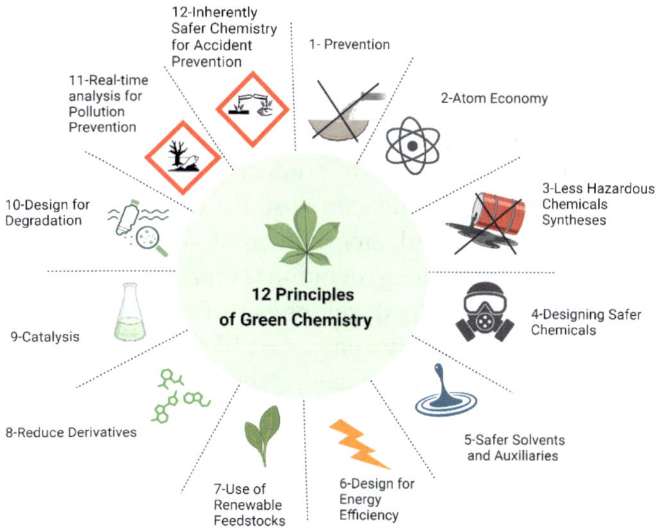

Figure 1.3. The 12 principles of green chemistry form the foundational tenets aimed at promoting sustainability in chemical processes.

Source: https://cdn.prod.website-files.com/621e95f9ac30687a56e4297e/65dfaee66fe1c9495893ff75_V2_1698514251379_fd26d014-6134-4442-bb0a-473e109ac9b0_HIGH_RES.png.

1.1.2. Examples of Each Principle in Practice

The application of the green chemistry principles may be realized through diversified industrial practices, in this case, from firms like BASF and Unilever, that have managed to include sustainable materials in their production processes. For instance, Unilever seeks to minimize waste by applying production processes that focus on atom economy, thus ensuring that the maximum possible percentage of raw material is being converted into its necessary product. Through the use of catalysts that allow for more efficient reactions, Unilever not only reduces the amount of waste created but also increases the overall yield of the cleaning and cosmetic products produced. This is founded on the same premise as that of green chemistry, which seeks to rationalize processes and promote environmental responsibility in industry, with proof of augmented reduction in the environmental footprint of manufacture (Ain, 2024). Within the pharmaceutical sector, the use of safer solvents has been implemented successfully by Roche, a leading producer in the field. Traditional organic solvents are not only environmentally and health dangerous but need to be replaced by green materials by Roche. For instance, the application of water as a solvent in certain reactions not only lowers the toxicity level but also enhances reaction efficiency. This approach is one of a global trend in green chemistry aimed at purifying safer chemical processes with no reduction in their effectiveness and fewer harmful emissions and by-products. These advances contribute to a greener pharmacy industry and demonstrate how new methodologies can balance scientific interests and environmental responsibility, producing an improved environment for future generations (Mark, 2023). Another significant feature of green chemistry is the development of renewable feedstocks, which are being actively investigated by companies like BASF. Emphasizing the use of biobased materials, BASF has made a step to source raw materials from renewable sources, ranging from plant-based biopolymers. The change aims at cutting dependence on fossil fuels and lowering carbon footprints in production processes. Besides this, their drive towards the utilization of biodegradable products also encourages additional waste management, eventually reducing environmental impacts. By transforming past linear consumption patterns to more circular modes, BASF demonstrates how green chemistry principles can revolutionize industry standards. The integration of these principles not only enhances sustainability but also demonstrates the viability of green chemistry principles in tangible, real-life applications, paving the way for more extensive implementations across diverse industries (Ain, 2024) (Figure 1.4).

Figure 1.4. By leveraging catalysts that promote more efficient reactions, Unilever not only reduces the amount of waste generated but also enhances the overall yield of their cosmetic and cleaning products.

Source: https://s7d1.scene7.com/is/image/CENODS/09902-feature4-opener?$responsive$&qlt=90,0&resMode=sharp2&fmt=webp.

1.1.3. Impact of Principles on Chemical Research

The principles of green chemistry radically redefine the research landscape of chemicals with a top-down focus on sustainability and environmental responsibility. Through the emphasis on eliminating harmful chemicals and chemical waste from chemicals processes, researchers are now more inclined to develop processes with minimal ecological footprint. Green chemistry procedures encourage the assessment of raw materials and procedures with respect to their impact on the environment, guiding chemists to employ safer substances and to create new pathways that ultimately result in safer end-products and efficiency. For instance, within the pharmacy sector, the application of green methods has seen the manufacturing of drugs that not only have lower toxicity but also lower the cost of manufacturing the drugs as a whole. Such changes in approach reflect a departure from conventional practices that tended to contribute to environmental deterioration and health hazards and hence aligning chemical research with contemporary demands of sustainability and public health (Purushottam et al., 2024). Advances in artificial intelligence (AI) are also amplifying the impact of green chemistry principles on chemical research. AI's capacity to analyze vast datasets, predict reaction outcomes, and optimize synthetic routes is compelling chemists to embrace new, sustainable techniques that align with the 12 principles of green chemistry. For instance, the application of AI in solvent selection can lead to the identification of eco-friendlier alternatives that minimize the use of volatile organic compounds. Furthermore, AI can also make catalyst development more efficient, potentially reducing waste and energy consumption in reactions by a significant extent. Such technology integrations not only streamline research steps but also elevate the sustainability level by favoring the alignment of advanced methodologies with green approaches. While scientists still struggle with such concerns as unification of data and model interpretability, the potential of AI in advancing green chemistry is enormous (Singh et al., 2024). The implementation of green chemistry principles in chemical science is the dawn of a new revolution towards a more responsible and innovative scientific world. Scientists are beginning to realize the merits of aligning their work with sustainable development, not merely for compliance reasons but also on ethical grounds concerning public health and environmental preservation. This is being reflected in new patterns of synthesis of chemicals, management of materials science, and even the development of solutions for energy that naturally predispose towards ecological well-being. With its emphasis on renewable resources, waste reduction, and energy conservation, green chemistry is encouraging a shift in paradigm that is encouraging cross-disciplinary collaboration to enable innovative solutions. Last but not least, as the implications of such ideas shape research agendas further, the growth of chemical science will construct an enduring and better future while maintaining the integrity and efficacy of its applications across various industries (Purushottam et al., 2024) (Figure 1.5).

Figure 1.5. Green chemistry principles encourage the assessment of raw materials and processes for their environmental impacts, guiding chemists to utilize more benign substances and to innovate pathways that ultimately enhance product safety and efficiency.

Source: https://www.reagent.co.uk/wp-content/uploads/2022/07/sustainable-chemistry-scaled-1.webp.

1.2. ENVIRONMENTAL IMPACT OF TRADITIONAL CHEMISTRY

The traditional practices in chemistry have significantly contributed to environmental degradation, largely as the emission of toxic chemicals to the environment. The manufacturing processes employed traditionally in the conventional production of chemicals normally result in huge amounts of waste, most of which are toxic and problematic to dispose of. Industrial activities continuously emit volatile organic compounds (VOCs) and greenhouse gases, which contribute to air quality issues and contribute to global warming. As observed in the Working Group II contribution to the Intergovernmental Panel on Climate Change (IPCC) Sixth Assessment Report, human need is to comprehend the intricate interaction of human activities, including chemical production, and their environmental implications. This awareness is crucial in formulating strategies that promote sustainable practices in the chemical industry, hence minimizing adverse effects on ecosystems and human health (Intergovernmental Panel on Climate Change, 2023). Utilization of non-renewable resources to support traditional chemistry not only depletes critical natural resources but also enhances the environment burden of chemical processes. Conventional processes tend to use petroleum-based feedstocks, resulting in higher carbon emissions during extraction, processing, and later use. The unsustainable loop adds to the burden on the environment and calls for a revolution toward sustainability. The environmental cost of large-scale chemical synthesis evokes serious ethical concerns about the responsibility of chemists and industries to create sustainability. As advances in green chemistry yield novel approaches that are efficient and minimize waste, it is crucial that the discipline shifts to practices that consider sustainability and the stewardship of the planet (Emily et al., 2021). The call to embrace green chemistry principles is a pressing response to increasing awareness of the environmental costs of conventional chemical practice. The transition toward sustainable operations is a stepping stone toward creating a circular economy in which reusing, recycling, and reducing the use of materials during their lifetime are the main priorities. The integration of environmental considerations into chemical design can serve to significantly eliminate the production of harmful chemicals and enhance overall worker and societal safety. Scientific efforts in accordance with the green chemistry principles seek not only to reduce waste but also to create less toxic chemicals by nature to human health and the environment. In this perspective, climate-resilient development in accordance with sustainable strategies in chemistry is a key means of attaining a livable future for humanity and biodiversity (Intergovernmental Panel on Climate Change, 2023) (Figure 1.6).

Figure 1.6. Industrial processes frequently emit volatile organic compounds (VOCs) and greenhouse gases, which exacerbate air quality issues and contribute to climate change.

Source: https://www.thecmmgroup.com/wp-content/uploads/2020/01/industrial-operation-releasing-VOC-emissions.png.

1.2.1. Pollution and Waste Generation

Waste generation and pollution are major global problems that strongly affect environmental sustainability, public health, and economic viability. Pollution is typically directly linked with industrial processes where waste generation is a major byproduct. Following the strategy of green chemistry, it is important to mention here that traditional manufacturing methods tend to overlook the impact of their waste products, which has harmful ecological consequences. It is vital to comprehend the character of industrial waste in order to formulate measures for lowering the overall pollution rate. Scholars have formulated diverse methods to evaluate and minimize industrial waste, but it usually causes confusion because various goals and methods overlap. This suggests the need for a concerted plan that pulls together different waste management strategies toward the same goal of zero waste production and ultimately positions green chemistry at the forefront of this drive (Bjelkemyr et al., 2013). Where pollution intersects with ocean ecosystems, it adds another dimension of complexity that demands urgent action and innovative solutions. When pollution reaches the oceans, it poses a threat to marine ecosystems, wildlife, and human health. Recent initiatives focused on pressing issues like combating marine littering and overfishing, which are both directly linked to waste generation. Increasing ocean habitat degradation has called for a Blue Revolution in which technology is being used to promote sustainable aquaculture and fishing enterprises. By facilitating transparency and traceability from ocean to coast, the stakeholders are able to adopt sustainable measures that have long-term effects of reducing ecological footprints. This emphasis on stemming pollution through strategic intervention is consistent with the objectives of green chemistry, which demands the invention and implementation of processes that intentionally limit hazardous waste. To actually prevent pollution and the production of waste, there must be an integrated strategy that represents education, research, and policy change. Academic fields must promote interdisciplinary studies that not only investigate the scientific causes behind waste generation but also entail policy-making and industrial operations. Interdisciplinary exchange between researchers, industrialists, and policymakers can unlock innovative interventions that prioritize sustainable solutions and social participation. Furthermore, public information campaigns have an important role in promoting an environmental awareness culture that urges citizens and companies to rethink their consumption and dumping habits. With the integration of the green chemistry philosophy in curriculums and industry standards, future generations will be able to manage the issues of pollution more effectively, leading to a sustainable and environmentally conscious society. Integration of various means of waste management still lends added strength to this potential, leading us one step closer towards a sustainable future free from pollution (Bjelkemyr et al., 2013) (Figure 1.7).

Figure 1.7. Pollution and waste generation are pressing global challenges that significantly affect environmental sustainability, public health, and economic viability.

Source: https://www.innovationnewsnetwork.com/wp-content/uploads/2024/03/shutterstockMOHAMED-ABDULRAHEEM_448263469.jpg.

1.2.2. Resource Depletion and Sustainability Issues

The high rate of depletion of resources is a major concern that highlights the need for sustainable practices across industries. As natural resources are depleting quickly because of misuse and environmental degradation, the pressure builds up on the ecosystem. Depletion has severe implications for human society as well as biodiversity, according to the IPCC. To address these issues, there needs to be an extreme shift towards the use of sustainable resource management. A sustainable approach not only strives to conserve current resources but also concentrates on the use of renewable alternatives. This transition is most applicable in the field of green chemistry, wherein work is being done towards minimizing waste and environmental impacts of chemical reactions, resulting in a synclastic coexistence between human activities and nature's integrity (Intergovernmental Panel on Climate Change, 2023). In addition to this, green chemistry metrics have also gained importance in measuring the sustainability of chemical processes. Indicators such as the E factor and atom economy are applied to quantify waste production and raw material efficiency and yield and inform us regarding the environmental footprint of chemical engineering processes. But as suggested by recent literature, such mass-based interventions will have to be complemented with life cycle assessments (LCA) to fully quantify the sustainability of chemical processes (Roger, 2017). Straightforward attention to the generation of waste does not take into consideration the full impact of the usage of resources on Earth. By integrating several metrics, industries and scholars can approximate their production process environmental efficiency in developing a circular economy, which is an emphasis on reducing waste and enhancing the efficiency of resources. Consequences of resource depletion extend beyond availability; they extend to economic stability, public health, and global equity. With decreasing resources, extraction and production cost increases the likelihood of economic disparity among countries. Besides, non-renewable resource dependency raises more environmental risks, including climate change, with severe threats to ecological and social systems. The IPCC highlights the interconnectedness of the problems and calls for climate-resilient development that encompasses adaptation and emission savings (Intergovernmental Panel on Climate Change, 2023). Here, green chemistry emerges as an innovation path of paramount importance as it provides avenues to sustainable alternatives that minimize pressure on resources while ensuring ecological sustainability and social responsibility. The convergence of green chemistry practices with overarching sustainability goals is a powerful step towards a more resilient future.

1.2.3. Health Risks Associated with Conventional Chemical Processes

The possible health risks with conventional chemical methods are a critical concern within conventional production and research procedures. These methods often utilize hazardous substances that pose serious risks to human health and the environment. Chemical substances such as solvents, heavy metals, and active ingredients could induce acute effects such as respiratory problems, skin lesions, and chronic effects such as cancer or reproductive toxicity. The limitations of risk assessment models in estimating such health impacts also exacerbate the issue, resulting in insufficient protection of workers who engage in hazardous working environments. The basis of the importance of changing to green practices in the chemical industry underscores the necessity to embrace principles from green chemistry to limit these health hazards. Also, the life cycle of conventional chemical processes reveals a range of possible hazards that extend beyond the first phase of manufacture. Waste disposal from toxic byproducts of such processes has the potential to pollute the environment, which in turn has an effect on public health by polluting air, land, and water bodies. Terrains surrounding chemical plants often experience increased instances of health-related events, like chronic diseases that result from prolonged exposure to poisons. The increasing focus on such environmental health disparities highlights the need to incorporate sustainable practices that provide priority to human safety. Applying green chemistry principles can allow industries to avoid exposure to toxic chemicals, hence reducing health risks not only to employees but to society in general.

Incorporating nanotechnology into traditional chemical processes both poses opportunities and challenges, in particular regarding health risks and toxicity. Nanomaterials have been advocated because of their unique properties and potential for application in improving therapeutic delivery systems. Their production and application are, however, problematic from a safety point of view. Conventional chemical synthesis methods often yield manufactured nanoparticles that could have unforeseen toxicological profiles. Apparently, the diverse categorization and complexity of nanomaterials necessitate stringent measures of risk assessment to determine their impact on human health and the environment (Jeevanandam et al., 2018). Lacking specific regulations that address these health issues, there is a need for frameworks that adopt safer alternatives, in line with the basic principles of green chemistry to evade the undesirable effects of conventional as well as advanced chemical processes (Figure 1.8).

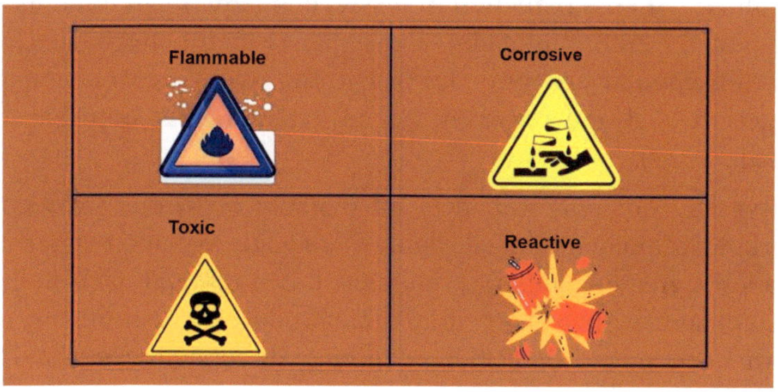

Figure 1.8. Chemicals such as solvents, heavy metals, and reactive agents can lead to acute symptoms including respiratory distress, skin irritation, and long-term effects like cancer or reproductive toxicity.

Source: https://lh7-rt.googleusercontent.com/docsz/AD_4nXeCd7WE_mSc-mXl918chpAfovRm7E7a5altZJZOgvs9zXqScE8ONbuvBWGr1VWFQA-1NoeHyWWRoSMPFSmRz8njG2LuamlYEeIDisBAGjg23-KAG-0nwzK8nC-tx4N-9SklIPewmxkqhiYYatk07jShfjMDCA?key=bCVk6aNOXQlzKZd_Vk-Kt6w.

1.3. BENEFITS OF GREEN CHEMISTRY

The increased demand to address environmental challenges has positioned green chemistry as a basic technique in contemporary scientific practice. Green chemistry focuses on the design of chemical products and processes that minimize the use of hazardous materials, thus significantly reducing the potential for pollution, thereby causing no harm to human health and the environment. With society grappling with the implications of climate change, the green chemistry philosophy presents a sustainable model of innovation.

As concluded by the Intergovernmental Panel on Climate Change, effective climate resilience demands steep reductions in greenhouse gas emissions, and green chemistry is an apt solution in balancing chemical manufacturing with such development goals (Intergovernmental Panel on Climate Change, 2023). Green chemistry facilitates safer alternatives to chemicals and reduced toxic byproducts, resulting in more sustainable companies, ultimately for a cleaner and more equitable future.

Apart from maximizing environmental protection, green chemistry is highly economically valuable. Transformation of production methods into sustainable methods can result in reduced cost of operation for companies as more effective processes are likely to use less raw materials and energy. By using renewable resources along with the reduction of waste, companies can not only minimize their impact on the environment but also enhance their bottom lines. Moreover, the tenets of a circular economy are in harmony with green chemistry, as the material gets reused and recycled, hence extending its lifecycle and usage (Ncube et al., 2023).

This synergy not only enhances sustainable economic practice but also stimulates innovation in product design, which leads to market opportunities. As business adopts green chemistry principles, it can rebrand itself in the market, respond to consumers' needs for sustainable products, and ultimately contribute to long-term economic resilience. In addition to its economic and environmental advantages, green chemistry encourages greater societal commitment to sustainability and responsible consumption. By teaching future generations the value and importance of green practices, the culture of green chemistry has the potential to create a shift towards environmental responsibility. The inclusion of green chemistry in school curricula serves to ensure that the pivotal role of chemistry in addressing issues the world is currently facing, such as pollution and resource depletion, is emphasized.

Promotion of green chemistry initiatives also aligns with the Sustainable Development Goals, especially those aimed at achieving sustainable consumption and production patterns (Ncube et al., 2023). By invoking such commitment from industries and students as well, green chemistry not only creates avenues for practical application but also evokes a sense of responsibility for the world, emphasis on the interconnectedness of human activity and environmental health.

1.3.1. Reduction of Hazardous Substances

Green chemistry relies on an underlying concept of reducing harmful substances to minimize the negative environmental and health impacts of chemical processes. Traditional processes are inclined to rely on the use of poisonous chemicals that endanger human well-being and ecosystems. With the inclusion of the principles of green chemistry, chemists are able to develop safer substitutes that not only work in the same ways but also minimize the harmful emissions and waste. For instance, formaldehyde, which is an ubiquitous industrial chemical implicated in many health complications, can be minimized by appropriate material selection and process design. Formaldehyde has been shown to cause risks through indoor exposure to furniture volatiles, and there have been demands for it to be regulated from use, especially in products meant for susceptible populations, such as infants (Neumann et al., 2008). Further, new strategies, such as the development of chrome-free tanning of leather, demonstrate real-world implementation of green chemistry guidelines of thumb to reduce dangerous materials. Wet-white tanning, based on tannic acid with Laponite nanoclay, avoids harmful Cr(VI) and formaldehyde successfully without compromising valuable physical properties of the leather. This methodology not only meets the criteria for furniture leather but also reflects an immense reduction in environmental impact, like diminished human toxicity potential and global warming potential as opposed to conventional tanning processes (Lin et al., 2016). Rationalizing such processes, green chemistry enables industries to mirror closer towards sustainability goals while meeting consumer demands for more environmentally safe and friendly products. In the end, the practice of minimizing harmful substances through green chemistry not only promotes public health but also supports innovation for different chemical industries. Through an emphasis on sustainable methods, the chemical industry is able to maximize its social responsibility as well as make a positive impact on environmental stewardship. As more firms embrace these values, it is evident that the pursuit of lower-risk alternatives is not just a regulatory imperative but a competitive advantage in an open market. The transition to safer chemistry practice has the potential to drive new areas of research, energizing future innovation with focus on health and sustainability. As understanding of these issues grows, so does the demand for continued advances in reducing hazardous chemicals, further cementing the importance of green chemistry in contemporary chemical science.

1.3.2. Economic Advantages and Cost Savings

The use of green chemistry principles in industrial applications has important economic advantages and savings for companies and the environment. By employing environmentally friendly practices, companies can minimize wastage and cut the utilization of harmful substances. In addition to lowering the expense of regulation compliance, this also reduces the financial burden of waste management. For instance, processes that can generate fewer by-products can lead to significant savings in the aspect of disposal. Companies adopting such green processes are usually more operationally efficient and can pass on the resultant lower production costs to the market, improving their competitiveness.

Furthermore, as green chemistry increasingly becomes the partner for what the consumer demands in the aspect of sustainability, companies are able to differentiate themselves, with a likelihood of increased revenues and loyalty. Furthermore, green chemistry shift ensures maximum resource utilization with additional economic benefits. Utilizing renewable sources of energy and feedstocks helps to decrease the use of fossil fuel by firms, easing market price fluctuations that disrupt operating budgets. The carbon neutrality technologies evident in recent proposals emphasize the importance of low-carbon technologies and energy-efficient systems. These innovations not only contribute to sustainability but also provide means to ensure long-term cost stability. Firms can use renewable technologies to build value along their supply chain, thus making greater profit without compromising on environmental sustainability. Reducing energy consumption in the course of production processes is another method by which embracing green chemistry is able to directly influence a firm's bottom line. The integration of green chemistry in industrial processes also opens new avenues for innovation and employment, and even more so, making the economy stronger. The ever-growing demand for green products and processes necessitates qualified practitioners who can handle and implement these green technologies effectively. This shift in human resource requirements spurs learning opportunities and provokes the generation of new jobs in future fields of sustainability, such as battery recycling and renewable energy systems. With industries becoming more integrated to encompass battery reuse and recycling strategies, as established in recent discourse on the world battery industry, they not only help solve environmental challenges but also create economic sustainability through novel business models. Ultimately, the expanded application of green chemistry not only fosters cost reductions but also supports a robust, innovative economy founded on resilience and sustainability, as dictated by the necessity to be carbon neutral over the next few decades (Chen et al., 2022; Zhao et al., 2021) (Figure 1.9).

Figure 1.9. By focusing on sustainable practices, organizations can minimize waste and reduce the consumption of hazardous materials.

Source: https://www.gep.com/prod/s3fs-public/blog-images/sustainable-e-waste-management-nurturing-a-greener-future-1920x1274.jpg.

1.3.3. Enhancement of Public Health and Safety

The relationship between green chemistry and public health is ever more vital, particularly in terms of enhancing levels of safety in different sectors. By focusing on the design of chemical products and processes to minimize the use of hazardous materials, green chemistry effectively aims at mitigating exposure risk from toxic substances. The mitigation is imperative both for chemical production employees and consumers who come into contact with products of everyday application. The public health implications extend beyond immediate safety; they encompass long-term health in the form of fewer poisonous chemicals entering the environment meaning healthier environments. Further, cleaner production can foster trust among communities, balancing industrial practice and public health agendas and enabling collective solutions for better health. Thus, green chemistry principles inadvertently create a healthier public health setting, requiring steady investment in sustainable practices. Apart from neutralizing chemical hazards, green chemistry promotes the use of renewable materials that, in turn, can prove extremely effective in optimizing public health and safety. The utilization of non-renewable fossil fuels not only raises environmental degradation but also endangers human health through air and water pollution. By employing bio-based feedstocks, green chemistry encourages cleaner production methods that limit toxic emissions. This new development not only reduces the health risks of traditional manufacturing but also tackles climate change issues related to fossil fuel use. Moreover, the promotion of sustainable agricultural methods, in accordance with green chemistry, can also ensure safer food production through lowered pesticide content and chemical fertilizers and thereby further improve public health. Therefore, the use of renewable materials in chemical processes is a fundamental element toward a healthier society as it addresses both short-term safety concerns and long-term environmental impact. Lastly, the One-World Medicine concept is the perfect example of the integration of traditional medicinal practices with green chemistry as a strategy for public health enhancement. By protecting traditional knowledge through initiatives like the Nagoya Protocol, green chemistry attempts to value indigenous practices but in a way that ensures their safe and sustainable use. Public health is enhanced through the promotion of sustainable herbal products that fall under strict quality controls and testing for efficacy. This approach not only supports the safety of complementary and alternative medicine but also extends to bridge the broader healthcare inequalities of developing nations, where accessible, effective care often is beyond reach. As countries implement regulations supporting both traditional and conventional medicine, they can encourage an integrated model of medicine that respects cultural heritage while keeping it safe. Here, green chemistry not only advocates innovation but also places utmost emphasis on its moral obligation towards the improvement of overall public health results worldwide (Abdelfatah et al., 2019; Curran et al., 2014) (Figure 1.10).

Figure 1.10. By opting for bio-based feedstocks, green chemistry fosters cleaner production methods that limit harmful emissions.

Source: https://media.licdn.com/dms/image/v2/D4E12AQEaAWcikMoutA/article-cover_image-shrink_720_1280/article-cover_image-shrink_720_1280/0/1721179906790?e=2147483647&v=beta&t=6IZ3z0SuC8GFgLOZc3xUC3dnjlVWTc8s3IItvYS5HAA.

1.4. APPLICATIONS OF GREEN CHEMISTRY

The applications of green chemistry are extensive, spanning industries and essentially revolutionizing the way chemical processes and products are carried out. A practical application is in the development of alternative solvents that reduce toxicity and environmental impact. Traditional solvents are usually significant air and water pollutants, so substituting them with more environmentally friendly versions is an objective for the majority of industries. For example, the design of bio-based solvents produced from renewable feedstocks is a good representation of how green chemistry applies principles grounded on sustainability. By trying to reduce hazardous chemicals within chemical processing, companies are not only able to comply with regulations but also enhance their reputation and appeal to environmentally-conscious consumers. This shift is particularly crucial in fields such as pharmaceuticals, where the need for effective yet non-toxic solvents can have serious consequences on drug development and efficacy. Further, the integration of green chemistry into the theory behind the Circular Economy shows a commitment to sustainability that proves to be successful on both social and economic levels. The Circular Economy is concerned with redesigning processes in order to maximise resource effectiveness, facilitating an industrial system where waste is kept to a minimum and materials are continually reprocessed.

This is entirely in line with the fundamental principles of green chemistry, which prefer process efficiency and reduction of environmental effect. As has been pointed out in more contemporary research, including research into Chinas adoption of the Circular Economy as part of national development strategies, this model increases the discussion of sustainability through linking economic activity with environmental well-being (Haynes et al., 2015). The benefits extend to competitiveness and innovation in those industries being increasingly tasked with becoming more sustainable, but another sign of the way green chemistry can take a leadership position in transforming the old economic model. Despite promising actions taken towards implementing the use of green chemistry, skepticism continues to be raised regarding companies' sustainability measures. Criticism tends to concentrate on the activities of green-washing in which companies deceptively present themselves as environmentally friendly in order to capitalize on consumer fads. This skepticism can limit actual progress in sustainable practice since consumers and activists demand openness and accountability from corporations. Corporations must chase actual practices in order to implement green chemistry efficiently, which not only validates green products but also maintains the principles of environmental stewardship (Wagner et al., 2014). This harmony between corporate sustainability initiatives and marketing is crucial in building communities' trust and a successful implementation of the principles of green chemistry for the common good. Therefore, ongoing discussion and criticism will be the means to overcome the challenges of business sustainability while encouraging genuine implementation of green chemistry (Figure 1.11).

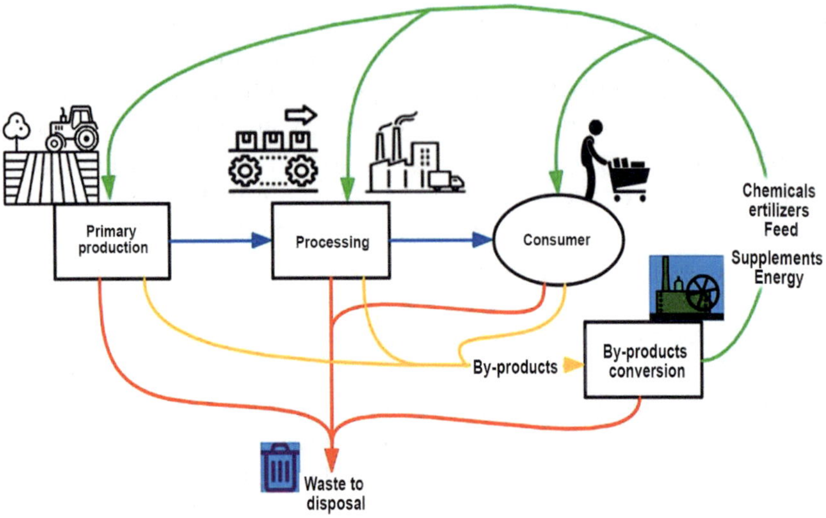

Figure 1.11. The circular economy emphasizes redesigning processes to maximize resource efficiency, promoting an industrial framework where waste is minimized, and materials are continually reprocessed.

Source: https://pub.mdpi-res.com/energies/energies-16-01752/article_deploy/html/images/energies-16-01752-g001.png?1676960078.

1.4.1. Green Solvents and Their Uses

Increased environmental consciousness has brought about a sudden shift in the chemical industry towards the adoption of green solvents. These solvents are either biological in origin or synthesized using environmentally friendly chemical processes, and they possess characteristics that reduce toxicity and environmental impact of chemical reactions. Volatile and hazardous traditional organic solvents produce serious health issues and lead to air and water contamination. Therefore, the development of green solvents, such as bio-based solvents from renewable materials, is tasked with promoting sustainable operations in the green chemistry industry. Utilization of alternatives not only aids in making chemical processes safer but also aligns with regulatory frameworks aiming to reduce hazardous waste and emissions, eventually paving the way for the sustainable chemical industry. Green solvents demonstrate their capability and usefulness in a wide range of applications as alternatives in most chemical processes. For instance, they are increasingly being utilized in natural product extraction of active compounds, demonstrating how they can dissolve a variety of substances with fewer adverse environmental effects. Moreover, products treated with green solvents can possess better properties and therefore improved product performance. Their application is not limited to extraction since they are also vital in catalysis, where green solvents can enhance the rate and selectivity of a reaction. As researchers expand their explorations for new applications of green solvents in different chemical industries, their ability to optimize efficiency while minimizing ecological footprints becomes even more evident, underlining their critical role in the paradigms of green chemistry. Apart from their benefits in operation, the use of green solvents in chemical processes supports the overall pursuit of sustainability and resource conservation. Mechanochemistry shows, for instance, how one can use mechanical energy to catalyze chemical reactions using non-toxic solvents, promoting more sustainable processes (Chetry, 2025). Besides, recent advances in the green synthesis of molecules, including pyrazole derivatives, emphasize the importance of green solvents towards developing operationally simple and atom-economical methodologies (Kachhadiya et al., 2025). With growing scientific understanding of green solvent use, it becomes clear that alternatives not only combat health and environmental challenges of today but also drive a change in the paradigm to which we react for the manufacture of chemicals. The ongoing advancements in green chemistry, with green solvents at its center, provide a hopeful path toward a greener future in chemical practice.

1.4.2. Biodegradable Materials and Their Significance

The production of biodegradable materials represents a significant milestone in the quest for environmentally friendly alternatives under the green chemistry principles. Compared to conventional plastics, which have taken centuries to be degraded in nature, biodegradable materials are designed to be degraded within a shorter period into harmless components through natural means. The degradation is done using microbial action, which decomposes the material into constituent units that go back to the environment, thereby minimizing long-term environmental impact. The importance of such materials is not just that they can prevent plastic contamination but also that they are relevant to resource preservation as they are typically produced from renewable biological sources. As industries shift increasingly toward environmentally friendly means, the adoption of biodegradable materials provides a way to contain the horrifically destructive effect of plastic waste, injecting the intersection of innovation and sustainability into contemporary material science. In pharmacy applications, biodegradable materials demonstrate their significance through the development of advanced drug delivery systems. By the application of principles of nanotechnology, biodegradable polymers can be formulated to encapsulate therapeutic agents and deliver them in a controlled manner to definite sites within the body. Not only does site-specific delivery enhance the efficacy of treatments but also minimizes system side effects that are usually associated with traditional dosing regimens. The employment of biodegradable material in drug delivery serves to mark a move toward the development of drugs and materials following the principles of sustainability through the aspect of ensuring biocompatibility while reducing pharmaceutical interventions-induced waste.

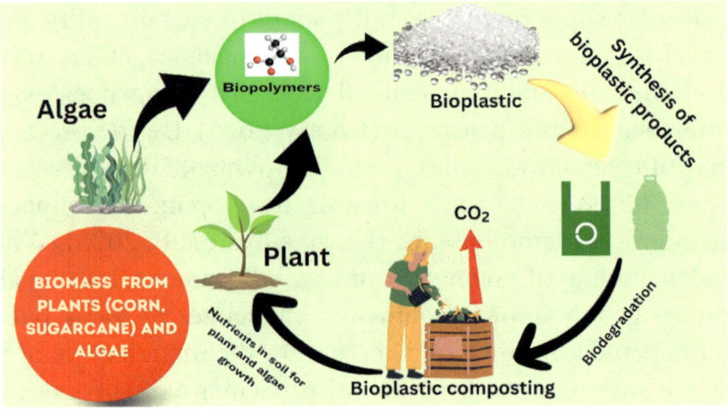

Figure 1.12. Biodegradable materials are designed to break down more rapidly into non-toxic components through natural processes.

Source: https://www.mdpi.com/processes/processes-11-03445/article_deploy/html/images/processes-11-03445-g001.png.

As the healthcare sector seeks new solutions, the application of biodegradable materials will redefine the dynamics of drug development and delivery, and sustainability in life sciences will become a priority (Patra et al., 2018). The greater uses of biodegradable materials intersect into a model of sustainable economic development where they provide the groundwork for a biobased economy. This economic system strives for the optimal utilisation of renewable resources with zero or negligible waste and environmental damage, supplementing green chemistry principles to a large extent. Specifically, advancements in the area of biocatalysis have played a pivotal role in optimizing the production process of biodegradable materials through phenomenal advances in the science of enzyme engineering (Roger et al., 2017). By streamlining the production of biodegradable plastics and other materials, biocatalysis not only makes the process more efficient but also facilitates cost-effectiveness, enabling shift towards a more sustainable material economy. As industries move to accept such sustainable processes, the need for biodegradable materials will further drive innovation, which in turn will instill a sense of responsibility to maintain ecological harmony without compromising the needs of modern society (Figure 1.12).

1.4.3. Renewable Energy Sources and Chemical Processes

The integration of renewable resources in chemical operations is a critical move towards sustainable development with the banner of green chemistry. By using energy from renewable sources such as solar, wind, and biomass, chemical manufacturing can be reduced to less reliance on fossil fuels and therefore lower emissions of greenhouse gases and ecological harm. For instance, bioenergy derived from organic material is a major contributor to this transformation. The International Energy Agency (IEA) report on Dutch bio-energy production, biomass utilization and bioraffinage usage illustrates how nations are adopting biomass not only as a fuel but as a viable feedstock for chemical manufacture. Utilization of renewable energy radically alters the energy dynamics of chemical processes with greater adoption of circular economy principles that reduce waste and enhance resource utilization. Transformation of chemical processes to make use of renewable energy resources also is optimized by next-generation design strategies, including life cycle assessment (LCA). LCA is a quantitative measure for evaluating environmental consequences for every stage of a product's life cycle, from manufacture to end-of-life disposal. Suggested in current literature, use of LCA has become increasingly prevalent in agro-chemical companies, maintaining its supreme status in deciding if processes are sustainable. This approach not only helps decide where to improve but also facilitates decision-making in selecting sustainable practice. Particularly, studies on biodiesel production from *Jatropha curcas* highlight the utility of LCA in evaluating the environmental trade-offs with renewable chemical processes for energy so that industries can optimize economic viability and ecological responsibility. Moreover, renewable energy enables the design of more efficient and greener green chemical processes. Advances in technology continue to allow for the direct application of renewable energy to chemical reactions, which can lead to less energy consumption and decreased production of noxious by-products. As more industries apply these cleaner technologies, they are part of an ongoing shift towards chemistry sustainability. The combination of LCA with these processes brings increased insight into the consumption of resources and encourages more stewardship of natural resources. This combined strategy of renewable energy and chemical processes is a quintessential example of the aims of green chemistry in which attention is not only paid to innovative chemistry but also to environmental responsibility with a focus on ecosystem and community welfare (Figure 1.13).

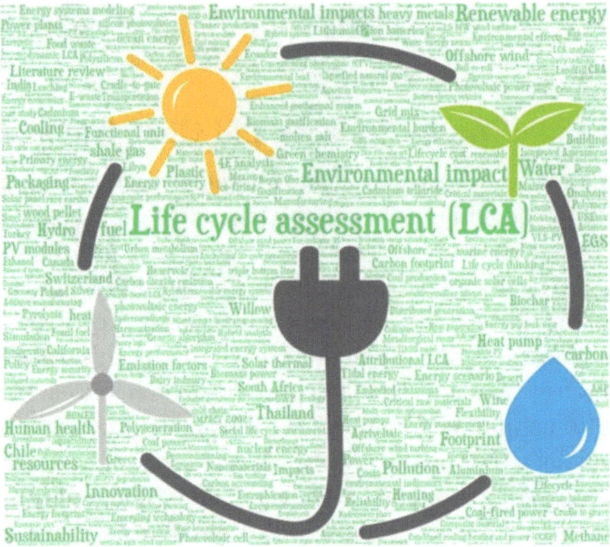

Figure 1.13. Transitioning chemical processes to utilize renewable energy sources is also optimized through innovative design methodologies, including life cycle assessment (LCA).

Source: https://www.mdpi.com/environments/environments-11-00147/article_deploy/html/images/environments-11-00147-g002-550.jpg.

1.5. CHALLENGES IN IMPLEMENTING GREEN CHEMISTRY

The use of green chemistry is hampered by numerous challenges that deter it from being used extensively, especially in the third world. One of the challenges that discourage its use is the availability of limited funding that discourages research and development activities essential in designing green chemistry processes. Schools and organizations are hesitant to include modern methods or renovate older plants due to scarce monetary resources required to attain the green mark. Moreover, resource constraints worsen this issue by restricting the availability of advanced technologies and materials necessary for such innovative solutions. Knowledge regarding green chemistry principles also lacks among a majority segment of society, excluding support of society in putting forward initiatives for encouraging sustainable practices. Hence, the convergence of financial, resource, and knowledge deficits poses a significant barrier to successful application of green chemistry, particularly in regions where environmental and societal issues are most widespread (Akinsipo et al., 2025). Additionally, the need for comprehensive overhaul of learning institutions' curriculum is a principal impediment to successful practice of green chemistry. Ancient methodologies continue to dictate modern chemistry education, with minimal focus on novel sustainable practice approaches.

Figure 1.14. Effective policy measures can provide the necessary regulatory framework to encourage industries to adopt sustainable practices.

Source: https://redingtongroup.com/wp-content/uploads/2023/06/How-Businesses-Can-Set-the-Standard-for-Sustainable-Business-Practices_shutterstock_2162001981.png.

This deficiency means a workforce that cannot properly engage with green chemistry principles, strangling innovation and uptake in an industrial setting. Education institutions should therefore put top priority on integrating the twelve principles of green chemistry into their curriculum in order to develop a mindset that gives prominence to sustainability from an early point. Academic-industry collaborations are also required for the creation of practical uses of green chemistry, but these are typically underdeveloped, which necessitates the enactment of systemic reforms. Collective efforts towards reforming pedagogical structures and promoting synergistic inter-stakeholder dynamics will be crucial in surmounting such challenges and propelling green chemistry initiatives (Ogodo et al., 2025). Government policy also plays a significant role in deciding the paradigm for the application of green chemistry. Effective policy interventions can provide proper regulatory support to the industries so that they are encouraged to adopt environmentally sound approaches of functioning. Unfortunately, in the majority of the world, inefficient laws and lack of proper incentives deter companies from transforming to greener alternatives due to apprehensions regarding increased operating expenses and risk of resistance. Green chemistry can be successful only if government agencies establish transparent regulations, economic benefits, and robust laws promoting sustainable means without stifling innovation. Additionally, promoting international cooperation will help in the dissemination of best practices and innovative technology tailored to individual contexts and also in the effectiveness of green chemistry measures. As nations grapple with pressing environmental issues, creating a supportive policy environment for green chemistry is essential in facilitating its effective application across various sectors (Akinsipo et al., 2025) (Figure 1.14).

1.5.1. Technological Limitations and Barriers

The motivation for green chemistry is typically met with a myriad of technological barriers and limitations that stymie its effective use. A major obstacle is the use of advanced materials and processes that are still in the process of development for most sustainable procedures. Research has shown that green chemistry transition would necessitate heavy investment in new technologies and laboratory methods that have the capability to minimize waste and enhance efficiency. However, the majority of developing countries lack the necessary capital and infrastructure to upgrade such processes. This deficit hence leads to a step-by-step adoption of green practices, as noted by Abbott et al. (2025). Meeting these challenges requires a concerted effort from government officials, research centers, and private organizations to provide level ground in terms of access to technological advancements. Some scientific limitations also present impossible hurdles in the area of green chemistry. Optimization and efficient design of sustainable chemical processes involve in most instances a complex knowledge of material property and behavior at the molecular scale. Traditional material discovery methods are resource- and time-intensive. For instance, while density functional theory (DFT) can potentially forecast material behavior in applications such as energy storage and photovoltaics, its commercial use is limited by computational cost and technical skills required for its effective application (Jain et al., 2016). These factors not only delay the timely development of green technology but also create a huge technology gap between research potential and commercialization. Further, the integration of new green technologies into conventional industrial environments also presents challenges. Companies will likely meet opposition to the change spurred by the astronomical cost of moving away from conventional techniques to novel, green ones. In addition, regulatory frameworks lag behind technological advances, and this uncertainty discourages investment in new green processes. This policy lag can cause an industry hesitation to adopt greener methods and thereby increase the current technological barriers. In addressing these challenges, it is crucial that policymakers create a conducive environment that will facilitate research, development, and implementation of green chemistry projects. It is only by breaking down these technological barriers that green chemistry science can get closer to achieving its ultimate goal of sustainability and environmental responsibility.

1.5.2. Economic Factors and Market Acceptance

The economic climate largely influences the adoption of green chemistry practices across various industries, particularly the manufacturing and construction industries. Demands for too much up-front investment typically pose a significant barrier to the implementation of green technology because stakeholders will weigh the potential future saving against immediate cash outlay. This is particularly true in sectors that have previously relied on proven processes and materials, in which the move towards sustainable alternatives will not necessarily seem economically viable. As Wu et al. (2025) infer, the lack of strong institutional incentives and high operating costs from low-carbon buildings deter developers from embracing such practices. Hence, there should be a properly designed financial system with subsidies, grants, or tax credits to make the market acceptable and induce transition towards more eco-friendly methods. Beyond economic constraints, market willingness to adopt green chemistry is heavily reliant on societal consciousness and public opinion regarding environmental issues. While there is still controversy surrounding sustainability, the public is not well cognizant of the actual advantages of green practice. Inadequate awareness of environmental advantages and potential long-term savings in cost generally leads to low consumer demand for green products. In addition, as Hoareau et al. (2024) further state, policy interventions aimed at promoting the use of green materials in construction have faced varying degrees of success due to public indifference towards sustainability. Therefore, enhancing public awareness through education outreach programs is important to build momentum for continued acceptance and demand for green alternatives.

Public awareness can change market forces, compelling companies to use sustainable practices in harmony with the values of consumers. In addition, the dependency of government policy and market forces sets the direction of adoption of green chemistry. While regulations can effectively promote eco-friendly practices, the effectiveness of such policies largely depends on effective enforcement and economic incentives that are aligned with market conditions. As Hoareau et al. (2024) explains, great margins generally exist between policy intent and actual industry practices, where most companies refrain from practicing green technology as a result of economic uncertainties as well as market impediments. In response to this challenge, governments must forge a robust framework that not only encourages sustainable practices but also ensures it becomes economically feasible for industries to adopt these strategies. By combining economic incentives with clear policy objectives, a conducive environment could be established to induce widescale adoption of green chemistry principles in industries and ensure a shift toward a more sustainable future (Figure 1.15).

Figure 1.15. High initial investment costs often pose a substantial barrier to the adoption of green technologies, as stakeholders weigh the potential for long-term savings against the immediate financial outlay.

Source: https://www.rapidops.com/mm-images/legacy-system-modernization-z7r69qyc.jpg.

1.5.3. Regulatory and Policy Challenges

The integration of green chemistry principles into existing regulatory frameworks presents an endless set of challenges that can severely hinder its uptake and efficiency. These are often exacerbated in developing nations where thin budgets and a lack of resources create gargantuan barriers to introduction. In the Nigerian context, for example, highlight group sessions among experts identified the rudimentary need for more funding and awareness to make green chemistry projects feasible. Lack of robust government and institutional support renders it difficult for industries and education sectors to embrace green practices, particularly in the wake of heightened complexities surrounding regulatory compliances and environmental management. This renders coordination with stakeholders between academia, industry, and policy-makers imperative in order to address such systemic issues and create an enabling environment where green chemistry can thrive (Akinsipo et al., 2025). Moreover, the integration of regulative processes and the adoption of new green technologies reveals significant impediments. Regulative processes governing chemical materials and manufacturing methods are usually lagging behind technological advancements, in a way that inhibits effective incorporation of greener methods to industrial operations. As seen in Malaysian construction, a review of the environmental management and development approval systems reveals a gap that typically impinges on sustainability. Policy makers must contend with how regulatory demands can be reconciled with new green technologies, such as renewable energy and smart building practices. Approval processes are at the center of this calculation, as bottlenecks or delays in the system can deter innovation and investment in green projects. Addressing these regulatory bottlenecks is needed to ease the process of shifting to environmentally more sustainable practices in various industries, thus promoting greater application of the principles of green chemistry (Siganul et al., 2025). Finally, the highest-level challenge of enforcement of regulations must be addressed so that the principles of green chemistry are not ideological but practically adhered to. Enforcement loopholes are responsible for great variations between policy objectives and real practices at the ground level, ultimately posing a danger to the effectiveness of green chemistry initiatives. The situation is very critical in developing nations where there may be insufficient regulatory oversight, leading to negligent utilization resulting in environmental deterioration. Improved regulatory mechanisms, along with stipulated enforcement systems, are needed in an effort to sustain levels of sustainability. Through guaranteed uniformity in regulation implementation, stakeholders can promote accountability culture and challenge industries to make green chemistry the ideal route toward economic development as well as environmental sustainability. The assiduity towards the formulation of suited regulatory strategies that adapt to regional conditions will eventually boost the success of green chemistry efforts globally (Akinsipo et al., 2025; Siganul et al., 2025).

1.6. CASE STUDIES IN GREEN CHEMISTRY

The application of green chemistry is best illustrated through various case studies that reveal the challenges of development and advancements in the field.

One such interesting feature of the landscape is how synthetic materials, and more specifically synthetic rubber, were created by war needs as well as market demands. In the German context, Peter Morris demonstrates how national pride pushed investments in the production of synthetic rubber even if there were questionable economic advantages. The case study not only points out the role political factors played in technological innovation but also suggests the importance of incorporating sustainability into industrial processes. By considering motives and consequences on synthetic rubber, we better understand the way green chemistry is able to rework existing technology to fit new environmental concerns, an active cross-over between scientific development and societal need (Krasnod, 2023). The consideration of nitrogen products in the First World War is another fascinating green chemistry case study, illustrating the power of crisis to generate innovation. Sandra Fehrs comparison of wartime Germany reveals that while the military's nitrogen deficit was all right solved by available Haber-Bosch technology, its lasting impact on agricultural practice was vast. That such a duality of near and far effects does exist illustrates the heart of a key green chemistry principle: sustainability through longer-term solutions rather than short-term fixes.

In recognizing that under compulsion made adjustments can perpetuate environmental problems, we have a requirement for a holistic approach to green chemical practice that involves consideration of product and process life cycle. These ideas provide valuable lessons on emerging green chemistry advances and serve as a warning that induces greater holistic thinking regarding ecologic impacts in innovation strategy (Krasnod, 2023). In the pharma sector, the shift towards continuous flow chemistry emphasizes the potential of sustainable methods in the highly regulated pharmaceutical market. As outlined in recent review articles, this approach has considerable advantages over conventional batch processing, at the very least through improved efficiency and reduced waste generation. Continuous flow systems possess the ability to reduce the detrimental environmental impacts that are often associated with pharmaceutical manufacture, making cleaner synthetic routes that support overall sustainability. However, there are obstacles to their widespread adoption, most significantly identifying alignment of the interests of chemists and chemical engineers in academia and industry. The on-going push towards the optimization of flow chemistry is a step change in the direction of more sustainable synthesis, with an open invitation for further investment and collaboration to facilitate more greener manufacturing methodologies in the pharmaceutical sector. These case histories demonstrate the convergence of green chemistry theory with experimental applications to achieve a brighter, sustainable industrial tomorrow (Figure 1.16).

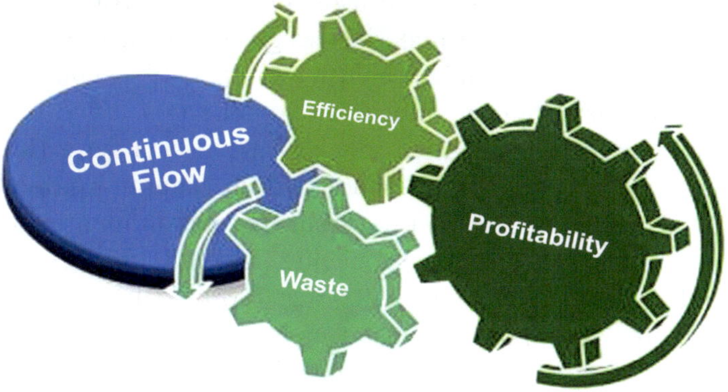

Figure 1.16. Continuous flow systems can mitigate the adverse environmental impact commonly associated with pharmaceutical manufacturing, allowing for greener synthetic routes that enhance overall sustainability.

Source: https://ars.els-cdn.com/content/image/1-s2.0-S2452223617300263-fx1.jpg.

1.6.1. Successful Green Chemistry Innovations

Among all the talks about successful green chemistry innovations, the appearance of next-generation battery technologies stands tall. With the need for sustainable energy storage materials increasing day by day in accordance with world energy demands and ecological pressures, researchers have given sufficient attention to beyond Li-ion technologies. The technologies are sodium-ion and potassium-ion batteries, which are focused on evading the challenges presented by traditional lithium-ion batteries, namely resource limitation and skyrocketing prices. These innovations are part of a broader trend in green chemistry as innovations seek to address practical energy storage requirements and minimize environmental impact. With new materials and chemistries, such innovations demonstrate the possibility to offer high energy density with less economic barriers for mass market deployment, eventually making way for a greener energy future (Tian et al., 2020). Apart from technological breakthroughs in batteries, circular chemistry has also emerged as a game-changing concept in green chemistry. Circular chemistry challenges the traditional linear model of production—where resources are extracted, made, and subsequently discarded—by encouraging the utilization of a closed-loop process where wastes are reduced and resources are continuously cycled back. This paradigm bursts from single-minded chemical manufacture to optimizing the entire lifecycle of products to achieve improved resource efficiency across value chains. Effective circular chemistry applications involve new recycling technologies and biodegradable material design, reducing the environmental footprint and allowing the chemical industry to operate sustainably. With the integration of circular principles, much can be done in a waste-free chemical industry, highlighting the innovative scope of green chemistry (Keijer et al., 2019). Green chemistry principles integrated in natural product synthesis and herbal medicine are an innovative front that has immense implications. By applying sustainable solvents, reducing harmful waste, and working with renewable feedstocks, chemists are transforming the herbal compound extraction and formulation. Improvements in these techniques not only enhance the efficacy of extraction processes but also the safety and efficacy of herbal preparations, making them increasingly attractive alternatives to chemical-based drugs. Further, the application of biocatalysis and enzyme-based processes is a greener approach that aims for lower energies and less environmental impact. These accomplishments demonstrate the richness of green chemistry, the wide scope of application to other areas, and the potential for systemic innovation that enhances human health and environmental sustainability.

1.6.2. Comparative Analysis of Traditional vs. Green Methods

The evolution of chemical strategies has prompted a critical comparison of traditional versus green chemistry, and the emphasis here is on sustainability. Traditional chemistry has a tendency to stress high-yielding and low-cost processes at the expense of environmental health. Such traditional techniques are a regular staple of strong solvents, waste-intensive, and pollution-prone. They consequently lead to environmental disruption, indicating an immediate need for alternatives. Green chemistry tries to address these adverse effects through the use of guidelines that encourage reduced toxicity, safer reactants, and waste minimization. The transition to green practices reflects society-wide trends demanding sustainable growth and civic responsibility in scientific endeavors. An extremely pertinent illustration of this transition can be found in the treatment of common diseases by different medical methodologies. Fu's subcutaneous needling (FSN) is a sustainable practice because it integrates traditional acupuncture with modern medical understanding, thereby ensuring patient safety and minimizing invasive interventions. Trials have revealed that FSN alleviates knee osteoarthritis (KOA) pain, increases mobility, and enhances quality of life, thus confirming that an integration of conventional practices can yield improved outcomes with fewer side effects (Liu et al., 2025). This approach not only remains pertinent but is also in accordance with green principles as it reduces the environmental effects of medical procedures. Furthermore, advances in analytical chemistry highlight the efficacy of green approaches to practical use. For instance, the development of a green analytical procedure for the analysis of statins indicates the significance of sustainable practice in pharmaceutical research. The blending of Partial Least Squares regression with the Firefly Algorithm made possible highly accurate and selective analysis with lesser resources and producing less waste (Farooq et al., 2025). Such kind of methods are the best example of the advantages of adopting green chemistry practices, exhibiting better performance parameters than the traditional chromatographic methods. This cross-comparison of conventional and green methods indicates that the adoption of sustainable practices not only addresses the increasing demand for environmental responsibility but also fosters innovation and optimality within the scientific community.

1.6.3. Lessons Learned from Industry Applications

The application of green chemistry principles in various industries has yielded valuable lessons, particularly in process improvement and waste minimization. Pharmaceutical and manufacturing sectors have been leaders in adopting new approaches that can lead to minimization of waste and improved productivity. It has been demonstrated by case studies that through the incorporation of source reduction techniques, companies can significantly minimize the amount of hazardous waste created through manufacturing processes. Such practices usually include redesigning processes to minimize the use of materials and energy, a practice that not only develops environmental benefits but also enhances cost-effectiveness. Ultimately, such integration of green chemistry practices can improve compliance with environmental legislation and further enhance the public image of a company, justifying the value of sustainable practices in modern industry (Singh et al., 2023). Aside from waste minimization, interdisciplinarity among academia, industry, and government has also become an inherent green chemistry learning lesson. Successful collaborations offer new solutions and facilitate the development of sustainable technologies. Sharing knowledge and sharing resources help develop new methodology, which in turn develops a culture of sustainability. For instance, industries collaborating on activities that involve green product design can leverage academic research to embrace cutting-edge technology that ultimately results in better environmental performance. Its appreciation of cooperation as essential can lead to a climate that not only encourages green chemistry initiatives but also creates continuous improvement and innovation in the sector (Singh et al., 2023). The values that derive from successful examples of green chemistry extend beyond technical achievement to include the need for high credibility in scientific methodology. In computationally intensive industries that rely on computational fluid dynamics (CFD) and other advanced modeling techniques, prediction accuracy is most important. That CFD results are credible implies ensuring complete validation protocols, quality assurance, and collaborative frameworks with team-oriented orientations. Synthesis between teamwork and quality has been shown to produce higher quality design outcomes and minimize threats of environmental impacts. Such emphasis on maintaining high credibility standards in different industries proves that effective utilization of the principles of green chemistry relies on teamwork with emphasis on scientific integrity and engagement with stakeholders.

1.7. CONCLUSION

Finally, the principles of green chemistry are a basis for sustainable processes that are beneficial to human health and environmental safety in the long run. Through prioritizing the reduction of hazardous chemicals in chemical product design, manufacturing, and use, green chemistry enables the development of an innovation culture that improves both the process of chemicals and makes it safer. The recognition of the interplay between climate, ecosystems, and human health, as detailed by the IPCCs full reports, underscores the need to adopt these ideas. The incorporation not only diminishes adverse environmental impacts but also guarantees a more sustainable development paradigm that aligns with global sustainability goals (Intergovernmental Panel on Climate Change, 2023). Thus, embracing green chemistry is not an option but a requirement in the quest for a healthier world. In the face of the global effort against the threat of climate change, clean energy sources such as hydrogen are becoming more prominent in decarbonizing key sectors, including transportation and manufacturing. The application of hydrogen technology is an embodiment of the innovative spirit of green chemistry and offers efficient solutions to the reduction of greenhouse gas emissions and fossil fuel dependence. Its successful application, however, is dependent on research and development with backing from policies facilitating their incorporation into existing infrastructures. The increasing developments in hydrogen as a green energy source mark a critical point in our treatment of environmental issues, uncovering the power of green chemistry principles to drive such a shift (Staffell et al., 2018). Finally, the path to a greener future through green chemistry includes the sensitive inclusion of environmental, social, and economic considerations. It is imperative that scientists, policymakers, and industries unite to formulate an integrated approach that unites innovation and responsibility. By constant monitoring of the impact of our chemical processes and using green alternatives, society can tackle the immediate needs of climate change and environmental degradation. Public discussion and integration of green chemistry into curricula is essential in creating awareness and fostering a culture of sustainability. Accordingly, the transition to green chemistry is not a transformation of technology but a paradigm shift in the way we perceive the world and interact with it to ensure an environmentally sound future for mankind and the world (Intergovernmental Panel on Climate Change, 2023).

1.7.1. Summary of Key Points Discussed

In the conversation regarding green chemistry, several key principles arise that reveal the pivotal position of sustainable practice in the manufacturing of chemicals. Primary among the principles of green chemistry is waste reduction by designing less harmful chemicals and using renewable resources. Through a focus on these areas, practitioners not only enhance the safety of chemical processes but contribute to the general vision of reducing the role of human activities towards the environment. This is in agreement with the imperative for urgent actions toward reducing climate change, as distilled in the Synthesis Report (SYR) that summarizes the principal findings of recent reports of the most recent IPCC assessments (Calvin et al., 2023). The inclusion of these principles forms an augmented vision of how chemistry can harmonize with ecological sustainability, ultimately benefiting industrial stakeholders and society at large. The second major element being discussed involves creative means that promote the use of green solvents and catalysts, which perform a key function to optimize the efficiency and safety of chemical reactions. The transition to greener solvents reduces the footprint of chemical processing on the environment and enhances performance simultaneously. This particularly applies in an era where moral limitation on chemical production is subject to increased debate. The development of flexible sensor technology illustrates the way of this, illustrating how advances in technology address technical as well as non-technical problems in green practice (Luo et al., 2023). As companies begin to adopt these green methods, they are also encouraged to explore regulatory and moral standards that ensure appropriate development and commercialization methodologies, further placing emphasis on the interdisciplinarity in the practice of green chemistry. Finally, it is critical that educational institutions and collaborative research networks be encouraged to principles of green chemistry at large. Through the education of future chemists with the knowledge and skills necessary to prioritize sustainability, learning institutions foster environmental responsibility. This culture of teamwork is not confined to campus life, as it promotes interdisciplinarity among scientists, industry players, and policymakers that can lead to revolutionary answers and technology. Witnessing the case of flexible sensors, the collaboration among different communities can significantly contribute to scientific findings (Luo et al., 2023). Lastly, the common pursuit of sustainable chemistry processes is not an academic pursuit but a shared responsibility that will shape the future of chemical manufacture and environmental stewardship.

1.7.2. Future Outlook for Green Chemistry

The future of green chemistry is promising with opportunities driven by advances in a number of technological fields that are aligned with sustainable processes. With increasing environmental awareness, industries are increasingly likely to adopt green chemistry methods that guarantee reducing hazardous chemicals and waste. One area in which this is most evident is in the development of next-generation materials, such as flexible sensors, that are being made to circumvent the limitations of traditional rigid equivalents. These technologies not only offer greater performance but also help to minimize environmental pressures and hence fulfill key sustainability needs. By sensor performance and commercialization challenge identification and resolution, scientists are opening doors to larger adoption, eventually creating enhanced quality of life in the era of digitalization and big data with solutions to difficult environmental problems (Luo et al., 2023). Another example is battery technology development, which represents the green chemistry potential to influence energy solutions that are critical to address climate change. The transition from conventional lithium-ion batteries to beyond Li-ion technologies is a significant step towards sustainable energy storage solutions. With scientists considering alternatives such as sodium-ion and multivalent batteries, there is a concerted effort to resolve the issues pertaining to battery production and recycling that are currently not environmentally friendly.

This shift not only holds the promise of augmented energy density and decreased cost but is also oriented toward responsibly sourced raw materials, therefore meeting principles of green chemistry. Such a future informed by such leading-edge technology not only improves energy efficiency but also ensures a decrease in the environmental footprint of energy storage devices, illustrating how green chemistry can guide the energy industry (Tian et al., 2020). As we look out to the green chemistry horizon, it will be interdisciplinary collaboration that brings it to its full potential. Interdisciplinary research has the potential to yield paradigm-shifting changes that merge chemistry with engineering, biology, and environmental science in support of integrated solutions to monumental global challenges. To create shared goals and facilitate transformation toward sustainable approaches, networks that bridge academia, industry, and regulatory agencies must be established. Additionally, dealing with non-scientific issues such as business ethics, regulatory frameworks, and social issues will be critical to the mass adoption of green technologies. By incorporating these issues into the agenda of green chemistry, an integrated strategy can be formulated that not only works on the environmental sustainability aspect but also promotes social equity and economic feasibility in the international context.

1.7.3. Call to Action for Further Research and Adoption

As environmental concerns become increasingly urgent on a global level, the demand for more research and adoption of green chemistry is increasingly critical. The negative impacts of pollution—particularly of chemical processes—are stark. For instance, ocean pollution has become a silent killer with enormous economic loss and sudden collapse in marine ecosystems. In order to meet these challenges, it's required to integrate the principles of green chemistry throughout various sectors as these practices focus on sustainability and reducing dangerous materials. Policymakers and entrepreneurs must be made aware that the adoption of green chemistry solves not only short-term environmental concerns but also drives innovation and economic development. Research to determine the effectiveness of green chemistry methods in different industrial processes can provide insightful findings on their applications and prospects for scalability, culminating in more sustainable activities in other sectors. Besides industrial application, research on the educational front of green chemistry is also crucial. Universities and schools offer the foundation where the future scientists and business leaders are taught to be sustainable method champions. The incorporation of green chemistry in curriculum is a powerful approach that can produce a generation of researchers who are not only proficient at the science but also committed to its principles. Moreover, greater collaboration between academia and industry can accelerate dissemination and application of green chemistry innovations. As a result, there must be collective effort to develop interdisciplinary partnerships that draw on various skill sets, in order to investigate novel means of resolving problems ensuing from traditional chemical practice. By instilling research-based culture, institutions can prepare students to be agents of change toward a greener environmentally friendly future. The large-scale adoption of green chemistry ultimately lies in the hands of regulatory policy and research financing that encourages its use. Institutions and governments worldwide need to take on the central role of spearheading the change by implementing policies and incentives that encourage industries to make greener choices. Robust legislation has the ability to enhance the visibility of green chemistry practice and legitimize its worthiness in providing solutions to significant environmental issues. Research encouraging cooperation, such as research awards and green chemistry research funding, can also facilitate innovative product development that reduces chemical waste and toxicity. These measures have implications beyond environmental sustainability because they also guarantee public health and public welfare. As is evident from the reports highlighting the negative impacts of pollution, it is not only a choice but a necessity to become green through green chemistry for a healthier future for the world and mankind (Philip et al., 2020; Tjoa et al., 2020).

REFERENCES

1. Abbott, L., Gruère, G., & Narrod, C. (2025). Agricultural, food, and water nanotechnologies for the poor: Opportunities, constraints, and role of the Consultative Group on International Agricultural Research. https://core.ac.uk/download/pdf/6237698.pdf (accessed on 24th October 2025).

2. Abdelfatah, S., Abu-Darwish, M. S., Banerjee, M., Bhakta-Guha, D., et al. (2019). Biopiracy versus one-world medicine–from colonial relicts to global collaborative concepts. https://core.ac.uk/download/228178415.pdf (accessed on 24th October 2025).

3. Ain, N. U. (2024). A review of the progress and developments of green chemistry. *International Journal of Science and Research Archive.* https://www.semanticscholar.org/paper/afee7ede80db118ba2fd1b9b137cf5d6b2a53e8e (accessed on 24th October 2025).

4. Akinsipo, O., & Anselm, O. H. (2025). Challenges and opportunities for implementing green chemistry in Nigerian universities: Educational and policy perspectives. *Sustainability & Circularity NOW.* https://www.semanticscholar.org/paper/9dd441ee2d3b3a10561cfb7660a773dda506cfa5 (accessed on 24th October 2025).

5. Almutairi, F. M., Althobaiti, Y. S., Abduljabbar, M. H., Alzhrani, R. M., Alnemari, R. M., Aldhafeeri, M. M., Serag, A., et al. (2025). A green and sustainable UV spectrophotometric approach for simultaneous determination of rosuvastatin, pravastatin, and atorvastatin in pharmaceuticals leveraging firefly algorithm-enhanced partial least squares regression. *Analytical Methods: Advancing Methods and Applications.* https://www.semanticscholar.org/paper/6ceecb9f8b72b67be8a3269d226011a9743cb33e (accessed on 24th October 2025).

6. Annevelink, E., & van Ree, R. (2010). Country report The Netherlands 2010: IEA Bioenergy Task 42. https://core.ac.uk/download/pdf/29240740.pdf (accessed on 24th October 2025).

7. Banerjee, P., Kemmler, E., Dunkel, M., & Preißner, R. (2024). ProTox 3.0: A webserver for the prediction of toxicity of chemicals. *Nucleic Acids Research, 52*, W513–W520. https://doi.org/10.1093/nar/gkae303 (accessed on 24th October 2025).

8. Baumann, M., Smyth, M., Moody, T., & Wharry, S. (2021). Evaluating the green credentials of flow chemistry towards industrial applications. *Synthesis, 53*, 3963–3976. https://www.semanticscholar.org/paper/ede0ffe926b71fff53ccf621d71a71bef45842a4 (accessed on 24th October 2025).

9. Belaud, J. P., Gillani, S. T., Le Lann, J. M., Sablayrolles, C., et al. (2010). Review of life cycle assessment in agro-chemical processes. https://core.ac.uk/download/12042676.pdf (accessed on 24th October 2025).

10. Bender, E. M., Gebru, T., McMillan-Major, A., & Shmitchell, S. (2021). On the dangers of stochastic parrots. https://doi.org/10.1145/3442188.3445922 (accessed on 24th October 2025).

11. Bjelkemyr, M., Jönsson, C., Kurdve, M., Shahbazi, S., et al. (2013). Industrial waste management within manufacturing: A comparative study of tools, policies, visions and concepts. https://core.ac.uk/download/42143949.pdf (accessed on 24th October 2025).

12. Calvin, K., Dasgupta, D., Krinner, G., Mukherji, A., Thorne, P., Trisos, C. H., Romero, J., et al. (2023). IPCC, 2023: Climate change 2023: Synthesis report. Contribution of Working Groups I, II and III to the Sixth Assessment Report of the Intergovernmental Panel on Climate Change [Core Writing Team, H. Lee & J. Romero (Eds.)]. *IPCC, Geneva, Switzerland.* https://doi.org/10.59327/ipcc/ar6-9789291691647 (accessed on 24th October 2025).

13. Chen, L., Msigwa, G., Yang, M., Osman, A. I., Fawzy, S., Rooney, D. W., & Yap, P. S. (2022). Strategies to achieve a carbon neutral society: A review. *Environmental Chemistry Letters, 20*, 2277–2310. https://doi.org/10.1007/s10311-022-01435-8 (accessed on 24th October 2025).

14. Chetry, A. B. (2025). Mechanochemistry: A new frontier in chemical synthesis. *Journal of Chemical Research, 49.* https://www.semanticscholar.org/paper/b601e758a61cbc16926f466890f1def81ccafbc0 (accessed on 24th October 2025).

15. Curran, J., Gatliff, R., Haszeldine, R. S., Masters, S., et al. (2014). Independent expert scientific panel–Report on unconventional oil and gas. https://core.ac.uk/download/16458850.pdf (accessed on 24th October 2025).

16. Dewulf, A., Hovelynck, J., & Sips, K. (2010). Perspective study: Governance for C2C. https://core.ac.uk/download/pdf/29239612.pdf (accessed on 24th October 2025).

17. Giannozzi, P., Andreussi, O., Brumme, T., Bunău, O., Buongiorno Nardelli, M., Calandra, M., Car, R., et al. (2017). Advanced capabilities for materials modelling with Quantum ESPRESSO. *Journal of Physics: Condensed Matter, 29*, 465901. https://doi.org/10.1088/1361-648x/aa8f79 (accessed on 24th October 2025).

18. Hastrup, K., Grønnow, B., & Mosbech, A. (2018). Preface. *Ambio, 47*, 159–161. https://www.semanticscholar.org/paper/143b726f81f0f6b488af

4ff116ea36e5bf465828 (accessed on 24th October 2025).

19. Haynes, K., Murray, A., & Skene, K. (2015). The circular economy: An interdisciplinary exploration of the concept and application in a global context. https://core.ac.uk/download/151157363.pdf (accessed on 24th October 2025).

20. Hoareau, C. E., & Tam, N. (2024). Evaluating the impact of government policies on circular economy adoption in the construction sector. *Civil and Sustainable Urban Engineering*. https://www.semanticscholar.org/paper/d02609275098e08dda141c6930884fbf2d4a9bec (accessed on 24th October 2025).

21. Jain, A., Persson, K. A., & Shin, Y. (2016). Computational predictions of energy materials using density functional theory. https://core.ac.uk/download/294625349.pdf (accessed on 24th October 2025).

22. Jeevanandam, J., Barhoum, A., Chan, Y. S., Dufresne, A., & Danquah, M. K. (2018). Review on nanoparticles and nanostructured materials: History, sources, toxicity and regulations. *Beilstein Journal of Nanotechnology, 9*, 1050–1074. https://doi.org/10.3762/bjnano.9.98 (accessed on 24th October 2025).

23. Kachhadiya, R., Shah, D., Prajapati, A., & Patel, A. D. (2025). Green synthetic strategies for pyrazole derivatives: A comprehensive review. *Archive der Pharmacies, 358*(7), e70055. https://www.semanticscholar.org/paper/dab2813c258afcee2054c5cb53dcc9aeacfca139 (accessed on 24th October 2025).

24. Keijer, T., Bakker, V., & Slootweg, J. C. (2019). Circular chemistry to enable a circular economy. *Nature Chemistry, 11*, 190–195. https://doi.org/10.1038/s41557-019-0226-9 (accessed on 24th October 2025).

25. Khan, I. S., Ahmad, M. O., & Majava, J. (2021). Industry 4.0 and sustainable development: A systematic mapping of triple bottom line, circular economy and sustainable business models perspectives. *Journal of Cleaner Production, 297*, 126655. https://doi.org/10.1016/j.jclepro.2021.126655 (accessed on 24th October 2025).

26. Krasnodębski, M. (2023). Research between science, society and politics: The history and scientific development of green chemistry. *Ambix, 70*, 333–335. https://www.semanticscholar.org/paper/2075f32d85679cc1a7281a4baeb344f2adff5b6d (accessed on 24th October 2025).

27. Laddha, P. R., Sitaphale, G. R., Charhate, K. B., & Tathe, P. R. (2024). Green chemistry in pharmaceutical synthesis: Sustainable strategies for drug production. *International Journal of Advanced Research in Science, Communication and Technology*. https://www.semanticscholar.org/pa

per/0b10f3c32499241c0564c879bfc236d0d9c8ab4a (accessed on 24th October 2025).

28. Landrigan, P. J., Stegeman, J. J., Fleming, L. E., Allemand, D., Anderson, D. M., Backer, L. C., Brücker-Davis, F., et al. (2020). Human health and ocean pollution. *Annals of Global Health, 86*, 151. https://doi.org/10.5334/aogh.2831 (accessed on 24th October 2025).
29. Lin, W., Puig Vidal, R., Sang, J., et al. (2016). A comprehensive evaluation of physical and environmental performances for wet-white leather manufacture. https://core.ac.uk/download/81578849.pdf (accessed on 24th October 2025).
30. Liu, L., Sun, J., Li, H., Zheng, Z., Lu, J., Lyu, Q., & Bai, T. (2025). Fu's subcutaneous needling approach versus electroacupuncture for knee osteoarthritis: Protocol for comparative effectiveness and safety randomized controlled trial. *Frontiers in Medicine, 12*. https://www.semanticscholar.org/paper/9783c49a77667dae2f4ff93b775cb117703136 3b (accessed on 24th October 2025).
31. Luo, Y., Abidian, M. R., Ahn, J. H., Akinwande, D., Andrews, A. M., Antonietti, M., Bao, Z., et al. (2023). Technology roadmap for flexible sensors. *ACS Nano, 17*, 5211–5295. https://doi.org/10.1021/acsnano.2c12606 (accessed on 24th October 2025).
32. Mehta, U. B. (2025). The aerospace plane design challenge: Credible computational fluid dynamics results. https://core.ac.uk/download/pdf/42819894.pdf (accessed on 24th October 2025).
33. Murphy, M. A. (2023). Professors Trost and Sheldon's promotion of catalytic technologies, atom economy, and the E-factor metrics in synthetic organic chemistry and the fine chemical and pharmaceutical industries, to speed the early evolution of "green chemistry." *Substantia*. https://www.semanticscholar.org/paper/b9e067007ddf240fd680e6e921ecfa9b112e5f9b (accessed on 24th October 2025).
34. N/A. (2023). Climate change 2022–Impacts, adaptation and vulnerability. https://doi.org/10.1017/9781009325844 (accessed on 24th October 2025).
35. Ncube, A., Mtetwa, S., Bukhari, M., Fiorentino, G., & Passaro, R. (2023). Circular economy and green chemistry: The need for radical innovative approaches in the design for new products. *Energies, 16*, 1752–1752. https://doi.org/10.3390/en16041752 (accessed on 24th October 2025).
36. Neumann, J., Gibson, R., & Madsen, T. (2008). Toxic baby furniture: The latest case for making products safe from the start. https://core.ac.uk/download/71355414.pdf (accessed on 24th October 2025).

37. Ogodo, U. P., & Abosede, O. (2025). The role of chemistry in achieving sustainable development goals: Green chemistry perspective. *International Research Journal of Pure and Applied Chemistry.* https://www.semanticscholar.org/paper/07af873696e6d8aad8a3273d34bfe3dd60458101 (accessed on 24th October 2025).

38. Patra, J. K., Das, G., Fraceto, L. F., Campos, E. V. R., Rodriguez-Torres, M. D. P., Acosta-Torres, L. S., Díaz-Torres, L. A., et al. (2018). Nano based drug delivery systems: Recent developments and future prospects. *Journal of Nanobiotechnology, 16.* https://doi.org/10.1186/s12951-018-0392-8 (accessed on 24th October 2025).

39. Proppe, J., & Reiher, M. (2018). Mechanism deduction from noisy chemical reaction networks. http://arxiv.org/abs/1803.09346 (accessed on 24th October 2025).

40. Saint Jean, M., & Oltra, V. (2025). Incrementalism of environmental innovations versus paradigmatic change: A comparative study of the automotive and chemical industries. https://core.ac.uk/download/pdf/6297038.pdf (accessed on 24th October 2025).

41. Sheldon, R. (2015). CHAPTER 2: Sustainability, green chemistry and white biotechnology. https://www.semanticscholar.org/paper/e1fa1d216d15d8eec8a7b1b1cc8839683dcf164c (accessed on 24th October 2025).

42. Sheldon, R. A. (2017). Metrics of green chemistry and sustainability: Past, present, and future. *ACS Sustainable Chemistry & Engineering, 6*, 32–48. https://doi.org/10.1021/acssuschemeng.7b03505 (accessed on 24th October 2025).

43. Sheldon, R. A., & Woodley, J. M. (2017). Role of biocatalysis in sustainable chemistry. *Chemical Reviews, 118*, 801–838. https://doi.org/10.1021/acs.chemrev.7b00203 (accessed on 24th October 2025).

44. Sholpan, K., Maral, K., Roza, M., & Assel, S. (2024). Formation of environmental literacy through the study of green chemistry. https://core.ac.uk/download/620316947.pdf (accessed on 24th October 2025).

45. Siganul, A. P., & Puspitasari, S. D. (2025). Challenges and opportunities in implementing green building materials in Malaysia. *Civil and Sustainable Urban Engineering*. https://www.semanticscholar.org/paper/905554d224b75c1440060f6b49d351845ba38af6 (accessed on 24th October 2025).

46. Singh, G., & Khullar, T. (2023). Green chemistry strategies for waste minimization: From waste to wealth. https://core.ac.uk/download/660318997.pdf (accessed on 24th October 2025).

47. Singh, G., Rajasekhar, E. S. K., Mounika, K., Tulasi, K. R. S. K.,

Dondapati, T., Himasaila, M., & Pulipati, S. (2024). Artificial intelligence in green organic chemistry: Pathway to sustainable and eco-friendly chemistry. *Asian Journal of Chemistry*. https://www.semanticscholar.org/paper/eb2daf33866f478f81177892f14dd72bd371e5c1 (accessed on 24th October 2025).

48. Staffell, I., Scamman, D., Velazquez Abad, A., Balcombe, P., Dodds, P. E., Ekins, P., Shah, N., et al. (2018). The role of hydrogen and fuel cells in the global energy system. *Energy & Environmental Science, 12*, 463–491. https://doi.org/10.1039/c8ee01157e (accessed on 24th October 2025).

49. Tian, Y., Zeng, G., Rutt, A., Shi, T., Kim, H., Wang, J., Koettgen, J., et al. (2020). Promises and challenges of next-generation "beyond Li-ion" batteries for electric vehicles and grid decarbonization. *Chemical Reviews, 121*, 1623–1669. https://doi.org/10.1021/acs.chemrev.0c00767 (accessed on 24th October 2025).

50. Tjoa, E., & Guan, C. (2020). A survey on explainable artificial intelligence (XAI): Toward medical XAI. *IEEE Transactions on Neural Networks and Learning Systems, 32*, 4793–4813. https://doi.org/10.1109/tnnls.2020.3027314 (accessed on 24th October 2025).

51. Wagner, W. E. (2014). Imagining corporate sustainability as a public good rather than a corporate bad. https://core.ac.uk/download/211333149.pdf (accessed on 24th October 2025).

52. Wiebe, A., Gieshoff, T., Möhle, S., Rodrigo, E., Zirbes, M., & Waldvogel, S. R. (2018). Electrifying organic synthesis. *Angewandte Chemie International Edition, 57*, 5594–5619. https://doi.org/10.1002/anie.201711060 (accessed on 24th October 2025).

53. Wu, S., Wang, Y., & Ren, H. (2025). Research on identification of factors influencing low carbon buildings in Hebei Province. *Scientific Journal of Economics and Management Research*. https://www.semanticscholar.org/paper/e75d14f5ab7aa67fcef473bfb14b3854e35fa6f7 (accessed on 24th October 2025).

54. Zhao, Y., Pohl, O., Bhatt, A. I., Collis, G. E., Mahon, P. J., Rüther, T., & Hollenkamp, A. F. (2021). A review on battery market trends, second-life reuse, and recycling. *Sustainable Chemistry, 2*, 167–205. https://doi.org/10.3390/suschem2010011 (accessed on 24th October 2025).

CHAPTER 2

PRINCIPLES OF SUSTAINABILITY IN CHEMISTRY

Contents

INTRODUCTION

The chemical principles of sustainability form a central intersection between science-based innovation and nature stewardship. The global environment confronts the challenge of sustainable action, and the chemical industry stands at a fork in the road. Challenges of resource limitation, climate change, and pollution have prompted the requirement to reconsider the design and operation of chemical processes. A recognition of these dilemmas has prompted professionals in the industry to pursue frameworks which integrate sustainability principles into mainstream chemical practice. This approach not only aims to minimize environmental impacts but also hopes to induce a paradigm shift in the learning and operational paradigms of the industry and pave the way towards a more sustainable future. Against this backdrop, there exists a paramount need for new strategies that balance the objectives of chemical manufacturing with sustainability parameters. Existing debate emphasizes the pivotal role of chemical engineers in the delivery of new solutions that enhance environmental integrity and serve industrial needs. Emphasizing the Triple Bottom Line approach, founded upon social, environmental, and economic imperatives, is essential in designing future chemical engineering curricula. As emphasized in recent literature, the chemical manufacturing industry has come under heightened pressures with respect to its sustainability strategies, necessitating a change in pedagogical curricula intended to guide future engineers to meet these challenges. The development of such curricular schemes is of utmost importance in imparting skills to future professionals that enable them to withstand the intricacies of sustainable chemistry (Allen et al., 2010). The quest for sustainability in chemistry is complemented by coming technologies, primarily nanotechnology, that have the potential to revolutionize energy solutions. With growing global energy requirements combined with depleting reserves of fossil fuels, new nanotechnologies offer multidimensional approaches towards sustainable energy systems. All these rely on novel scientific principles for energy transformation applications with the guarantee of safety and protection of the environment. The emphasis on sustainable materials and processes ensures that there is shared motivation towards better solutions that satisfy energy needs without compromising environmental integrity (Kelarakis et al., 2014). Through embracing this improvement, the chemical industry can play a major role in moving towards a sustainable future, reasserting its vital role in addressing some of the world's most important challenges.

Definition of Sustainability in the Context of Chemistry

The sustainability principle in chemistry is a complete approach that integrates conservation of the environment, economic viability, and social responsibility. At its core, chemistry sustainability entails creating chemical products and processes that guarantee lower environmental impact throughout the life cycle. This calls for creative solutions as well as a transformation of culture in how researchers and chemical engineers undertake their work. By redefining traditional practices to encompass sustainable principles, the field strives to reduce waste, conserve resources, and optimize the efficiency of chemical reactions. The change promotes the ethical responsibility of chemical engineering to contribute positively to society with reduced negative effects on ecosystems and human health, thereby shaping future directions in chemical research and education (Allen et al., 2010). In uncovering sustainability, one of the most important frameworks is the Circular Economy that aims to rethink production and consumption patterns in the best interest of resource efficiency and environmental health. The concept turns the traditional linear model around and targets rethinking material design, use, and regeneration in chemical processes.

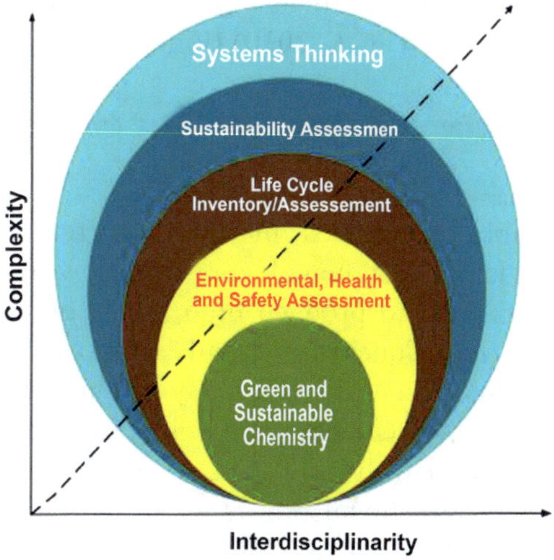

Figure 2.1. Sustainability in chemistry promotes the development of chemical processes and products that minimize environmental impact throughout their life cycle.

Source: https://ars.els-cdn.com/content/image/1-s2.0-S2589004221014607-fx1.jpg.

By advocating closed-loop systems that allow for the recycling of materials back into production, the Circular Economy offers the solution for a sustainable practice that avoids waste and environmental deterioration. While this model promises tremendous potential for advancing sustainable practice in the chemicals industry, it also raises significant questions about ethical issues and social equity in sustainable development. As stressed, the model should take into consideration broader social contexts to ensure that interventions for sustainability impact not only the environment but also the community's well-being (Haynes et al., 2015). Finally, in the chemical sense, sustainability is about innovation in process engineering and material science since it recognizes the necessity for solution-driven solutions. It can be initiated through insuring that the chemical sector assumes a proactive role in considering sustainability elements and possessing mechanisms for transparent reporting, particularly by means like the Triple Bottom Line approach addressing environmental, economic, and societal impacts. The integration of these sustainability principles is imperative not merely for regulatory compliance needs but also for establishing corporate culture in terms of long-term ecological health and social health. By incorporating sustainability into the underlying principles of chemistry instruction and professional practice, the discipline can lead the changes needed to make the production and application of chemicals more sustainable, addressing the dual challenge of a growing population and a finite terrestrial resource base (Allen et al., 2010) (Figure 2.1).

Importance of Sustainable Practices in Chemical Research and Industry

The use of sustainable techniques in chemical research is crucial for addressing the pressing environmental concerns now faced by the world. As industrial processes tend to produce high levels of waste and pollution, employing sustainable techniques can greatly reduce the adverse effect of chemical production on the environment. By placing greater emphasis on green chemistry practices, researchers and scientists can engineer processes that conserve the use of harmful substances and maximize the utilization of resources. Here, water treatment processes are an essential area in which sustainable practices can be adopted. Simulation-based methods, if utilized correctly, can enhance these processes so that they become efficient and less environmentally destructive. The role of simulation in identifying and evading potential environmental impacts underpins the urgent necessity of sustainable practice in chemical engineering and signifies the revolutionizing potential of creative imagination within the field (Obiuto et al., 2024). Of particular interest also is the application of flexible sensor technology in the chemical industry, which is an emerging trend towards sustainable innovation. Sensors also play a crucial role, as they allow real-time monitoring of chemical processes and therefore improve safety and efficiency. However, to enable the application of these technologies in the broadest sense possible, several impediments must be overcome, including optimization of function and development of compatible biocompatible interfaces. Environmental concerns for the disposal of sensor material as well as sustainability must also be considered, indicating the requirement for sustainable growth practices within this field. By tackling these issues from a sustainability point of view, scientists can design smart flexible sensors that cater to the evolving requirements of the industry while minimizing negative impacts on the environment (Luo et al., 2023). This emphasis on sustainability not only fuels innovation but also ensures that technology is utilized wisely in chemical science. The collective effort towards sustainable practice in the chemical industry brings significant developments in environmental protection and public health. Collaborative action across disciplines from chemical engineering to environmental science is needed to establish such success. Interdisciplinary solutions, including the integration of simulation-driven techniques for water management and advancing enhanced sensing technologies, can enhance the effectiveness of sustainable practice. Also, emphasizing non-technological issues such as regulatory matters and ethics promotes the sustainable development of the chemical industry. Promoting the full understanding of sustainability, the sector is enabled to implement responsible innovation with concern for human health while preserving the environment. Establishing these sustainability guidelines in chemistry not only benefits the current generation but also accounts for future generations to thrive in a greener, cleaner world (Figure 2.2).

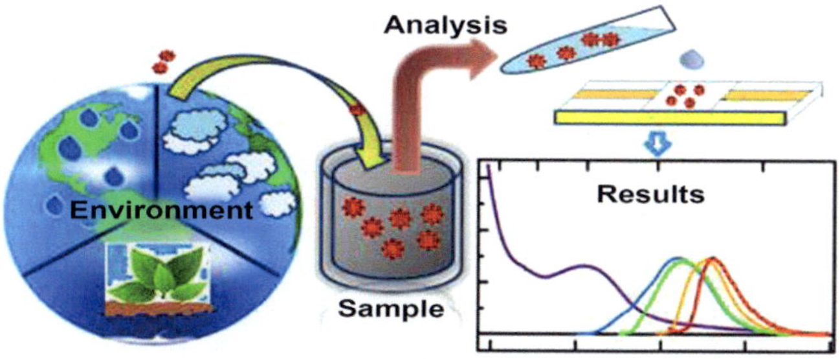

Figure 2.2. Sensors play a critical role by enabling real-time monitoring of chemical processes, thus improving safety and efficiency.

Source: https://ars.els-cdn.com/content/image/1-s2.0-S0165993622002345-ga1.jpg.

2.1. HISTORICAL CONTEXT OF SUSTAINABILITY IN CHEMISTRY

The history behind chemistry sustainability traces its roots back to early recognition of the environmental impacts associated with the use and manufacture of chemicals. The industrial revolution was a key landmark, as it catalyzed rapid advancements in chemical operations without much regard for environmental impacts. Such neglect was intensified by early environmental efforts, which focused on highlighting the adverse impacts of toxins on the ecosystem and human health. Since the late 20th century, the growing concern for the environment brought about a shift in paradigm among the chemistry community, espousing more environmentally friendly practices. Scientists started focusing on the importance of sustainable practices, highlighting the importance of chemical processes that reduce waste and energy usage. This created the foundation for what later would become the green chemistry principles, which aim to redesign chemical processes in a manner consistent with stewardship of the environment. During the late 20th century, the concept of sustainability within chemistry was becoming popular with overall environmental movements and legislation.

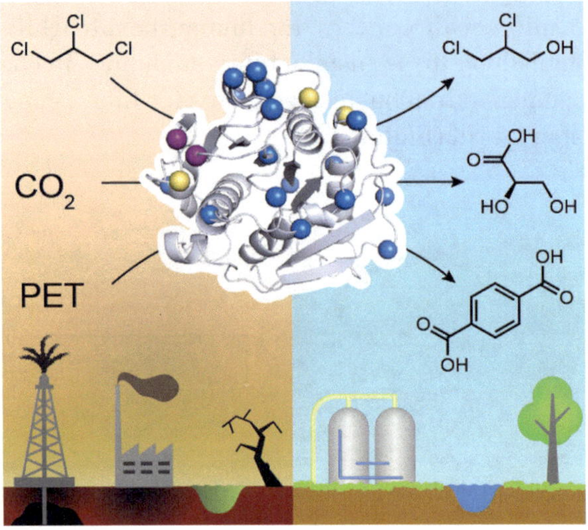

Figure 2.3. Emerging from the fields of molecular biology and biotechnology, biocatalysis leverages enzymatic reactions to facilitate chemical transformations with minimal environmental impact.

Source: https://onlinelibrary.wiley.com/cms/asset/54d420e6-e917-4b0b-a921-2393cf8c7a30/anie202309305-toc-0001-m.jpg.

Key milestones in history, such as the publication of Green Chemistry: Theory and Practice by Paul Anastas and John Warner in 1998, made sustainable methods official in the industry. The publication introduced twelve principles that provide guidance toward reducing harmful substances and enhancing efficiency in synthesis chemical processes. Also, increased awareness about global warming and resource depletion sped up the development of interest in the life-cycle footprint analysis of chemical products. Integration of sustainability into chemical research resulted in the development of superior technologies that had minimized environmental effects, leaving a paradigm shift from conventional practices. Shift towards sustainability not only transformed the face of chemistry but also had an instrumental role to place it alongside bigger social goals of safeguarding natural resources for generations to come. Biocatalysis's evolution embodies the path toward sustainable chemistry, closing the gap from earlier accomplishments to present-day applications. Spawned from the fields of molecular biology and biotechnology, biocatalysis leverages enzyme catalysis to power chemical reaction at little environmental cost. The past two decades have witnessed significant breakthroughs in protein engineering that facilitated designing enzymes for operation under mild conditions and minimal levels of byproducts. In this way, biocatalysis embodies the principles of green chemistry in that it offers processes that are sustainable by nature, especially in the pharmaceutical industry where manufacturing processes must be efficient and clean. The prospect of reuse of immobilized enzymes also has the added benefits of enhancing stability as well as commercial viability, which further promotes the use of such technology. Hence, biocatalysis not only indicates the historical trajectory of sustainability in chemistry but also signifies its influential role in developing more sustainable production methods in the economy of today (McKenney et al., 2018; Roger et al., 2017) (Figure 2.3).

2.1.1. Evolution of Chemical Practices Over the Decades

The evolution of chemical practice throughout the decades has been heavily influenced by the growing awareness of principles of sustainability, driving the transition towards greener methods from the traditional approach. The branch of chemistry first concentrated predominantly on maximizing the efficiencies of production and enhancing the value of the product at the costs of the environmental consequences. But with an increasing international awareness of environmental pollution, the phrase "green chemistry" was born, promoting methodologies that not only minimize dangerous chemicals but also minimize waste production. Applying the concepts of sustainability to chemical practices has thus initiated a paradigm shift based on the necessity for secure and effective chemical processes. This shift has led to an increased exploration of renewable resources and new materials that emphasize the issue of a circular economy, eventually transforming the manner in which chemists perform research and development in their respective areas. Throughout the latter part of the twentieth century and the early twenty-first century, there has been a distinct increase in research on sustainable management practices in chemistry, particularly in the area of treating metals and other hazardous chemicals. A bibliometric analysis reveals that globally, there was a meteoric rise in publications in this area between 1993 and 2017, with 63% of all the articles published over a period of five years before 2017. Furthermore, journals that led this discussion, such as Journal of Cleaner Production and ACS Sustainable Chemistry and

Engineering, reflect a specific emphasis on topics like decontamination of metal and waste management processes towards recycling and reuse. This new practice in research is a coming together of fields, and it is a testament to the need for cross-cutting practices in the chemicals industry that are responsive to environmental, economic, and social dimensions of sustainability, as evident in the collaborative networks instituted among leading nations in the field like China, the United States, and Germany. However, chemical practice change also represents the spirit behind the role of urban areas in shaping sustainability outcomes. Urban chemistry, as an emerging field, shows how chemical thinking can be applied to address the particular problems posed by urbanization, including management of water quality, waste treatment, and pollution control. As urban areas expand and advance, some of these pressure forces from human endeavors necessitate innovative chemical solutions that are adapted to these changing dynamics. For instance, historical urban drainage transformations highlight the evolving strategy to respond to stormwater and sewage, prioritizing a paradigm shift toward sustainable urban infrastructure. Recognizing water as a primary driver of urbanization, scholars prescribe flexible practices, which can transform cityscapes, to maintain ecological equilibrium and enhance the quality of life in urban agglomerations. This intersection of urban planning and chemistry shows the broader applications of sustainable principles in the discipline and, in the end, upholds the ideas of sustainability in chemistry as the foundation of progressive chemical practice over the course of the decades (Figure 2.4).

Figure 2.4. Urban chemistry, as an emergent field, demonstrates how chemical principles can be applied to address the unique challenges posed by urbanization, including water quality management, waste treatment, and pollution control.

Source: https://compton.chem.ox.ac.uk/lab/mainlab1.jpg.

2.1.2. Key Milestones in the Development of Green Chemistry

Throughout its history, the development of green chemistry has been punctuated with several milestone events that have collectively propelled the discipline in its direction today toward a sustainability framework. One such epochal moment was when Paul Anastas and John Warner articulated the Twelve Principles of Green Chemistry in 1998. These principles constitute a roadmap instrument that endeavors to reduce waste, conserve energy, and achieve non-toxicity in chemical reactions. The official sanctioning of these principles encouraged researchers and industry specialists alike to rethink conventional synthetic strategies, paving the way for innovation through cleaner technology. This groundbreaking endeavor emphasized the call for protecting the environment in chemical research, which becomes increasingly significant in the environmentally friendly world of the present times.

The incorporation of such principles works to align chemical practice with broader sustainability goals, a fundamental shift in the practice of chemistry both in educational and industrial environments. Another milestone was achieved during the early 21st century when regulatory systems were developed based on green chemistry principles. Such legislation as the Pollution Prevention Act of 1990 and numerous programs by the U.S. Environmental Protection Agency (EPA), promoted reduction of harmful chemicals and stimulated the adoption of sustainable practices across industries. Additionally, funding for green chemistry research through grants and public-private collaborations further stimulated innovation. Such infrastructure assistance resulted in innovations in numerous sectors, including green solvent design and biodegradable materials that significantly minimized environmental impacts. These regulatory advances not only legitimized greener alternatives but also motivated the scientific community to drive solutions proactively that successfully addressed environmental concerns and market needs, maintaining the harmonious balance of chemistry and sustainability. Another milestone in the history of green chemistry is the evolution of reticular chemistry, particularly in synthesizing new materials with reduced environmental footprint. With approaches such as Metal-Organic Framework (MOF) and Covalent Organic Framework (COF) designing, researchers have ushered in never-before-seen ways of synthesizing highly effective materials with tunable properties, basically pushing the frontier of green chemistry applications. Such frameworks possess unrivaled surface areas and discriminate adsorption, making them ideal for application in gas storage, catalysis, and pollution cleanup. The advancement in reticular chemistry not only enhances the efficiency of chemical processes but also promotes the values of sustainability in waste reduction and energy consumption. This intersection of fields engages the potential that is always there for chemistry to evolve and advance, in the long-term promoting even more sustainable practices that are the basis for our activities in solving environmental issues and a sustainable future for the chemical sciences (Calvin et al., 2023; Freund et al., 2021).

2.1.3. Impact of Environmental Movements on Chemical Regulations

The rise of green movements has had a substantial impact on chemical regulations, which have prompted reconsideration of the production, sale, and use of chemicals. Pressure groups have played an essential role in putting people in focus with the harmful impact of chemicals and demanding stronger controls that will minimize environmental and health damage. For instance, the push towards legislations such as the Toxic Substances Control Act (TSCA) in the US illustrates growing public alarm regarding chemicals' safety. Such campaigns have mobilized the public and policymakers to make demands for open practice and alternatives to hazardous chemicals, which has a profound effect on regulation systems. As such, increasing regulation of chemical agencies and agency practice demonstrates a more holistic convergence of environmental issues and regulatory action, resulting in a safer, cleaner environment in harmony with the principles of sustainability in chemistry. Environmental movements beyond consciousness-raising have also promoted integration among scientists, commercial interests, and regulatory bodies to facilitate sustainable tendencies. Integration is vital in defining safer chemical substitutes and fresh solutions. The integration of inter-disciplinary studies has led to breakthroughs in areas such as biomimicry and green chemistry, where the efforts revolve around making chemical processes greener. Moreover, initiatives led by these movements have gravitated towards the development of guidelines that push for the usage of non-toxic materials in products, especially in industries such as cosmetics and agriculture. By giving a priority to green practices, activists have not just guided regulation policies but also ignited a cultural transition towards sustainability, which forces industries to seek out chemical innovations that reduce ecological footprints while safeguarding public health. Environmental movements have also increased discussions around the ethical consequences of chemical use in industry and society. The application of sustainable practices calls for an integrated approach that responds to the long-term consequences of chemical production and its consequence on the environment. As highlighted in the discussion on flexible sensors and their challenge with commercialization, the understanding of the broader context of chemical regulation requires recognizing aspects such as compatibility with biological systems and environmental concerns related to waste management. These are necessary not only in terms of regulation but also to create public trust in science and technology. With increasing wider-ranging regulations being driven by movements, these emphasize the necessity of incorporating ethical, regulatory, and scientific inputs into the construction of sustainable chemical strategies that consider both human health as well as the health of the environment, leading society eventually toward a more sustainable direction (Malik et al., 2023; Luo et al., 2023) (Figure 2.5).

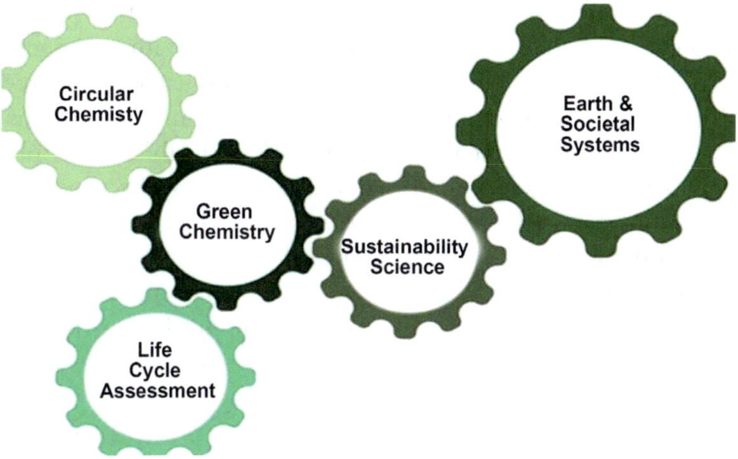

Figure 2.5. The integration of interdisciplinary research has led to advancements in areas such as biomimicry and green chemistry, which prioritize sustainability in chemical processes.

Source: https://pubs.acs.org/cms/10.1021/acs.jchemed.9b00390/asset/images/medium/ed9b00390_0007.gif.

2.2. PRINCIPLES OF SUSTAINABILITY IN CHEMISTRY

The green chemistry principles constitute a model that guides the endeavor towards sustainable operation in chemistry. The principles encourage waste prevention at the outset, demanding cautious design of chemical reactions with fewer byproducts. With the addition of efficiency in chemical synthesis, green chemistry aims to heighten safety for human and environmental well-being. The emphasis on the use of renewable resources is also an essential aspect that forces chemists to employ biologically sourced materials instead of draining fossil resources. The synergy between chemical production and sustainable methods is thus central in balancing society's requirement for chemical products with nature's requirement to conserve natural resources. One of the best examples of the green chemistry principles is the design of biodegradable materials, which itself is waste minimization and resource recovery. In the development of new chemical processes with renewable resources, such as plant-based polymerics, researchers are making alternatives with much less environmental impact. For example, green chemistry promotes the use of catalytic processes so that waste can be converted into useful materials. Recycling and reuse of chemical by-products aid in compatibility with sustainability goals and shows how green chemistry can be applied to obtain a circular economy. With such technologies, the employment of biodegradable solutions in general use not only reduces the burden of long-term pollutants but also instills environmental responsibility in industrial processes (Sanghratna, 2025). In addition to this, the application of green chemistry is even wider than material science as observed through the creation of environmentally friendly consumer products. In recent research, creating a natural mosquito repellent is an easy reference to green principles in practice, with essential oil extracts as the working components. The operations involved, e.g., cold pressing and steam distillation, are low-carbon operations that seek to keep emissions to a minimum and sustainability at a maximum. Not only does this project offer a safer alternative for consumers, but it also cuts down on chemical waste typically produced by the production of traditional products. This study demonstrates the potential of green chemistry to create new products that are capable of addressing health and environmental issues, thus underscoring the significance of responsible development in the manufacturing of everyday commodities (Choudhury, 2025).

2.2.1. Waste Prevention and Minimization

The place of waste prevention and minimization in the field of chemistry cannot be overstated as these dictates mirror the very spirit of sustainability. By centering on reducing waste at the origin, chemists are provided with a stimulus to redesign processes and materials linked to their activities. Such a progressive practice not only conserves resources but also prevents pollution and the overall environmental burden of chemical manufacture. Methods such as green chemistry principles promote the use of harmless solvents and conditions, hence minimizing waste and adopting a sustainable model. With these principles, scientists have the capability of finding opportunities in their methods where by-products can be minimized, leading to a sustainable means that also suits industry and the environment. Industrial symbiosis is an alluring concept for avoidance of wastes by resource sharing between companies. The idea encourages companies to work in partnership and exchange materials, energy, and by-products as a means of maximizing resources. Through physical proximity and synergies, firms can minimize waste while at the same time minimizing input expenses. For instance, effluent from one manufacturing process can act as the input for another, thus enabling successful recycling and reuse of resources (Marca et al., 2015). This system approach not only minimizes the release of toxic emissions but also enables higher sustainability in industrial processes, showcasing how interdependence can lead to improved resource efficiency. In addition, the principles of resource productivity provide a structured roadmap for companies to successfully utilize in formulating waste minimization strategies. The principles cause companies to analyze critically their activities and identify areas of utmost importance to address. By responding to questions related to the use of resources and by-product management, companies can formulate a holistic sustainability program dependent on their individual operations (Harvey et al., 2001). Such a system will lead firms in prioritizing priorities with waste prevention activities, process redesign, and eventually greater productivity. As firms embrace the waste minimization philosophy, they not only become participants in environmental stewardship initiatives but also become economically more sustainable, thereby contributing to the establishment of a sustainability culture that will be required for future advancements in the chemistry field (Figure 2.6).

Figure 2.6. The importance of waste prevention and minimization in the field of chemistry cannot be overstated, as these practices embody the core principles of sustainability.

Source: https://media.springernature.com/lw1200/springer-static/image/art%3A10.1007%2Fs44274-024-00033-5/MediaObjects/44274_2024_33_Fig1_HTML.png.

2.2.2. Use of Renewable Feedstocks

The transition from renewable feedstocks is one of the driving maxims of sustainability in the world of chemistry, with it bringing into focus the transition away from depletable fossil fuels to durable biological compounds. By employing biomass from plants, crop by-products, and waste materials, the chemistry sector will be in a position to reduce the discharge of greenhouse gases and maintain its independence from petroleum-based derivatives. The increased use of renewable feedstocks is in accordance with global sustainability goals, as it fosters a circular economy based on renewing resources and limiting waste. Advances in biorefinery technologies will play a leading role in facilitating this change, facilitating the recovery of high-value chemicals alongside contributing to environmental sustainability and social-economic benefit. These innovations are central to the potential of bio-based economies as envisioned in programs like those highlighted in the Roadmap for Bio-refinery development in the Netherlands (Annevelink et al., 2009).

The incorporation of renewable feedstocks into chemical manufacturing processes not only plays the role of environmental solution but also augments the economic viability of the sector even more. The utilization of sustainable resources can lead to better energy efficiency and decreased costs of production, primarily by valorizing waste matter. Furthermore, bio-based products are less toxic and environmentally persistent and therefore can reduce pollution. This paradigm requires a shift away from traditional engineering strategies in the chemical industry to allow a new paradigm where sustainability is the driver of design and operational decisions. Consequences of this realignment extend beyond the chemical engineering curriculum into models of education, necessitating curricula for chemical engineering to embrace its relevance and promote a sustainability culture among new professionals (Allen et al., 2010). Successful adoption of renewable feedstocks will necessitate an interdisciplinary approach that mingles chemistry, biology, and engineering expertise. Interdisciplinary integration among the disciplines will be essential to surmounting feedstock supply issues, processing effectiveness, and product competitiveness. Industry stakeholders and researchers need to focus on scaling up bioprocessing technologies while enhancing the economic feasibility of bio-based production systems. Furthermore, the application of open book sustainability reporting methods, such as the Triple Bottom Line approach, will help ensure that the impacts of these innovations are properly assessed and communicated. Lastly, the proactive use of renewable feedstocks is an innovative approach that not only addresses current environmental concerns but also facilitates long-term sustainable innovation in the chemical industry (Allen et al., 2010) (Figure 2.7).

Figure 2.7. The transition to renewable feedstocks embodies one of the core principles of sustainability in the field of chemistry, highlighting the shift from finite fossil resources to sustainable biological materials.

Source: https://www.reagent.co.uk/wp-content/uploads/2022/07/biogas-plant-behind-a-rape-field.jpg.webp.

2.2.3. Energy Efficiency in Chemical Processes

Energy efficiency in chemical processes is an important field to achieve sustainability in the discipline of chemistry as a whole. The principle behind it is minimizing energy use and maximizing output to decrease the environmental footprint of chemical manufacture. Process optimization is an important part of this endeavor where innovative approach methodologies such as the integration of artificial intelligence (AI) and traditional chemical engineering become the standard. AI can be utilized to assist real-time monitoring and optimization of chemical processes, effectively identifying wasted energy and improving productivity. Through AI-based optimization models, industries can more closely meet sustainable processes that not only conserve energy but also reduce waste, thereby setting the best standards for sustainability in chemistry (Taiwo et al., 2025). The role of fluid mechanics in enhancing energy efficiency cannot be overemphasized, particularly in chemical engineering applications. Fluid behavior governs many aspects of operation such as mixing, transport phenomena, and heat transfer, which are fundamental to maximizing chemical processes. Understanding fluids' properties and flow regimes enables engineers to design systems that minimize energy loss during operations. Advanced computation techniques like computational fluid dynamics (CFD) allow for precise modeling of fluid flow, which can lead to innovation in process design and operation. By focusing on momentum transfer principles and non-Newtonian fluid mechanics, engineers can improve processes, ultimately resulting in improved energy efficiency and promoting sustainable chemical production methods (Nandiyanto et al., 2025). Incorporation of the most recent technological developments also improves energy efficiency in chemical processes. The hybridization of hybrid AI-CFD methods is the transformative strength of the combination of artificial intelligence with sophisticated fluid simulation methods. These fusions can streamline simulations, enabling improved fluid flow and energy consumption predictions in chemical processing. Moreover, such breakthroughs provide reduced operating expenses and promote a greener approach to producing chemicals. As firms increasingly adopt these new ways, they embrace the culture of sustainability inherent in chemical engineering. This transformation not only provides them with a competitive advantage in the market but also addresses the urgent need for green practices in chemical production (Taiwo et al., 2025; Nandiyanto et al., 2025) (Figure 2.8).

Figure 2.8. Energy efficiency in chemical processes represents a pivotal area for enhancing sustainability within the broader field of chemistry.

Source: https://media.licdn.com/dms/image/v2/D5612AQEqoUuULpXDnQ/article-cover_image-shrink_600_2000/article-cover_image-shrink_600_2000/0/1721193876207?e=2147483647&v=beta&t=lPHF2_OA1dNPkSCUzQSmVBoJbEHHw8BKw9b4S6Fo8hs.

2.3. ROLE OF CHEMISTRY IN SUSTAINABLE MATERIALS

The role of chemistry in creating sustainable materials is critical in resolving the environmental problems facing us today. With the use of chemical principles, researchers create materials that are not just functional but also environmentally friendly. For instance, biodegradable polymers were synthesized much more effectively, allowing for the creation of materials that will naturally degrade and reduce plastic waste. Secondly, advances in material chemistry involve creating sustainable composites through the use of agricultural wastes to cut back on fossil fuel dependence. Such advances subsequently represent where chemistry intersects with sustainability because they directly tie into minimizing environmental footprints without compromising consumer demands for performing materials. An understanding of chemical processes is thus crucial to push materials towards sustainable development targets. Moreover, the chemistry of sustainable materials equates to efficient energy processing during product and manufacturing life cycles. Green chemistry practices promote practices that minimize hazardous materials and maximize resource efficiency in the processes of manufacturing. For example, using environmentally friendly solvents and catalysts can significantly lower energy requirements and waste for chemical reactions. This is part of overall societal objectives for sustainability through limiting detrimental emissions and the advancement of cleaner technologies. With more industries focusing on sustainable processes, chemists are challenged to not just create new materials but also simplify processes to minimize damage to the environment. The ethical responsibility of chemists towards embracing sustainability is underscored by the persistent skepticism about the intentions of corporate interests towards environmental sustainability, as seen in cases of green-washing (Wagner et al., 2014). No less necessary is global collaboration in chemistry that facilitates knowledge and resource sharing to enhance sustainable processes. While along with material science breakthroughs, so too comes the need for global access to technological advances in various regions of the globe. Implementing mentorship programs that bridge the divide between the developed and developing world has the potential to scale up research in sustainable materials towards a more inclusive scientific clan. Scaling up mentorship around the world, as the existing debates call for, emphasizes the imperative of building talent worldwide in addressing issues of global environmental interest (Cordova et al., 2015). Such an interdisciplinary strategy not only supports chemistry innovation but also addresses an interdependence of scientists to further sustainability. The transition towards sustainable materials is thus an interdisciplinary collective effort that is strongly reliant on shared knowledge and ethical science practice.

2.3.1. Development of Biodegradable Materials

The production of biodegradable products is a turning point of chemistry and sustainability that addresses one of the most pressing concerns of society: plastic waste. Traditional plastics derived from fossil fuel last for centuries to degrade, contaminate the environment, and lead to the death of animals. Biodegradable ones, which are designed to be broken down by bacteria into inert byproducts, cause much less damage to the environment. To make them more practical, the chemists are examining various polymer structures that guarantee biodegradability without compromising ideal mechanical properties. The clever application of renewable resources and creative synthesis methods allow for the manufacture of materials that not only serve functional purposes but also satisfy the requirements of sustainability in chemistry through ensuring minimal reliance on non-renewable resources as well as waste minimization. In addition, sustainability of biodegradable products is very interrelated with the product lifecycle. The entire life cycle from feedstock selection to the end-of-life disposal must be taken into account in a holistic way. For instance, integration of self-healing capability in biodegradable polymers can improve their durability and roundabout influence on the environment by reducing the need for new materials over time.

Figure 2.9. Traditional plastics, derived from fossil fuels, persist in the environment for hundreds of years, contributing to pollution and harming wildlife.

Source: https://media.assettype.com/indiawaterportal%2Fimport%2Fsites%2Fdefault%2Ffiles%2Fstyles%2Fimage_1200x675%2Fpublic%2F2023-05%2Fclimate-change-recycling-plastic-waste-wallpaper-preview%20(1).jpg.

By extending the service lifetime and enhancing the recyclability of the materials used, chemists are able to address a number of sustainability challenges simultaneously. To that end, there needs to be a thorough evaluation of the performance of the materials under various environmental conditions.

This ensures that the materials continue to perform while still remaining capable of biodegrading in suitable circumstances, as discussed in the context of sustainable development. The synergy of material attributes and biodegradation is a key driver for the successful design of future biodegradable materials (Kenneth et al., 2024). Finally, White Analytical Chemistry (WAC) principles form the basis upon which the sustainability of biodegradable materials can be quantified in terms of a balance between environmental impact and function. As WAC highlights, effective sustainable methods do not value greenness above performance, but instead uphold a more inclusive evaluation of analytical and functional aspects. This is critical in the development of biodegradable materials as it strongly focuses on environmental benefits and material performance. By adhering to the twelve principles of WAC, scientists can develop biodegradable materials that not only degrade in a controlled process but also meet the stringent demands of numerous applications. Hence, the balancing act between chemistry principles and sustainable methods endorses growth and encourages material development that contributes favorably to environmental management and plastic waste elimination, presenting the overall goals in the sustainability sector (Paweł et al., 2021) (Figure 2.9).

2.3.2. Innovations in Recycling and Upcycling Processes

Upcycling and recycling technologies play a significant role in the creation of sustainability in the science of chemistry. The rate of waste generated by mankind's activities continues to increase, and therefore it is crucial that there are efficient recycling processes that do not cause much damage to the environment. There have been new technologies that have come up that not only enhance the efficiency of the existing recycling operations but also cause upcycling, which turns the waste into a product of higher value. For instance, chemical recycling processes have been developed to break down plastics into their original monomers, facilitating the production of new, high-quality plastics. This groundbreaking fix addresses the ancient issue of plastic pollution, and in the process, recirculates materials back into the production cycle, facilitating a circular economy. The existing research aims to develop biodegradable alternatives to non-recyclable plastics, thereby alleviating the consumption of fossil fuel and ensuring a sustainable use of resources (Hofmann et al., 2023). In conjunction with chemical advancements, the integration of sensor technologies has been crucial in improving the recycling and upcycling of materials. Flexible sensors have emerged as a method to measure and regulate waste streams effectively through the provision of real-time data acquisition and analysis. Sensors can be integrated into recycling plants to track material flow and identify contaminants, hence improving the overall efficiency of operations.

Moreover, breakthroughs in sensor-biology interfaces present new possibilities for waste separation and management, ultimately resulting in more efficient recycling. Such technological innovations not only increase more sustainable waste management but also have an integral role in meeting the increasing demand for resource efficiency in today's digitally connected world. There is a need to tackle challenges in commercializing these technologies since it opens up room for more sustainable methods in various industries (Luo et al., 2023). With the global struggle to deal with waste and environmental degradation, innovative recycling and upcycling methods offer the change we need. Placing sustainability at the forefront of chemical processes enables us to create a frugal culture that extends beyond traditional waste management. This needs the input of scientists, industries, and policymakers to outline standards for creating and introducing such innovations. In addition, the societal implications of these technologies cannot be undervalued; education and public engagement are key to promoting participation in recycling efforts and encouraging environmental stewardship. At the end of the day, recycling and upcycling technologies not only provide sustainable material use but also enhance economic viability and social responsibility, combining the foundations of sustainability in chemistry with community goals (Figure 2.10).

Figure 2.10. Flexible sensors have emerged as a means to monitor and manage waste streams effectively, allowing for real-time data collection and analysis.

Source: https://ars.els-cdn.com/content/image/1-s2.0-S2949750724000385-gr4.jpg.

2.3.3. Sustainable Alternatives to Traditional Plastics

The requirement to fight environmental degradation due to plastic pollution has fueled the quest for renewable and sustainable alternatives to traditional plastics. Traditional plastics, which are predominately petrochemical-based, release copious amounts of greenhouse gases and trigger ecological trauma. Bioplastics, derived from renewable biomass such as starch, polylactic acid, and lignocellulosic biomass, represent a viable pathway to sustainability. The recent advances in biorefineries target the utilization of crop waste and the transformation of lignocellulosic biomass into various bioproducts ranging from biofuels to bioplastics. The literature indicates that all these advances have a dual benefit: in addition to having reduced fossil resource dependency, they also turn wastes into new products and thus promote a circular economy for the chemical sector (Mujtaba et al., 2023). Such alternatives' strategic innovation is essential to minimize the ecological impact of conventional plastic production and consumption. In the perspective of sustainability, the development of flexible, biodegradable materials is a breakthrough in the destruction of the bounds of conventional plastics. Flexible sensors and other high-performance materials, though meant to be used in high-performance applications, also need to address environmental challenges by making a change towards biodegradable alternatives. These kinds of materials can minimize the plastic waste burden while providing essential functions for modern technology. Research says that transition towards green sensor solutions not only fulfills technical demands but also aligns with greater environmental goals (Luo et al., 2023). As the technology on sustainable materials continues to improve, the challenge is to maximize their performance capabilities as biodegradable alternatives so they are acceptable for mass market penetration and spark industry standard changes and consumer behavior. Change from unsustainable to sustainable alternatives ultimately requires interdisciplinarity involving scientific research, regulatory backing, and consumer awareness. The manufacture of bioplastics using renewable raw materials must be followed up with efforts to enlighten stakeholders about the benefits of such products. For instance, sensitization on the properties and applications of bioplastics may encourage a higher degree of consumption as well as industry acceptance. In addition, addressing nontechnical concerns such as business viability and regulatory issues is paramount to the commercial success of these environmentally friendly products. Researching collaborative research practices among disciplines will enhance the effectiveness of research in the field and enable implementation of sustainable practices in the chemical industry. The current debate on sustainability in chemistry underscores the value of interdisciplinary collaboration towards a sustainable future, and hence the diffusion of sustainable alternatives to traditional plastics.

2.4. CHEMICAL SAFETY AND ENVIRONMENTAL IMPACT

The integration of chemical safety and environmental impacts is increasingly essential to contemporary chemistry, particularly as the world tries to cope with the consequences of industrial processes on the environment. The principles of sustainability demand that chemists do the safety of their products and processes keeping in mind, minimizing toxicological risks to human health and the environment. In practice, this involves the utilization of safer solvents, the reduction of toxic chemicals, and the control of waste management practices. Through such endeavors, chemists are able to minimize the environmental footprint of chemical synthesis and usage significantly. In addition, sensor technology advancements are crucial as they can facilitate real-time monitoring of toxic substances and contaminants of the environment, further enhancing chemical safety features in various industrial procedures. The integration of these technologies must address not just the technical concerns but also the environmental concerns, reinforcing the call for a systematic approach to chemical safety within the context of sustainability (Luo et al., 2023). The second major advantage of sustainable chemistry is the necessity for creating metrics to accurately measure the environmental impact of chemical processes. One of the new approaches proposes employing the blue applicability grade index (BAGI), a new analytical tool for quantifying the usefulness of analytical methods. Metrics like BAGI allow chemists to quantify not only the efficiency of a chemical reaction but also its safety and sustainability, which will guide decisions in chemical method design and application. By implementing such practices, scientists are able to gain improved comprehension of trade-offs in selecting chemicals and techniques. Such adherence to sustainability principles not only encourages corporate responsibility but also facilitates the development of greener alternatives, and in turn reduces reliance on harmful substances. Ultimately, embracing quantifiable and physical performance measurements is an essential step towards the attainment of a more sustainable chemical industry that optimizes security and environmental stewardship (Manousi et al., 2023). Beyond technical issues, difficulties in implementing sustainable procedures within chemistry also encompass regulatory and moral aspects. With the advancement of new technologies such as intelligent flexible sensors, there are significant barriers to commercialization of such innovations in real application. Issues of business viability, compliance with environmental regulations, and ethical issues of deploying sensors must be properly addressed. For instance, ensuring safe production of sensor-biology interfaces with minimum environmental disruption is essential in guaranteeing responsible innovation (Luo et al., 2023). Thus, interdisciplinary collaboration is essential to navigate these challenges because policymakers, engineers, and chemists must work together to establish guidelines for safe development. By realizing the broader implications of their actions and encouraging innovation with the needs of society, chemists can contribute to developing an integrated framework of safety and sustainability that serves human beings and the environment (Figure 2.11).

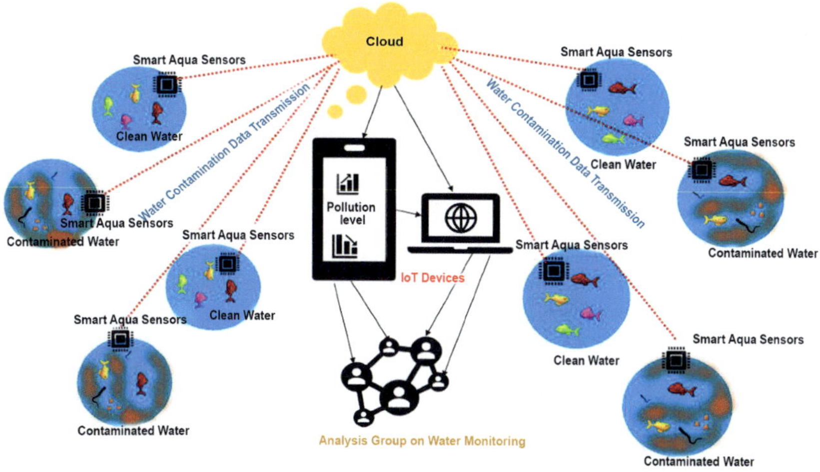

Figure 2.11. Advancements in sensor technology are essential, as they can facilitate real-time monitoring of hazardous materials and environmental pollutants, enhancing chemical safety protocols in various industrial applications.

Source: https://www.mdpi.com/sensors/sensors-20-03113/article_deploy/html/images/sensors-20-03113-g001.png.

2.4.1. Assessment of Chemical Hazards and Risks

The assessment of chemical hazards and risks is an integral part of encouraging sustainability in chemistry as a science. Systematically identifying and evaluating the risks to human health and the environment that can be posed by chemicals used in various products prevents chemical scientists from adverse effects. For instance, a report by the Campaign for Healthier Solutions reveals shocking findings of products available in dollar stores; it had over 81% of products tested for hazardous chemicals at concentrations above what they should be found to contain them. This finding emphasizes the necessity of intensive chemical hazard tests, particularly on consumer goods that fly under the radar. The inclusion of large-scale evaluation can lead to more effective regulations and guidelines that protect vulnerable groups, thus balancing product safety with sustainable chemistry principles. Secondly, the procedures applied in chemical risk assessment have become more advanced to incorporate quantitative and qualitative approaches for evaluating potential effects of chemical exposure on the public. The building of firm structures, such as applied in new industries like aquaculture, highlights the importance of cross-disciplinary collaboration. As an example, in the tilapia sector, professionals employed a trisectoral structure, integrating data from the public's health, technical needs, and marketing to quantify the potential impact on consumers and the environment.

Figure 2.12. Engaging stakeholders, including consumers, business owners, and regulatory bodies, fosters a culture of safety and informed decision-making.

Source: https://cdn-ileafpk.nitrocdn.com/ZaBJDsgdcwKtdTgMXICGMSQdP-tHvUweX/assets/images/optimized/simplystakeholders.com/wp-content/uploads/2020/08/SSTK-Blog-Post-Covers-20.png.

These integrated assessments offer improved understanding of occupational and environmental hazards resulting from chemical and biological inputs employed in agriculture. This integrated assessment is a classic example of how holistic risk assessment not only maintains product safety but also guarantees sustainable methods in the chemical industry. In achieving sustainability, the most important factor should be the transparency of chemical hazard assessment and growing awareness among the public in relation to the threats represented by chemical exposure. Engagement of stakeholders like consumers, business leaders, and regulatory agencies fosters a safety culture and informed decision-making. Appropriate use of public health impact assessment methods, as demonstrated in aquaculture research, can serve as a model for other sectors working with toxic chemicals. Sectors can collaborate to minimize health risk and environmental impact by exchanging information and stimulating discourse. Finally, the creation of an authoritative and visionary framework for assessing chemical hazards is critical to progress toward sustainability in chemistry and creating safer places for today's and tomorrow's generations (Figure 2.12).

2.4.2. Strategies for Reducing Toxic Emissions

Reduction of toxic emissions in chemistry is crucial in order to make chemistry sustainable and also protect human health as well as the environment. Efficacious methods to achieve this goal involve a total method that encompasses resource efficiency in its use and waste minimization. Laboratories, in particular, are big contributors of toxic emissions, and thus strict waste management has to be adopted. For example, in a national shellfish biotoxin monitoring program, innovative strategies were employed to become more resource-efficient that resulted in 69% reduction of one-use plastics by using sustainable substitutes such as paperboard and glass. This is the greatest example of how strategically changes can be made to facilitate a more sustainable practice of chemistry while resulting in substantial reductions in emissions of hazardous materials and plastics (Kilcoyne et al., 2022).

The incorporation of the principles of green analytical chemistry is a vital strategy in minimizing toxic emissions during sample preparation. By prioritizing the use of renewable, recycled, and reusable materials, the laboratory not only reduces the amount of wastes generated but also minimizes the energy requirements of their operations. The application of these principles in terms of using safe solvents and reagents leads to enormous reduction of toxic chemical waste. For instance, one lab was able to reduce its chemical waste by some 23% by undertaking quality evaluation of procurement processes and adopting more environmentally friendly solutions. Furthermore, the use of green metrics to assess the sustainability of sample preparation protocols allows researchers to continually refine their practice, which is consistent with overall sustainability goals (Ángela et al., 2022). Lastly, instilling a culture of sustainability in chemistry entails continuous improvement and innovation. Practices such as switch off campaigns and optimizing equipment management can also cut energy consumption in laboratories. A good example is the efforts towards managing fume hood and cold storage equipment that resulted in a whopping 30% cutback of energy usage. This not only reduces toxic emissions but also achieves long-term cost reductions for laboratories. By implementing these multi-faceted strategies, the scientific community can effectively reduce toxic emissions, demonstrating that it is possible to practice sustainability in chemistry through collective action and unwavering adherence to green practices (Kilcoyne et al., 2022).

2.4.3. Importance of Life Cycle Analysis in Chemical Products

Life cycle assessment (LCA) is a key method of evaluating the environmental performance of chemical products, and a necessary tool of sustainable chemistry. Through analysis of the entire life cycle of a product—from extraction of raw materials, manufacture, and consumption to disposal—LCA provides an overall picture of the resource consumption and emissions associated with these chemical processes. This decomposition enables producers and researchers to visualize where efficiencies may be made, hence cutting waste and optimizing returns. In doing so, LCA encourages more sustainable chemistry because it reveals hidden environmental costs that might not otherwise become evident when passing through the discrete production stages. Through systematic evaluation, stakeholders can prioritize strategies that optimize both economic feasibility and environmental responsibility, thereby, ultimately contributing to long-term sustainability in product design and chemical manufacturing. The role of LCA extends beyond compliance requirements and addresses worldwide imperatives such as climate change and resource depletion. As the building sector accounts for a significant percentage of CO_2 emissions worldwide, material design innovations that are stimulated by LCA can significantly influence minimizing such impacts. For instance, through life cycle assessment of building materials, such as engineered wood products, researchers can determine the scope for innovation and functionality in conformity with the demands of sustainability. Through chemical modification of conventional materials, it becomes feasible to create building elements that, aside from using fewer resources, can support carbon-negative building design. Such strategic consideration follows the ideals of chemistry-based sustainability by looking at the convergence between environmental integrity and product lifecycles in an effort to improve the quality of life while protecting natural ecosystems. LCA is also a bridge too between scientific data and actual use in the chemical sector, guiding decisions towards a more sustainable direction. LCA is increasingly utilized by corporations to decide whether or not their products are sustainable, including analyzing supply chains, manufacturing processes, and wastage. This combined process brings greater transparency of environmental impact, allowing consumers to make more informed decisions. Furthermore, the use of LCA during product development can result in competitive alternatives to less sustainable products. With the use of information obtained from life cycle analysis, chemical manufacturers can be sustainability leaders, thus driving industry norms and customer practices. This evolution reflects the importance of LCA as not only a regulatory requirement but as a pillar of visionary strategies that guarantee long-term sustainability in the chemical industry, ultimately resulting in a healthier planet for generations to come (Figure 2.13).

Figure 2.13. Life cycle analysis (LCA) serves as a critical tool in evaluating the environmental impacts of chemical products, playing an essential role in sustainable chemistry.

Source: https://library.fiveable.me/_next/image?url=https%3A%2F%2Fstorage.googleapis.com%2Fstatic.prod.fiveable.me%2Fsearch-images%252F%2522Key_concepts_of_life_cycle_assessment_LCA_stages_environmental_impacts_chemical_engineering_sustainability_analysis%2522-fenrg-08-00015-g001.jpg&w=640&q=75.

2.5. EDUCATION AND AWARENESS IN SUSTAINABLE CHEMISTRY

One of the most important elements in progressing towards green chemistry is promotion of education and awareness among future chemists and teachers. Effective education in this field not only provides students with fundamental chemical information but also sound awareness of the environmental and moral implications of chemical processes. With green chemistry, students will need to learn about synthesizing chemical processes that produce little dangerous waste and reduce waste to a minimum. This instructional focus reflects the urgent demand for growing a generation of green chemical experts who can address issues in the environment using green and sustainable methods. As noted, one viable strategy is integrating green chemistry principles into curriculum, thereby enhancing students consciousness and subsequent application of sustainability principles in their professional practice (Jhun et al., 2024). It is necessary to have instructors well learned in these subjects for the sake of improving sustainability on all levels of education. The integration of sustainable chemistry education challenges existing schooling systems, particularly where awareness of green actions is still limited. For instance, a research study investigating Senior High School science educators' insights in the Philippines highlights a pressing need to increase green chemistry principles awareness among educators (Jhun et al., 2024). Such lack of familiarity may be detrimental to teaching principles of sustainability effectively, highlighting the imperative for professional training programs that prepare teachers to be able to communicate such principles confidently. By increasing teacher training with sustainable chemistry methods and pedagogical approaches, schools can instill innovation-think in potential chemists. With increased teacher expertise, there is potential to not only enhance student comprehension but to encourage a long-term commitment to sustainability across discipline, encouraging the understanding that education has an active role in ensuring change to societal attitudes toward chemistry and care for the environment. In order to further enhance the scholarly environment of sustainable chemistry, it is also important to place priority on experiential learning and practical uses in the curriculum of the course. Providing students with access to experiential experiments to shed light on the principles of greener methodologies familiarizes the students with the subject. Investigating the ten green principles of sample preparation, for example, can provide students with an insight into safe solvents, renewable materials, and waste reduction strategies and thus develop their practical skills. Also, being able to apply green metrics to estimate the greenness of analytical methodologies enables students to evaluate and optimize their work critically. Through integration with practice, education courses can establish a learning environment where students are effective change agents in chemistry. This experiential approach not only demystifies complex sustainability challenges but also equips students with the ability to develop innovative sustainable solutions in their subsequent professional lives.

2.5.1. Incorporating Sustainability Into Chemistry Curricula

Embedding sustainability in chemistry teaching involves a pedagogical shift of focus that emphasizes the applicability of chemical principles to environmental stewardship. Traditional teaching of chemistry emphasizes pure science and offers limited scope for applications pertinent to present sustainability challenges. By being incorporated into the core curriculum, teachers can generate a generation of chemists who understand not just chemical reactions but also their implications in the global ecological system. Such a shift requires the incorporation of inter-disciplinary courses such as green chemistry, which promotes the creation of processes and products that minimize negative environmental impacts. Such promotion of new practices is in harmony with the education reforms to prepare students for meeting the more advanced demands of the job market, particularly industries with a focus on sustainable processes. Additionally, a sustainability-oriented chemistry curriculum must engage students with problem-solving and critical thinking in the real world. Fixed encounters with case studies of environmental chemistry, waste management, and sustainable use of resources can complete the circle between theory and practice.

Figure 2.14. Incorporating sustainability into chemistry curricula necessitates a transformative approach to teaching that underscores the interconnection between chemical principles and environmental stewardship.

Source: https://cdn.rit.edu/images/program/2025-02/environmental-sustainability-health-and-safety-bs-1200x675b.jpg.

Project work on studying local environmental issues, such as pollution or conservation, enables students to apply their learning to a related context. Such learning through experience has the ability to instill ownership and accountability for environmental problems, hence convincing students to devise sustainable solutions. Aligning chemistry education within sustainability frameworks is the way institutions can better prepare graduates who are capable of addressing some of the most ambitious world challenges. Curricula enhancement in this aspect not only enhances the applicability of chemistry education but also its contribution to society's welfare. Lastly, successful incorporation of sustainability into chemistry curricula relies on cooperation between academia and industry. Cooperation with environmental organizations, government, and private sector organizations can provide valuable insights into the applications of chemistry. Cooperations can facilitate the design of innovative, demand-driven curriculum models mapping the needs of various industries, thereby creating an industry-suitable workforce with the capacity to tackle sustainability challenges. For instance, Maredia et al. (2025) highlights the imperative of education systems that are adaptable to the evolving demands of agriculture and environmental sectors. With such recommendations in place, schools can strengthen their curricula in order to benefit the new job market as well as enhance sustainable development agendas. An interconnection among education, industry, and policy can further support the sustainability principles of chemistry to wider social repercussions beyond the classroom (Figure 2.14).

2.5.2. Role of Outreach Programs in Promoting Sustainable Practices

Outreach programs have an important role to play in the development of sustainable practices through bridging the gap between theoretical knowledge and involvement at the local community level. The programs exist to disseminate information about principles of sustainability, particularly that of chemistry, to the broader community. By creating communities through workshops, seminars, and community meetings, outreach programs sensitize individuals and organizations about the importance of sustainable practices, such as reducing waste, conserving energy, and using green materials. The efficacy of such demonstrations can go a long way in raising public sensitivity and promoting behavioral changes conducive to sustainable practices, and developing a culture of sustainability among local communities. As these courses are taken up by different populations, they also facilitate the population to adopt sustainable practices in daily life, with a multiplier effect extending beyond the education sector and into society in general, which is a primary engagement component identified in the Gettysburg College Sustainability Plan elaborated in Boone et al. (2011). Besides, outreach programs also focus on integrating sustainability into local industries, particularly relevant for chemistry applications. Outreach programs can establish opportunities for partnerships between corporations and academic institutions, promoting sustainable practices supported by scientific research. By bringing to the forefront effective case studies and practical applications of sustainability in local industries, outreach programs accentuate the economic and environmental benefits of utilizing sustainable practices. For instance, outreach projects for local farmers in chemical-free agriculture highlight the prospects of chemistry as a game-changer to sustainable agriculture. Furthermore, as has been established by research on gender-sensitive outreach in microfinance in Uganda (Almeyda et al., 2025), flexibility and openness to inclusivity in outreach initiatives can enhance overall impact, so that sustainable solutions are accessible and worthwhile to everyone in the community. Finally, the effectiveness of outreach programs in ensuring a sustainable future lies in whether or not they are able to foster cooperation among sectors. Through stakeholders' engagement, including students, teachers, local governments, and NGOs, outreach programs can build networks that increase the sharing of knowledge and the distribution of resources. The creation of sustainability committees, as proposed in the Gettysburg College Sustainability Plan, is a case in point of cooperation by fostering the shared responsibility of the community. Effective outreach activities can play a role in generating creative solutions to community sustainability problems, such as carbon footprint minimization and environmental stewardship. As is observed both in urban and rural settings, for instance, in Uganda's microfinance environment (Almeyda et al. 2025), effective outreach can empower various segments of society to become advocates of sustainability, hence making sure the concepts of sustainable chemistry are not merely taught but practiced and respected actively within the community (Figure 2.15).

Figure 2.15. By engaging stakeholders, including students, faculty, local governments, and NGOs, outreach programs can create networks that facilitate knowledge sharing and resource allocation.

Source: https://aworld.org/wp-content/uploads/2024/02/shutterstock_2258443385-scaled-1.jpg.

2.5.3. Collaboration Between Academia and Industry for Sustainability

This collaboration between academia and industry is increasingly seen as the most important driver of sustainability, especially in the field of chemistry.

Universities and research institutions have the capability to facilitate innovation and the know-how, while industry possesses the means and logistical know-how for implementation. This collaboration allows more efficient green chemistry technologies and practices to be developed. Notably, initiatives such as the American Chemical Society's Green Chemistry Institute Pharmaceutical Roundtable accentuate such collaboration. Through its outright partnership with over 50 pharma corporations, the Roundtable provides a forum for integrating sustainable practices into the processes of drug manufacturing. Later curricular resources, such as the Green Chemistry and Engineering Learning Platform (GChELP), offer valuable training modules that can bridge the gaps between theory knowledge and real-world implementation, enhancing workforce preparedness for sustainability challenges in the chemical industry (Babij et al., 2025). Effective collaboration can also address immediate training education gaps that hinder the advancement of sustainable processes in chemistry. Graduate students and industry chemists also get limited exposure to green chemistry principles and practices during their course of study, and this impairs them from contributing meaningfully to innovations in the subject. Promoting collaboration through such initiatives can bridge this gap by creating targeted training materials and programs with a focus on sustainable chemistry.

For instance, partnerships between academia and industry associations can facilitate the creation of workshops and hands-on exercises that introduce students to real-life scenarios where they practice implementing green chemistry principles. These are critical in ensuring that future chemists have the competencies required to balance rising industry needs with a focus on sustainability (Babij et al., 2025). In addition to chemical innovations, the academy-industry interface can also involve the development of sustainable software programs that are the underpinning of scientific inquiry and industrial practice. The Third Workshop on Sustainable Software for Science highlighted the importance of funding as well as community participation in sustaining essential software tools used in scientific investigation. As both academic researchers and industry partners look for ways to facilitate sustainable practice, developing user-friendly, efficient software is key to maximizing research productivity while minimizing the environmental impact. By working in conjunction with a network of professionals dedicated to sustainable software engineering, academia and industry can collaborate to co-create solutions to sustainability in every aspect of science. Lastly, these couples not only encourage the philosophy of sustainability in chemistry but also foster a wider context involving chemical and software sustainability, developing a full-fledged platform for sustainable development.

2.6. CASE STUDIES IN SUSTAINABLE CHEMISTRY

From the case studies of green chemistry, it can be observed that the use of green principles has a significant impact on the production processes in various chemical industries. The pharmaceutical industry, for instance, has been embracing more sustainable methods of manufacturing centered on green chemistry in a bid to mitigate the environmental load created by traditional methods. Case studies verify that business companies utilizing such green technologies gain not only from cost savings through increased efficiency but also from increased competitiveness within their markets since they position themselves in consonance with regulatory programs that ensure sustainability. This shift towards greener production stresses the potential for economic benefits alongside addressing environmental concerns, a two-sided imperative becoming increasingly crucial within the present industrial age (Muthoni, 2025). Further research into sustainable methods reveals the importance of Green Analytical Chemistry (GAC) indices in quantifying the environmental footprint of analytical procedures. The indices, including the Analytical Eco-Scale and the Green Analytical Procedure Index (GAPI), enable researchers to quantify and enhance the greenness of analytical processes effectively. By adopting the use of such measurement tools, industries can identify areas for enhancement, bringing their processes in line with sustainability principles. Case studies prove efficient applications of GAC metrics in bioanalytical methods, substantiating their role in furthering environmentally friendly practices within the chemical analysis environment. Besides the preservation of natural resources, such focus on sustainability guarantees greater accuracy and integrity in analysis processes through encouraging the use of renewable and less harmful products (Gandu et al., 2025). The continuous evolution of sustainable chemistry is defined by an ongoing interplay between compliance and innovation. The more industries expand to utilize greener approaches in compliance with green chemistry, the more there arises a demand for robust regulatory systems. These must balance progressing sustainability with ensuring safety and efficiency in chemical processes. Case studies analyzed show that overcoming financial and regulatory hurdles is necessary to facilitate wider use of sustainable innovations. Future trends will be founded on fostering cooperation among stakeholders, the integration of pollution prevention principles, and giving importance to eco-design. By noting best practices in case studies, sustainable chemistry as a discipline can benefit from collaboration in solving environmental issues as well as advancing science (Muthoni, 2025).

2.6.1. Successful Implementation of Green Chemistry in Industry

The successful implementation of green chemistry principles to industry relies on a multidisciplinary effort that reengineers traditional chemical processes. By focusing on reducing or eliminating hazardous materials, industries not only can comply with regulatory standards but also cater to the increasing green consciousness among consumers. One such big strategy is adopting new materials and techniques, such as metal-organic frameworks (MOFs) and covalent organic frameworks (COFs).

These new materials allow for precise control of chemical function and structure, which is maximally efficient with minimal waste generation. The application of reticular chemistry is a demonstration of the way industries are shifting more towards science innovation in order to create sustainable solutions that address world sustainability goals. This renders the transition to green chemistry not just theoretical but indeed transforming practice at a concrete level, which in time brings in higher acceptance from varied industrial fronts (Freund et al., 2021). The integration of green toxicology into the green chemistry platform also solidifies its place in business. The principle attempts to unravel the toxicological implications of chemicals as much as possible while promoting safer alternatives through better test protocols.

This includes the use of non-animal testing, where *in vitro* and in silico technology is utilized for toxicity prediction, enhancing product safety without compromising sustainability. As industries embrace such principles, cross-disciplinary communication between toxicologists and chemists is a must. Such communication not only aids in the development of harmless products but also addresses the complexities involved in chemical safety assessment. By integrating green toxicology into chemical development, industries can make it possible to pioneer solutions that minimize harmful exposure and environmental degradation (Krebs et al., 2020). Finally, the road map towards the successful application of green chemistry in industry is fraught with challenges that call for strict scrutiny and innovative minds.

Even though possible benefits are huge, economic constraints, the need for trained labor, and resistance by entrenched manufacturing procedures are profound barriers. Furthermore, it is also necessary that industry leaders embrace research and development aimed at sustainability. This should involve not only the adoption of cleaner technologies, but also the continuous questioning of what they have done to ensure that they are in line with green chemistry principles. With more discussion about sustainability in chemicals, innovative contributions towards the development of infrastructure and technology that facilitates environmentally friendly operation will need to be made to drive long-term change in the industry ecosystem (Freund et al., 2021) (Figure 2.16).

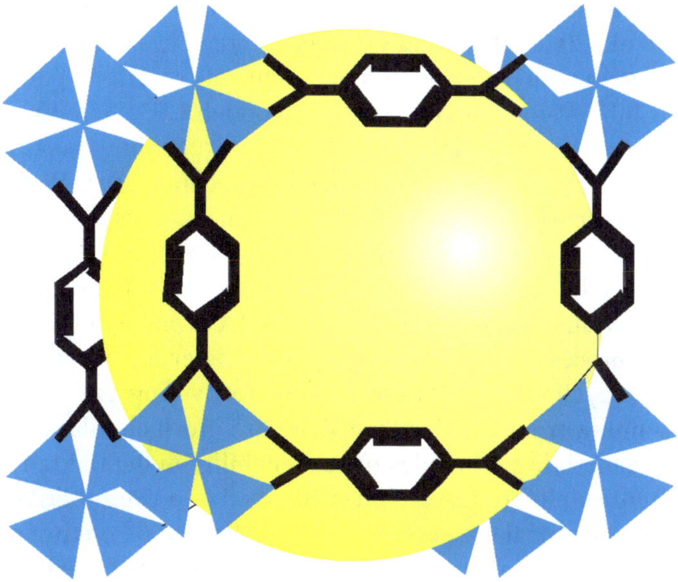

Figure 2.16. The integration of reticular chemistry exemplifies how industries are increasingly relying on scientific advancements to create sustainable solutions that align with global sustainability goals.

Source: https://pub.mdpi-res.com/ijms/ijms-23-07980/article_deploy/html/images/ijms-23-07980-g004.png?1658297326.

2.6.2. Research Breakthroughs in Sustainable Chemical Processes

In the last ten years, the chemistry discipline has made huge research strides towards having more environmentally friendly chemical processes. At its heart lies the inclusion of principles of green chemistry, which mandate the reduction of harmful materials and waste. Catalysis advances have been at the forefront, and researchers have developed catalysts that will operate under milder conditions and with greater efficiencies. These advancements not only reduce energy consumption but also assist in minimizing the environmental impact associated with typical chemical manufacturing processes. With environmental degradation continuing to raise crucial challenges, the role of chemists in formulating sustainable solutions plays an increasingly important role. Educational paradigms have also continued to support these advancements, with emphasis on a curriculum that embeds sustainability values and prepares upcoming scientists to solve global challenges related to chemical processes (Abubakar et al., 2024). The field of energy storage has also continued to show potential progress with the formation of sustainable practices, particularly in the recycling of lithium-ion batteries (LIB). Although LIBs revolutionized energy storage since their commercial release, concerns regarding their environmental impact have prompted research in recycling technologies. Present recycling methods are primarily labor-based and use hazardous chemicals, contrary to the principles of circular economy and sustainability (Piątek et al., 2020). However, research has begun exploring other methods that use the principle of upcycling material, minimization of waste, and less energy consumption during the recycling process. With the progress of cleaner technologies, researchers aim to have recycling of LIBs not just for recovering precious material but also for sustaining processes that contribute to a green circular economy, which supports the underlying principles of sustainability in the field of chemistry. In addition, interdisciplinary research has emerged as a central driver of progress in sustainable chemical processes. The complexity of environmental issues requires an interdisciplinary solution with the synthesis of chemistry with other scientific branches, such as biology and environmental science. Collaboration has led to the development of new approaches to pollutant treatment and the recovery of the environment. For example, bioremediation techniques involve the use of living organisms to facilitate the breakdown of harmful substances, providing a novel and environmentally friendly alternative to traditional chemical approaches. As scientists bridge gaps between disciplines, the likelihood that more sustainable answers become available increases, eventually leading to an understanding of how deeply connected our environmental issues are to chemical processes. This interdisciplinarity not only adds to the richness of the discipline of chemistry but also places emphasis on the key role scientists play in promoting sustainable measures (Abubakar et al., 2024).

2.6.3. Lessons Learned from Failed Sustainability Initiatives

The study of failed projects on sustainability has significant lessons about the basic principles that sustain sound practices in chemistry and environmental protection. One of the key lessons is stakeholder engagement. The majority of projects fail because they lack the incorporation of local people's aspirations and interests, scientists,' and politicians.' In the case of anti-corruption initiatives in the private sector, for instance, there was a lack of collective action and commitment among stakeholders that resulted in inefficient implementations, highlighting the necessity for collaboration and trust in achieving sustainable results (Dan, 2020). If sustainability initiatives prescribe merely top-down approaches, they are likely to overlook the important local knowledge that can enhance the practicability and usability of such projects. Engaging stakeholders early ensures that diverse views are applied to inform strategies, and therefore improving the chances of success and fostering a stronger sense of collective responsibility for sustainability outcomes. Another major lesson from analyses of defunct sustainability projects is that there is value in adaptive management and flexibility when addressing unforeseen problems. The COVID-19 pandemic illustrated how rigid policies can lead to adverse effects, especially in the disability services and public health sectors, where pre-determined frameworks did not consider vulnerable populations (Dan, 2020). This experience underscores the need for making plans for sustainability that enable adaptability based on evolving environmental conditions and social needs. In chemistry, this flexibility is essential because new research developments can require a change in sustainable approaches. Rather than following fixed methods, adopting an adaptable style encourages innovation and resilience to allow projects to evolve and succeed in the context of complexity and uncertainty. Finally, transparency and accountability in assessing sustainability projects should not be emphasized enough. Dead projects reveal systemic breakdowns in monitoring and reporting mechanisms that sap trust and impede progress. Organizations which eschew honest appraisal and fail to return results to the public guarantee breeding grounds for doubt, thus jeopardizing future endeavors. The Argentinean anti-corruption crusade is a prime example; without credible enforcement of collective action arrangements, hoped-for benefits proved elusive, causing disillusionment and charges of hypocrisy among stakeholders. Therefore, developing solid guidelines to evaluate the success of sustainability programs is critical not only for accountability but also for learning from past mistakes. By putting transparency and formal assessment in priority, stakeholders can optimize their credibility and foster an environment of continuous improvement in sustainability practices in the chemical sector.

2.7. CONCLUSION

In short, the chemistry of sustainability has emerged as core principles for addressing some of the most serious questions facing our world. The integration of sustainable practice into chemical industry and research not only attempts to reduce environmental impacts but also encourages innovations that can lead to improved utilization of resources. With the ever-increasing demand for chemicals worldwide, it is more vital that chemists adopt practices that are healthiest for the environment. Implementation of sustainable practices bridges the gap between scientific advancement and environmental stewardship and ultimately fosters a harmonious coexistence between technology advancement and the environment. This study into sustainability in chemistry also suggests exposing future chemists to the implications of their practice. The ongoing development of chemical education needs to center on cultivating a comprehensive perspective of sustainability principles, hence enabling students to design more eco-friendly solutions. Only through a dedicated educational system can the future generation of chemists come up with strategies that effectively address global challenges like climate change and resource depletion. Fostering cross-disciplinary dialogue and integrating sustainability into chemistry education can be a difference-maker in making chemical research more pertinent to addressing the advanced problems of society and ensuring scientific progress is linked with the ethical responsibility for saving the environment. Ultimately, advancing the sustainability tenets of chemistry is an intellectual undertaking; it's an ethical imperative for the scientific community. The future of chemistry depends on the adoption of sustainable practices whose utmost concern is the health of humans as well as the health of the world. As bearing witness by new technologies, such as those faced in noisy intermediate-scale quantum (NISQ) computing, there is potential for superior development that, when it comes sustainably, can drive revolutionary developments across the board. It is the task of chemists to carry out research that not only meets current requirements but does so in a way that supports long-term sustainability and minimum ecological disturbance, paving the way to a sustainable future for chemistry and beyond (McKenney et al., 2018; Preskill, 2018).

2.7.1. Summary of Key Points Discussed

The research of sustainability in chemistry focuses on the adoption of environmentally friendly approaches to the synthesis of materials. One of the prominent fields is the establishment of green synthesis processes, which are gaining favor in materials science as excellent alternatives to traditional processes. These processes leverage natural extracts in synthesizing nanoparticles and other materials with significantly reduced environmental impacts associated with traditional synthetic procedures. By employing biological materials and required phytochemicals such as flavonoids and alkaloids, green synthesis not only minimizes toxicity but also ensures greater stability for the resulting products. In doing so, it promotes the fundamental tenets of sustainability by promoting processes that are not just efficient but also environmentally friendly. This transition exhibits the chemistry community's commitment to finding sustainable solutions to meeting key environmental challenges, thereby proving a paradigm shift in materials chemistry (Singh et al., 2018). Another crucial element in the subject of sustainability in chemistry is the emphasis on applying innovative technologies that improve stewardship of the environment even more. Quantum computing has been shown to be a potentially game-changing tool in this area. While scientists are in the process of developing Noisy Intermediate-Scale Quantum (NISQ) machines, we have a promise of applications going as far as material design and optimization. While these quantum computers are currently constrained, they will be capable of carrying out complex calculations and simulations unfeasible with traditional computing methods. This will potentially change the development of sustainable materials and processes for the better by enabling chemists to model material properties and reactions better. As quantum computing is still in its early stages, its use in chemistry is a forward-looking approach that can potentially revolutionize the process of developing sustainable solutions in the field (Preskill, 2018). The overall sustainability agenda in chemistry emphasizes the necessity to develop methodologies and technologies that consider long-term environmental consequences. Change in the profession does not only rely on technical advancement; it also entails a cultural change towards more sustainable research practices. Chemists are more and more being required to evaluate the lifecycle of chemical substances, from production to destruction, ensuring that each process adheres to sustainability regulations. This involves adopting a culture of minimizing wastage and employing renewable resources. Integration of green chemistry principles into curricula remains to equip the next generation of chemists with the competencies required to innovate responsibly. As the incumbent and emerging technologies keep advancing, the principles of sustainable chemistry practice will continue to be pivotal in achieving balance between scientific advancement and environmental care (Singh et al., 2018).

2.7.2. Future Directions for Sustainability in Chemistry

The future of sustainable chemistry relies on the implementation of advanced technologies and methodologies that optimize the protection of both human health and the environment. One key potential direction is the path of flexible sensors, which are being developed to enhance our ability to observe conditions in a variety of settings. These sensors hold promise for addressing several grand challenges, such as detecting contamination and resource management, by offering superior performance compared to fixed, conventional counterparts. However, their application in the market remains constrained by various bottlenecks like performance limitations, biocompatibility, and challenges in powering and networking sensor networks. By offering integrated solutions to overcome such bottlenecks, researchers are able to enable extensive deployment of these smart flexible sensors, thereby promoting sustainable development of the chemical industry and its contribution towards ecological conservation and better quality of life (Luo et al., 2023). Another key future emphasis on sustainability in chemistry is the establishment of good data management practice in line with the FAIR Data Principles. With the rapid digitization of research workflows, sound data management becomes increasingly important. Initiatives like the National Research Data Infrastructure for Chemistry (NFDI4Chem) are examples of this trend, as it aims to foster good research practices through the guarantee that chemical data is well-structured, accessible, and reusable in the long term. Not only does this project emphasize strongly on data provenance and the documentation of metadata, but it also aims to build a culture of open science in the field. By closing the gap between existing research data initiatives and engaging different stakeholders, NFDI4Chem is a remarkable step towards developing an inclusive framework to support sustainable data management in chemistry (Herres et al., 2019). Further research into sustainable synthetic routes is another avenue through which sustainability can be fostered in chemistry. The application of green chemistry principles is one critical necessity to curb environmental impacts associated with chemical operations. The emphasis of future research has to be on fresh approaches that maximize efficiencies of reaction and minimize toxic waste by the application of renewable feedstocks and renewable energy. For instance, better catalytic technology and biocatalyst engineering have the potential to optimize the manufacture of many chemicals simultaneously while minimizing ecological harm. Facilitating interdisciplinary cooperation among chemists, engineers, and environmental scientists will turn out to be crucial to spur these innovations. Through maintaining sustainability as its number one concern in synthetic chemistry, the discipline is able to steer towards a more responsible and sustainable path, aligning its activities with the overall goals of sustainability and public well-being.

2.7.3. Call to Action for Chemists and Researchers to Prioritize Sustainability

The efforts of scientists and chemists drive the solution of the pressing challenges of sustainability. As the realization of the impact of chemical processes on the environment becomes increasingly clear, action by experts is needed to drive sustainable practices. This involves not only research and development of environmentally friendly materials and processes but also adoption of new approaches. For instance, the application of in silico methods to screen for chemical safety can be used to reduce animal testing and maximize regulatory adherence. Tools like ProTox 3.0 can predict several endpoints of toxicity, which can help chemists make informed decisions that override human and environmental health (Banerjee et al., 2024). All these advancements indicate that green chemistry cannot be an option but a reality for the future of chemical research and development. In addition, chemistry sustainability also extends to energy storage and management, as evident from the increasing reliance on lithium-ion batteries (LIBs). Increased usage of LIBs across industries serves to raise a call for chemists to adopt the entire product life cycle of such products, including recycling and reuse. With the incorporation of recycling technologies and involvement in the development of circular economy solutions, chemists are likely to develop solutions that will ensure minimal environmental impact. Recycling efforts for LIB can alleviate the strain on natural resources and lower greenhouse gas emissions. Through collaborative vision between researchers, new approaches can be explored to improve recycling techniques, leading to a sustainable future. The LIBs discussion gives room for research to be conducted, motivating scientists, policymakers, and entrepreneurs to gather and engineer more environmentally friendly energy systems (Zanoletti et al., 2024). Educational institutions and professional societies must promote a culture of sustainability in chemistry science based on these arguments. Through the integration of sustainability principles into curricula and research projects in academia, future scientists and chemists will be adequately equipped to address environmental problems. Professional societies can also lead the way by encouraging more green-lab procedures and sustainable partnerships with industrial partners. Not only is this going to enhance the profession's image, it is going to attract talent and funding dedicated to sustainable research into the profession. Finally, ensuring sustainability in chemistry is everyone's role that demands determined action and innovative thinking from scientists and chemists alike. It is through such action that the scientific community can effectively make significant strides toward developing a more sustainable and resilient world.

REFERENCES

1. Abubakar, M. Y., Haladu, M., Titilayo, O. A., & Adam, A. B. (2024). Effect and remediation of environmental pollution on the concept of chemistry–Review. *African Journal of Sciences and Traditional Medicine.* https://www.semanticscholar.org/paper/61b8ea89e7f9b121f569d6ef56c950a44a649146 (accessed on 24th October 2025).

2. Allen, D., Allenby, B., American Chemistry Society, Aragon-Correa, J. A., Armstrong, F., Batterham, R., Benoit, J. P., et al. (2010). Sustainability of the chemical manufacturing industry–Towards a new paradigm? https://core.ac.uk/download/339962.pdf (accessed on 24th October 2025).

3. Almeyda, G., & Seibel, H. D. (2025). Women and men in rural microfinance: The case of Uganda. https://core.ac.uk/download/pdf/6581473.pdf (accessed on 24th October 2025).

4. Annevelink, E., Broeze, J., van Ree, R., et al. (2009). Opportunities for Dutch biorefineries. https://core.ac.uk/download/pdf/29248632.pdf (accessed on 24th October 2025).

5. Azim, E., Boyd, K. A., Doi, L., Little, J., et al. (2012). Scoping a public health impact assessment of aquaculture with particular reference to tilapia in the UK. https://core.ac.uk/download/1438606.pdf (accessed on 24th October 2025).

6. Aznar Sánchez, J. Á., García Gómez, J. J., López Serrano, M. J., Velasco Muñoz, J. F., et al. (2018). The sustainable management of metals: An analysis of global research. https://core.ac.uk/download/286590215.pdf (accessed on 24th October 2025).

7. Babij, N., Baker, A., Briddell, C., Ichiishi, A., et al. (2025). GChELP: A vital resource for teaching practical green chemistry in industry and academia. https://core.ac.uk/download/661944912.pdf (accessed on 24th October 2025).

8. Banerjee, P., Kemmler, E., Dunkel, M., & Preißner, R. (2024). ProTox 3.0: A webserver for the prediction of toxicity of chemicals. *Nucleic Acids Research, 52*, W513–W520. https://doi.org/10.1093/nar/gkae303 (accessed on 24th October 2025).

9. Bateman, A., Martin, M., Orchard, S., Magrane, M., Agivetova, R., Ahmad, S., Alpi, E., et al. (2020). UniProt: The universal protein knowledgebase in 2021. *Nucleic Acids Research, 49*, D480–D489. https://doi.org/10.1093/nar/gkaa1100 (accessed on 24th October 2025).

10. Belt, K. T., Kaushal, S. S., Mayer, P. M., McDowell, W. H., et al. (2015). Urban evolution: The role of water. https://core.ac.uk/

download/215538902.pdf (accessed on 24th October 2025).

11. Boone, J. W., Carson, A., Hecklau, S. F., Hughes, S., et al. (2011). Gettysburg College sustainability proposal. https://cupola.gettysburg.edu/cgi/viewcontent.cgi?article=1022&context=student_scholarship (accessed on 24th October 2025).
12. Busquets Rubio, P., Hernández Gómez, M. Á., Horta Bernús, R., Segalàs Coral, J., et al. (2018). The EDINSOST project. Training sustainability change agents in Spanish and Catalan engineering education. https://core.ac.uk/download/158801938.pdf (accessed on 24th October 2025).
13. Calvin, K., Dasgupta, D., Krinner, G., Mukherji, A., Thorne, P., Trisos, C. H., Romero, J., et al. (2023). IPCC, 2023: Climate change 2023: Synthesis report. Contribution of Working Groups I, II and III to the Sixth Assessment Report of the Intergovernmental Panel on Climate Change [Core Writing Team, H. Lee & J. Romero (Eds.)]. *IPCC, Geneva, Switzerland.* https://doi.org/10.59327/ipcc/ar6-9789291691647 (accessed on 24th October 2025).
14. Cérdan, K., Thys, M., Costa Cornellà, A., Demir, F., Norvez, S., Vendamme, R., Van den Brande, N., et al. (2024). Sustainability of self-healing polymers: A holistic perspective towards circularity in polymer networks. *Progress in Polymer Science, 152*, 101816. https://doi.org/10.1016/j.progpolymsci.2024.101816 (accessed on 24th October 2025).
15. Choudhury, H. K. (2025). Synthesis of mosquito repellent product through green chemistry. *Journal of Advances in Science and Technology.* https://www.semanticscholar.org/paper/08779d3108baf96c127c86f2afe45cdffed89c02 (accessed on 24th October 2025).
16. Cordova, K. E., Furukawa, H., & Yaghi, O. M. (2015). The development of global science. https://core.ac.uk/download/323081301.pdf (accessed on 24th October 2025).
17. Dan, B. (2020). Long-term impact of COVID-19 on disabled children. *Developmental Medicine & Child Neurology, 62.* https://www.semanticscholar.org/paper/104e918425l8ac293dca7ee4601f5d3f63e51d7b (accessed on 24th October 2025).
18. Ding, Y., Pang, Z., Lan, K., Yao, Y., Panzarasa, G., Xu, L., Lo Ricco, M., et al. (2022). Emerging engineered wood for building applications. *Chemical Reviews, 123*, 1843–1888. https://doi.org/10.1021/acs.chemrev.2c00450 (accessed on 24th October 2025).
19. Freund, R., Zaremba, O., Arnauts, G., Ameloot, R., Skorupskii, G., Dincă, M., Bavykina, A., et al. (2021). The current status of MOF and COF applications. *Angewandte Chemie International Edition, 60*, 23975–

24001. https://doi.org/10.1002/anie.202106259 (accessed on 24th October 2025).

20. Gandu, S., Gandla, K., & Repudi, L. (2025). Green metric tools and white assessment in bioanalytical method development and validation: A comprehensive review. *World Journal of Advanced Research and Reviews.* https://www.semanticscholar.org/paper/a021fa05ccef89647d719b8221605e05d20c48af (accessed on 24th October 2025).
21. Harvey, D., Jimerson, L., Kimmel, M., Loupin, C., et al. (2001). A template for resource productivity/sustainability programs in forest products manufacturing facilities. https://core.ac.uk/download/36680648.pdf (accessed on 24th October 2025).
22. Haynes, K., Murray, A., & Skene, K. (2015). The circular economy: An interdisciplinary exploration of the concept and application in a global context. https://core.ac.uk/download/151157363.pdf (accessed on 24th October 2025).
23. Herres-Pawlis, S., Koepler, O., & Steinbeck, C. (2019). NFDI4Chem: Shaping a digital and cultural change in chemistry. *Angewandte Chemie International Edition, 58,* 10766–10768. https://doi.org/10.1002/anie.201907260 (accessed on 24th October 2025).
24. Hofmann, T., Ghoshal, S., Tufenkji, N., Adamowski, J., Bayen, S., Chen, Q., Demokritou, P., et al. (2023). Plastics can be used more sustainably in agriculture. *Communications Earth & Environment, 4.* https://doi.org/10.1038/s43247-023-00982-4 (accessed on 24th October 2025).
25. Idul, J. J. A., & Walag, A. M. P. (2024). Integrating green chemistry and sustainability principles to a secondary science curriculum: A mixed-methods needs assessment. *Journal of Chemical Education, 101*, 2765–2778. https://doi.org/10.1021/acs.jchemed.4c00341 (accessed on 24th October 2025).
26. Kasare, S. L. (2025). Green chemistry approaches for sustainable waste management and resource recovery. *International Journal of Scientific Research in Science and Technology.* https://www.semanticscholar.org/paper/5fc742f532b843b84086d71bde2065bacb8451c4 (accessed on 24th October 2025).
27. Katz, D. S., Choi, S. C. T., Niemeyer, K. E., Hetherington, J., et al. (2012). Report on the Third Workshop on Sustainable Software for Science: Practice and Experiences (WSSSPE3). https://core.ac.uk/download/74212159.pdf (accessed on 24th October 2025).
28. Kelarakis, A. (2014). Functional nanomaterials for energy and

sustainability. https://core.ac.uk/download/42136808.pdf (accessed on 24th October 2025).

29. Kilcoyne, J., Bogan, Y., Duffy, C., & Hollowell, T. (2022). Reducing environmental impacts of marine biotoxin monitoring: A laboratory report. *PLOS Sustainability and Transformation, 1*, e0000001. https://doi.org/10.1371/journal.pstr.0000001 (accessed on 24th October 2025).

30. Krebs, J., & McKeague, M. (2020). Green toxicology: Connecting green chemistry and modern toxicology. *Chemical Research in Toxicology, 33*, 2919–2931. https://doi.org/10.1021/acs.chemrestox.0c00260 (accessed on 24th October 2025).

31. La Marca, F. (2015). Treatment and valorization plants in materials recovery supply chain. https://core.ac.uk/download/54525197.pdf (accessed on 24th October 2025).

32. López-Lorente, Á. I., Pena-Pereira, F., Pedersen-Bjergaard, S., Gomes Zuin, V., Özkan, S. A., & Psillakis, E. (2022). The ten principles of green sample preparation. *TrAC Trends in Analytical Chemistry, 148*, 116530. https://doi.org/10.1016/j.trac.2022.116530 (accessed on 24th October 2025).

33. Luo, Y., Abidian, M. R., Ahn, J. H., Akinwande, D., Andrews, A. M., Antonietti, M., Bao, Z., et al. (2023). Technology roadmap for flexible sensors. *ACS Nano, 17*, 5211–5295. https://doi.org/10.1021/acsnano.2c12606 (accessed on 24th October 2025).

34. Malik, S., Muhammad, K., & Waheed, Y. (2023). Emerging applications of nanotechnology in healthcare and medicine. *Molecules, 28*, 6624. https://doi.org/10.3390/molecules28186624 (accessed on 24th October 2025).

35. Manousi, N., Wojnowski, W., Płotka-Wasylka, J., & Samanidou, V. (2023). Blue applicability grade index (BAGI) and software: A new tool for the evaluation of method practicality. *Green Chemistry, 25*, 7598–7604. https://doi.org/10.1039/d3gc02347h (accessed on 24th October 2025).

36. Maredia, M. K. (2025). Curriculum enhancement and reform to meet the needs of smallholder farmers in developing countries: Survey of literature. https://core.ac.uk/download/pdf/6546149.pdf (accessed on 24th October 2025).

37. McKenney, S., & Reeves, T. C. (2018). *Conducting Educational Design Research.* Routledge eBooks. https://doi.org/10.4324/9781315105642 (accessed on 24th October 2025).

38. Mujtaba, M., Fraceto, L. F., Fazeli, M., Mukherjee, S., Savassa, S. M., de Medeiros, G. A., Pereira, A. E. S., et al. (2023). Lignocellulosic biomass from agricultural waste to the circular economy: A review with focus on

biofuels, bio composites and bioplastics. *Journal of Cleaner Production, 402*, 136815. https://doi.org/10.1016/j.jclepro.2023.136815 (accessed on 24th October 2025).

39. Muthoni, K. L. (2025). Sustainable practices in pharmaceutical manufacturing. *NEWPORT International Journal of Public Health and Pharmacy.* https://www.semanticscholar.org/paper/310b95c646915d31d3ebb522d084fe7d0635335d (accessed on 24th October 2025).
40. Nandiyanto, A., Che Sidik, N. A. C., Kurniawan, T., Bilad, M. R., Gandidi, I. M., Sukrawan, Y., & Mahmudatussa'adah, A. (2025). How to understand fluid mechanics in chemical engineering: Principles, applications, research trends, and bibliometric insights. *Journal of Advanced Research in Fluid Mechanics and Thermal Sciences.* https://www.semanticscholar.org/paper/e48a2b07671fa1cc53b3fd5950a8fecb2bee1094 (accessed on 24th October 2025).
41. Nowak, P. M., Wietecha-Posłuszny, R., & Pawliszyn, J. (2021). White analytical chemistry: An approach to reconcile the principles of green analytical chemistry and functionality. *TrAC Trends in Analytical Chemistry, 138*, 116223. https://doi.org/10.1016/j.trac.2021.116223 (accessed on 24th October 2025).
42. Nwankwo, C. O., Nwakamma, N. E., Ani, E. C., Olu-lawal, K. A., & Ugwuanyi, E. D. (2024). Simulation-driven strategies for enhancing water treatment processes in chemical engineering: Addressing environmental challenges. *Engineering Science & Technology Journal, 5*, 854–872. https://doi.org/10.51594/estj.v5i3.942 (accessed on 24th October 2025).
43. Piątek, J., Afyon, S., Budnyak, T. M., Budnyk, S., Sipponen, M., & Slabon, A. (2020). Sustainable Li-ion batteries: Chemistry and recycling. *Advanced Energy Materials, 11.* https://www.semanticscholar.org/paper/f0ddf0afbc9bba7c8dd3f27f07762c63ca7e763f (accessed on 24th October 2025).
44. Preskill, J. (2018). Quantum computing in the NISQ era and beyond. *Quantum, 2*, 79. https://doi.org/10.22331/q-2018-08-06-79 (accessed on 24th October 2025).
45. Sheldon, R. A., & Woodley, J. M. (2017). Role of biocatalysis in sustainable chemistry. *Chemical Reviews, 118*, 801–838. https://doi.org/10.1021/acs.chemrev.7b00203 (accessed on 24th October 2025).
46. Singh, J., Dutta, T., Kim, K. H., Rawat, M., Samddar, P., & Kumar, P. (2018). 'Green' synthesis of metals and their oxide nanoparticles: Applications for environmental remediation. *Journal of Nanobiotechnology, 16.* https://doi.org/10.1186/s12951-018-0408-4 (accessed on 24th October 2025).

47. Taiwo, A. I., Isi, L. R., Okereke, M., Sofoluwe, O., Olugbemi, G. I. T., & Essien, N. A. (2025). Development of AI-powered optimization frameworks for enhancing chemical processes in sustainable and energy-efficient water treatment. *International Journal of Scientific Research in Science, Engineering and Technology*. https://www.semanticscholar.org/paper/2e225a9206f32285f8453a33868030c11b82de34 (accessed on 24th October 2025).

48. Taylor, S. (2015). A day late and a dollar short: Discount retailers are falling behind on safer chemicals. https://core.ac.uk/download/75780358.pdf (accessed on 24th October 2025).

49. Wagner, W. E. (2014). Imagining corporate sustainability as a public good rather than a corporate bad. https://core.ac.uk/download/211333149.pdf (accessed on 24th October 2025).

50. Zanoletti, A., Carena, E., Ferrara, C., & Bontempi, E. (2024). A review of lithium-ion battery recycling: Technologies, sustainability, and open issues. *Batteries, 10*, 38. https://doi.org/10.3390/batteries10010038 (accessed on 24th October 2025).

CHAPTER 3

ATOM ECONOMY AND PROCESS EFFICIENCY

Contents

INTRODUCTION

Atom economy has turned into a key principle in the quest for a greener, more efficient chemical process. It was driven by the need to cut down on waste and maximize the utilization of resources. Atom economy is the measurement of how efficiently atoms of the reactants are utilized in the final product. This standard not only indicates the need to reduce chemical waste but also encourages the synthesis methods to focus on incorporating more material into the intended chemical products. Focusing on high atom economy in products allows industries to move towards green techniques, thus achieving international standards for environmental protection. At its core is an appreciation that the achievement of a chemical process is not solely measured on yield alone but also by the degree to which it maintains the molecular wholesomeness of the raw materials utilized for production. In addition to atom economy, process efficiency evaluation is significant in sustainable chemistry development. Process efficiency considers the entire chemical manufacturing life cycle—raw material extraction to end-product distribution—and aims to minimize energy consumption and waste production at each stage. This holistic approach meets current-day sustainability goals because process-efficient processes can better minimize their footprint on the environment. Furthermore, the use of efficient methodologies in the chemical sector can have the impact of developing innovative technologies that replace traditional, resource-deficient processes. Difficult as the adoption of these changes might be, industries must find a way to adopt these new measures so that they can remain competitive and ethical in a changing market. Since industries aim to maximize atom economy and process efficiency, it becomes essential to utilize appropriate analytical tools for determining their "green" nature. Accessing productive tools to measure the sustainability of such techniques can be quite cumbersome, especially when considering the sheer amount of compounds, materials, and chemicals before analysis (Justyna et al., 2021). In this revolutionary context of sustainable practice, the study of alternative forms of energy such as hydrogen fuel also shows the importance of process efficiency in achieving a greener future (Zhu et al., 2019). By organically addressing the catalytic processes and materials used in sustainable energy generation, industries can optimize their entire process and facilitate an economically more sustainable hydrogen economy. This audio appreciation of atom economy and process efficiency is essential for fostering innovation while spearheading environmental responsibility in chemical manufacturing.

Definition of Atom Economy

In green chemistry, atom economy is a main indicator of the efficiency and sustainability of a process. Atom economy is a metric that measures the degree to which a synthesis route transforms all the materials entering the chemical reaction into the final product. At the center of atom economy is waste reduction. High atom economy implies that the majority of the reactants get transformed into the product of interest with more economic as well as environmental benefits. This does not only assist in reducing costs by putting an end to the handling and disposal of waste and procurement of raw material but also towards the general aim of sustainability. Hence, atom economy's adoption is essential in the design of both efficient and less environmentally damaging methods, particularly in the context of increased demand for sustainable products (Samir et al., 2022). Moreover, atom economy's applicability is not just in individual chemical reactions but also in entire production processes. As industries strive to adopt greener practices, understanding and optimizing atom economy become crucial for both process efficiency and resource management. For example, synthesizing biodegradable polymers can significantly benefit from high atom economy approaches, as these materials are becoming increasingly necessary for addressing environmental concerns surrounding plastic waste. Through targeting high atom efficiency responses, chemists can create materials that are not only extremely efficient in the desired function but also place as little as possible into lifecycle waste, thereby encouraging a circular economy.

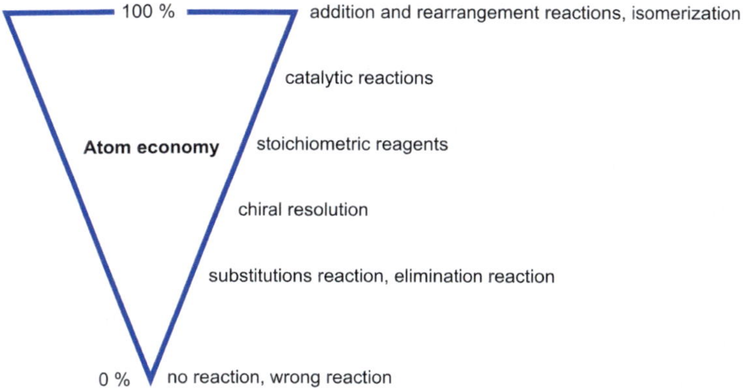

Figure 3.1. The concept of atom economy serves as a pivotal indicator of the efficiency and sustainability of chemical processes.

Source: https://upload.wikimedia.org/wikipedia/commons/a/a9/Atom_economy_V3_en.svg.

This clearly outlines the connection of atom economy to material design and sustainability development objectives, reinforcing the premise that effective chemical processes are an essential aspect of driving ecological objectives (Kabeyi et al., 2022). Most significantly, atom economy is a sophisticated concept that has relevance to various industries including pharma, materials, and energy. When industries transition toward sustainability, they must balance the consideration of atom economy with other factors including energy consumption and scalability of production. Integration of atom economy into process design necessitates an integrated strategy where environmental factors must be balanced against economic factors. For instance, when designing synthetic routes, chemists would have to evaluate not only atom efficiency but also the potential of working with renewable feedstocks and limiting greenhouse gas emissions. Such methods harmonize with the ideals of sustainable energy conversion and emphasize the need for technological development in an effort to make the most of resources while conserving the environment (Kabeyi et al., 2022). This holistic perspective on atom economy ultimately guides the changes towards cleaner industrial processes, providing a model for future research and development (Figure 3.1).

Importance of Process Efficiency

Process efficiency cannot be overstressed, particularly with regard to green practices in various industries. Efficient processes not only make operations easier but also greatly reduce wastage and consumption of resources. By maximizing the usage of resources and reducing the by-products, companies can make their processes more productive while contributing to the conservation of the environment. With the call for environmentally friendly practices, process efficiency is central to finding high atom economy. When processes are designed to generate minimal waste, the entire operation's carbon footprint is reduced, aligning with the world's sustainability initiative. The move towards the more efficient practices is also supported by developments in technology, which enable better comprehension of process inefficiencies and potential solutions to the inefficiencies. In addition to the environmental, improving process efficiency has extremely important economic profitability and competitiveness implications. Those industries that target efficient processes are likely to enjoy lower production costs and higher product quality, translating into better market positioning. For instance, the use of alternative carriers of energy such as ammonia is a good illustration of this trend. Scholars have addressed ammonias potential as a leading energy transport actor due to its carbon-neutral profile and high-density energy, the effectiveness of which is crucial in the low carbon economy (Valera et al., 2018). Companies that manage to exploit such innovations will be better positioned to thrive in increasingly competitive landscapes in accordance with evolving regulatory environments demanding higher levels of sustainability standards. Furthermore, biocatalysis innovation demonstrates a giant leap towards the improvement of process efficiency via novel technologies. The enzyme improvement through protein engineering facilitates a revolutionary approach in industrial processes, particularly in pharmaceutical synthesis (Roger et al., 2017). Not only does the technology enhance reaction efficiency and sustainability, but also the precise design of processes to meet desired parameters leads to more efficient production processes. Further, the ability to immobilize enzymes allows for stability and reuse, which greatly enhances economic efficiency and commercial viability. As industries strive to become more sustainable, the integration of biocatalysis is proof of the absolute importance of process efficiency in achieving not just economic but also environmental objectives in current production processes (Figure 3.2).

Figure 3.2. Efficient processes not only streamline operations but also substantively reduce waste and resource consumption.

Source: https://greenbankwastesolutions.com/wp-content/uploads/2023/12/Baled-waste-outside-scaled-3.webp.

Overview of the Relationship Between the Two Concepts

The combination of atom economy and process efficiency is pivotal in enhancing the practice of sustainable chemistry. Atom economy centers on transforming reactants to products with virtually no waste, essentially advocating use of atoms in chemical reactions. Process efficiency has a direct link with atom economy because process efficiency quantifies the overall performance of a reaction in terms of time, expenses, and impacts on the environment. The atom economy of target compounds does not only require maximum atom utilization but also utilization of lean and efficient process steps. Therefore, a highly atom economy reaction can be inefficient if it consumes extremely long reaction times or is marked by high energy consumption. The interplay between these themes highlights the importance of developing sustainable approaches that both enhance chemical conversions and minimize environmental footprint. Application of transfer hydrogenation methods is an ideal illustration of concurrence of atom economy and process efficiency in synthetic organic chemistry. Transfer hydrogenation is a valuable pathway towards selective reduction of complex organic molecules through high atomic efficiency via direct transfer of hydrogen atoms to the substrate. The use of catalytic systems, as with the example of hydrazine-mediated through (Michael et al., 2025), demonstrates the way high atom economy can be obtained through these reactions and also achieve unprecedented reductions in time and byproducts. With selective hydrogen transfer, the process streamlines chemical transformation and eventually abides by the principles of green chemistry. The efficiency of these processes not only enhances the sustainability of the reactions but also enhances the finding of cleaner chemical processes, which are fast becoming very popular in modern chemistry. In addition to hydrogenation reactions, multicomponent reactions (MCRs) are also a great example of how process efficiency and atom economy go together to assist in the synthesis of heterocyclic compounds. MCRs integrate a number of reactants in one process, thereby maximizing atomic utilization with minimal wastage and reaction protocols minimization. The strategic use of solid base catalysts in the reactions, as explained by Naik et al. (2025), in the case of this work, for instance, shows the connection by enhancing selectivity and reaction process minimization. By enabling catalyst recovery and reuse, MCRs follow green principles and emphasize the fundamental principles of sustainable chemistry. By enabling high atom economy and process efficiency, these protocols enable synthesis of crucial heterocycles of medicines and agrochemicals, reiterating the critical role of atom economy and process efficiency in contemporary chemical science.

3.1. HISTORICAL CONTEXT

The historical context of atom economy and process efficiency is not merely a backdrop but a turning point that has helped determine today's practices in manufacturing and chemistry. Practice in industry has made a gradual shift from the traditional, which generated a lot of waste, to more efficient techniques conserving resources. This transition gained momentum to the background of worsening environmental challenges, particularly greenhouse gas emissions from fossil fuels, which made a record 61.3% contribution to global electricity generation in 2020. International summits, such as Stockholm, Rio, and Johannesburg, placed great importance on sustainable energy development, paving the way for policies that ensure eco-efficient operations in the production of chemicals, serving as an example of the link between past happenings and ongoing advancements in atom economy (Kabeyi et al., 2022). When the demand for greener solutions increased, the approaches came to be modified to integrate not just environmental factors but also economic and social dimensions of sustainability. The traditional ideas of resource efficiency were hardly considered, and activities that generated a lot of material loss and environmental degradation prevailed. Recent developments have been guided by the principles laid out in sustainability discourses, advocating for a five-dimensional policy involving environmental, economic, social, technical, and political aspects of energy sustainability. This is vital for examining the efficacy of processes in contemporary chemistry, where atom utilization is optimized at the expense of by-products. Towards this end, the historical trend of recognizing the necessity of sustainability has provided the foundation for developing new technologies and strategies that optimize atom economy and overall process efficiency (Bogaerts et al., 2020). Furthermore, the history of atom economy and process efficiency is inextricably connected with scientific advances in catalysis and chemical engineering. The emergence of plasma catalysis, for example, is a significant leap in the quest for more efficient catalytic processes. The process has been of concern for the fact that the method has the capability to use greenhouse gases like CO_2 to make valuable chemicals, thus the necessity of innovation to address issues in the environment. Nonetheless, the effective use of such technology requires a comprehensive comprehension of the chemical processes and the interaction of the plasma environment. The history of catalytic processes evolving from ancient treatments through to sophisticated plasma-assisted processes indicates that science is dynamic and can respond to the challenges set by climate change (Bogaerts et al., 2020). This setting not only situates the advances in atom economy in their historical context but indicates that there must be a necessary interdependence between history of development and invention to come.

3.1.1. Development of Green Chemistry

Green chemistry has emerged as the central aspect of modern chemical practice, an answer to the demand for sustainability in responding to the urgent environmental issues of the day. Unlike traditional chemistry, which often prioritizes yield over environmental concerns, green chemistry aims to reduce unwanted substances in the design of chemical processes. The principles of green chemistry promote the use of renewable feedstocks, waste minimization, and enhanced energy efficiency, thereby making the shift towards practices that ensure ecological balance. With industries only now realizing the importance of sustainability, the principles of green chemistry have ushered in a change in mindset regarding chemical processes, promoting a paradigm that is as environmentally friendly as it is financially sound (Maïder et al., 2025). A principal focus in the design of green chemistry is the criterion of atom economy, with a focus on the selectivity of a chemical reaction to transform the reactants into desirable products. With the optimization of atom economy, chemists can construct synthesis routes that not only reduce waste, but also enhance the overall efficiency of chemical manufacture. This is vital to both economic considerations and environmental concern, as such processes that minimize the formation of by-products minimize the burden on waste treatment systems and guarantee the conservation of resources. Moreover, increased atom economy is in accordance with the basic principles of green chemistry, with an innovative system where sustainability can help drive technology progress (Maïder et al., 2025). The embrace of the precepts of green chemistry is also supported by an emerging regulatory system that incentivizes sustainable methods in the industry. Restrictions on harmful chemicals and incentive policies to favor cleaner processes of production have driven the shift to green chemistry. Such a regulation environment, therefore, provokes an interface between academia, the industry, and the government to facilitate a more open approach to innovation that weaves science advancement together with environmental responsibility. When industries adapt according to these guidelines, the reciprocal dynamic interplay between regulating systems and the development of green chemistry will become increasingly responsible for shaping the fate of chemical processes so that sustainability is made a part of technological advancement and economic growth (Maïder et al., 2025) (Figure 3.3).

Figure 3.3. Improved atom economy aligns with the core principles of green chemistry, fostering an innovative environment where sustainability can drive technological advancements.

Source: https://ttconsultants.com/wp-content/uploads/2023/12/Eco-Friendly-Alchemy-Redefining-Chemical-Innovation-for-Sustainability-1.webp.

3.1.2. Evolution of Atom Economy Concepts

The evolution of atom economy principles reflects a fundamental shift in the mindset of chemical synthesis and process optimization over the decades. The classical metrics within chemistry first emphasized yield and purity at the cost of the environmental impact of the chemicals themselves. As sustainability became more emphasized, though, researchers realized that it is important to be able to use every atom in a reaction completely, which gave rise to a new definition of efficiency. This change developed atom economy as a fundamental concept, which placed emphasis on keeping waste to a minimum but producing the desired product. Through a focus on strategic use of reagents and process pathway optimization, chemists have not only increased the efficiency of the process but also produced a more sustainable chemical manufacturing process aligned with current environmental requirements. Building of atom economy principles has also occurred simultaneously with progress in chemical technologies and techniques. Electrochemical reactions, for example, have been advocated for their potential to improve atom economy via the provision of lower-step reactions and less waste. Recent reviews have demonstrated how electrochemical technologies can be applied towards remediation of troublesome organic pollutants in wastewater, suggesting a promising avenue for achieving higher atom efficiency in addition to resolving environmental issues. Here, the marriage of atom economy principles and waste treatment technologies permits useful chemical intermediates to be generated from by-products, illustrating the capability of abstract principles to be brought into beneficial practice. Such a success clearly indicates that the implementation of atom economy can lead to substantial contributions to process efficiency and environmental sustainability. Secondly, the interaction between biomass-derived products and atom economy postulates is a necessary area of research as companies attempt to discover substitutes for fossil inputs. Biomass is a renewable, CO_2-neutral product, which gives it a central role in the formation of green chemistry. The new combination of biomass compound oxidation with hydrogen evolution reactions not only boosts yields of high-value chemicals but also keeps atom economy principles in mind through low energy requirements and waste generation. Experiments have analyzed reaction pathways and catalyst requirements critically in effectively converting biomass into valuable products, demonstrating a highly advanced understanding of how to extract maximum value from renewable feedstocks. Last but not least, these developments demonstrate the ongoing evolution of atom economy concepts as they adapt to this era's sustainable energy and utilization challenges, securing their place in the overall process efficiency paradigm.

3.1.3. Key Figures and Milestones in Process Efficiency

The concept of process efficiency has evolved significantly, as can be seen through the works of various great minds in diverse fields. Pioneers like Frederick Winslow Taylor, who came up with the concept of Scientific Management back in the early 20th century, devised basic principles that encouraged productivity through time and motion studies. The concepts served as a basis for modern attempts at attaining efficiency in manufacturing and business processes, calling for systematic optimization. Secondly, the introduction of Lean Manufacturing ideology, made famous by the likes of Taiichi Ohno at Toyota, focused on reducing waste and optimizing value addition.

Their methodology became definitive in reshaping methods of production, providing a platform that integrates performance well with the philosophy embedded in atom economy. This past milestone illustrates how previous success contributes to the ongoing quest for efficiency within industries, impacting economic outcomes as well as environmentally friendly processes. In recent environments, technological innovations have significantly influenced process efficiency. Advances in intelligent computing have been a revolutionary force in this aspect. Merging artificial intelligence with big data analysis has made organizations able to streamline their operation workflows and identify clogs with greater accuracy. Intelligent computing facilitates multiple paradigms of computing beyond the traditional, such as cognitive and autonomous intelligent systems. This paradigm shift has broad implications for process-efficiency industries, driving the technology of automation and predictive maintenance. Through these technologies, as firms take advantage of them, they can have their processes optimized for increased atom economy, leading to additional sustainability in chemical production and more.

The marriage of process efficiency and technological innovation is a watershed moment in the history of industrial practice. Furthermore, the development of high-performance materials, like silicon carbide (SiC), opens new avenues for process efficiency improvement in diverse applications. SiCs special electronic and photonic properties are revolutionizing the field of device fabrication, especially in high-power and high-frequency applications. This material's tremendous progress over the past two decades highlights its diversity, and it is a focal point of attention for scientists and practitioners engaged in development of photonics integration with electronics. The search for SiC as a viable material in quantum applications is also testament to the intersection between process efficiency and materials science. The progression of SiC technology mirrors the necessity for industries to adopt advanced materials that enhance not just product performance but also manufacturing efficiency. Therefore, these developments are yet another crucial step in the ongoing saga of process efficiency and atom economy because they combine classical and contemporary methods for enhanced outcomes (Figure 3.4).

Figure 3.4. Frederick Winslow Taylor.

Source: https://upload.wikimedia.org/wikipedia/commons/thumb/1/14/Fredrick_Winslow_Taylor_c1907_retouched.png/800px-Fredrick_Winslow_Taylor_c1907_retouched.png.

3.2. PRINCIPLES OF ATOM ECONOMY

The principles of atom economy substantially redefine industrial processes and chemists' approach to synthesizing chemical compounds. Atom economy calculates the efficiency of a chemical reaction in utilizing the raw materials towards yielding the product of interest with little waste. This metric has become increasingly significant as industries are compelled to be greener. Through an emphasis on responses facilitating atom use, industries can reduce raw material expenses and ecological footprint. Modern research reveals that development of advanced biorefineries focuses on the use of renewable and sustainable feedstocks to achieve maximum energy gains without consuming non-renewable materials, which is in direct alliance with overall goals of atom economy. This consistency emphasizes the importance of incorporating considerations of atom economy into the development of chemical synthesis pathways to make the entire process more efficient and eco-friendlier (Arndt et al., 2007). Furthermore, atom economy principles are not limited to economic need; they coincide with environmental and societal needs. Biochemical processes, such as producing biofuels from sustainable sources, show the application of atom economy principles in practice. The atom-efficient processes also yield more with fewer toxic byproducts that more traditional synthesis protocols would produce. For instance, the future microalgae and crop biomass biorefineries promise higher resource efficiency through synergistic process intensification technologies. These processes not only minimize energy input but also lower chemical intake and accompanying environmental expenses. Consequently, the research is directed towards methodologies that achieve maximum atom economy and process efficiency, thus demonstrating an integrated approach to sustainable chemical production (Arndt et al., 2007). Alongside this is the fact that new technologies provide compelling prospects for enhancing atom economy in most chemical processes. Direct synthesis of nitrogen oxides from nitrogen and oxygen through non-thermal plasma, for instance, is a novel method of overcoming conventional methods such as the Haber-Bosch process. The method reduces emissions as well as chemical transformation under reduced temperatures, which is a prime example of an effective method of maximizing atom economy. The role of catalysis in this process cannot be overstated, since the use of appropriate supports and reaction conditions has a direct bearing on atom use efficiency. Work illustrates that there are catalysts that have the potential to increase significantly target product concentrations by sacrificing decreased energy consumption. Lastly, by implementing these new technologies and abiding by atom economy rules, the industries can become more efficient and environmentally friendly in operations and thus effectively address the environmental challenges of the contemporary world.

3.2.1. Calculation of Atom Economy

It is crucial to comprehend atom economy calculation in assessing the efficiency of chemical processes. Atom economy describes the measure of the extent to which the reactants of a chemical reaction are utilized in creating the target product, based on the minimization of waste. It is calculated by the molar mass of the desired product divided by the total combined molar mass of the entire suite of reactants and then multiplied by 100. A higher percentage of atom economy indicates a more efficient process, consistent with those of green chemistry and sustainable processes. By favoring reactions with fewer by-products, chemists avoid wasteful use of resources and damage to the environment, resulting in a greener industry. So, atom economy calculation is not merely a helpful measure but also a design philosophy for more economical chemical processes. Lately, the significance of atom economy has picked up steam due to the growing environmental regulations and worldwide movement towards sustainable development. Industries are now interested in optimizing their processes in order to obtain higher atom economy values and so reduce the amount of toxic waste produced and bring down the cost. For instance, in renewable energy production of hydrogen, atom economy knowledge can impact the choice of reactions and catalysts that have high conversion efficiency. As discussed in the literature, advancements in electrocatalytic hydrogen evolution reaction (HER) have been central to achieving high atom economy values in hydrogen production processes, particularly with low-cost and high-performance catalysts. On this view, the drive towards enhancing atom economy benefits not just environmental but also economic effectiveness in industrial processes. Studies on alternative energy carriers, namely ammonia and methanol, demonstrate the practical significance of atom economy for prospective technologies. These carriers have the advantage of offering good alternatives to hydrogen transport and storage, addressing problems associated with transporting hydrogen directly. Ammonia, which possesses high atom economy and well-established infrastructures, is the best example of how atom economy can assist in transition to low carbon energy systems. By maintaining that energy carriers should be produced with high atom economy, industries can become more financially viable as well as environmentally friendly. Furthermore, pioneering studies on the efficient use of ammonia for power generation highlight the importance of integrating atom economy in the design of the future energy technologies, citing it as crucial in the attainment of sustainable energy solutions (Figure 3.5).

Figure 3.5. Atom economy refers to the measure of how well the reactants in a chemical reaction are used to produce the desired product, with a focus on minimizing waste.

Source: https://cards.algoreducation.com/_next/image?url=https%3A %2F%2Ffiles.algoreducation.com%2Fproduction-ts%2F__S3__0f3c582a-171c-4872-9244-ab3651a62ca9&w=3840&q=75.

3.2.2. Factors Influencing Atom Economy

In green chemistry, a number of factors are critical in atom economy, which has been defined as the efficiency in product formation from reactants. One of the key contributors to improved atom economy is catalyst selection. Catalysts enable chemical reactions without being consumed in the reaction so that more molecular conversion with very little byproduct can be made. By accelerating reaction rates and facilitating reactions under milder conditions, catalysts reduce the amount of energy input in synthesis, finally yielding enhanced levels of target products. This is in line with the principles of green chemistry, which promote the reduction of hazardous materials and maximization of resource utilization, thereby promoting environmentally friendly processes in chemical manufacturing (Sakhahari, 2025).

Figure 3.6. Catalysts promote chemical reactions without being consumed in the process, allowing for a higher degree of molecular transformation with fewer by-products.

Source: https://upload.wikimedia.org/wikipedia/commons/9/9d/Catalysts.JPG.

The reaction route selection and design is also a significant factor influencing atom economy. Choice of the synthesis route through careful selection can maximize the overall efficiency of a chemical process. Reactions that minimize side product formation maximize atom economy because they leave most of the reactants in the product. For instance, multicomponent reactions involving the simultaneous formation of multiple bonds in a single operation are characteristic of strategies with the potential for very high atom economy. By employing such practices, chemists can in fact reduce waste generation while recovering useful compounds to the fullest degree. Through methodology advancement and integration, by the way, shows promise for efficient synthesis routes that are consistent with sustainability and reachable with economic viability (Saffar, 2025). Moreover, physical properties of reactants and products also play a major role in atom economy. Molecular weight of reactants and product structure design can impact the yield of waste and reaction efficiency. For instance, reactions producing lightweight or smaller molecular-sized products could be more atom-economical than those that produce bulky or complex products. In addition, solubility and stability are also definite parameters that affect reaction conditions that could, in return, affect atom incorporation efficiency in target products. Such intricate interrelatedness facilitates the identification of devoted compounds and catalysts that maximize atom economy, thereby facilitating realization of more general goals of process efficacy and environmental sustainability (Sakhahari, 2025) (Figure 3.6).

3.2.3. Examples of High vs. Low Atom Economy Reactions

In organic chemistry, atom economy is a major efficiency benchmark of chemical reactions. Highly efficient atom economy reactions are highly minimized in waste and utilize maximum amounts of the reactants, an ideal best exemplified by enyne carboalkoxycyclization. This Pt-promoted catalyst-induced reaction is greatly selective, giving functionalized five-membered carbo- and heterocycles with very high efficiencies, showing enantioselectivity as high as 85% (Charruault et al., 2006). By making use of the best-suited combination of silver salts and an atropisomeric ligand, this approach not only ensures maximum product yield but also shows how cautious catalyst selection may be compatible with green chemistry, thereby ensuring minimum environmental impact while ensuring maximum utilization of the reactants. Alternatively, however, low atom economy reactions do generate a high proportion of waste materials to desired products. They typically take many steps and the creation of by-products that can counteract efficiency. One such example is the traditional use of halogen oxidation, where organometallic complexes of halogen metallic salts have functioned as catalysts. While effective, the technique tends to be plagued by extensive wastage, and is therefore inferior in terms of sustainability. However, newer techniques have been formulated to address such weaknesses through the implementation of cyclodextrin-coupled complexes. Such adaptations not only reduce catalyst amounts, but also inhibit by-product formation, leading finally to more favorable results in both efficiency and environmental concerns (Rossi et al., 2012). Lastly, the comparison between high and low atom economy reactions underscores the importance of process efficiency in organic synthesis. High atom economy is a modern trend that advocates for the utilization of advanced catalysts and methods to enable greener practice in the discipline. Conversely, addressing the inefficiencies associated with low atom economy reactions is important in further developing the field towards more environmentally friendly methodologies. Through constant demands for improvement, chemists are able to effectively influence chemical production chemistry in terms of bigger environmental goals as well as enhancing the economic viability of chemical operations. Atom economy is, therefore, a ground zero in promoting progress in the science of organic synthesis.

3.3. PROCESS EFFICIENCY IN CHEMICAL MANUFACTURING

In the quest to optimize process efficiency in chemical manufacturing, of crucial importance is the use of sustainable processes such as Green Chemistry. By reducing waste production and employing renewable materials, the chemical industry can significantly lessen its impact on the environment. For instance, the focus on atom economy—equilibrating the mass of reactants to that of desired products—can lead to fewer by-products and wastes. The drug industry, in turn, has embraced these principles with enthusiasm as a try to account for its own substantial portion of chemical contamination. As highlighted by recent literature, the tenets of Green Chemistry are not mere theory; they are actual practices which make catalysis and solvent choice paramount, render chemical transformations safer and more economized, and thus demonstrate the efficacy of innovation in environmentally friendly production practice (Achieng, 2025). Catalysts are crucial in enhancing process efficiency by reducing chemical reaction energy barriers, accelerating reaction rates, and enhancing yields. The trans-methylation reaction provides insight into how Brønsted-acid catalysts can enhance conventional approaches to more efficient processes. Using highly sophisticated computational techniques such as density functional theory (DFT), researchers have been able to establish an understanding of such catalytic processes, highlighting how catalysts could facilitate reaction kinetics significantly. As witnessed by the dramatic reduction of energy barriers—over 40 kcal/mol in the majority of cases—these findings underscore the importance of further research and application of these catalysts in industry. Tuning reaction mechanisms and activating molecular reactivity, catalytic reactions can produce crucial compounds more effectively at environmental norms (Ang et al., 2025). Lastly, where process efficiency and atom economy converge, is a paradigm shift in chemical manufacturing. The green synthesis of the widely used chemicals benzene, toluene, and xylene not only benefits many industrial processes but also presents challenges to environmental sustainability. Adoption of greener synthesis pathways and enhanced catalytic efficiencies presents the manufacturer with an opportunity to align with circular economy principles. This merger guarantees a reduction of toxic by-products while guaranteeing maximum use of resources. By promoting a culture of ongoing improvement and innovation, the chemistry sector can step towards processes that not only yield high-value products but also prioritize ecological balance, thus guaranteeing process efficiency together with environmental stewardship (Achieng, 2025) (Figure 3.7).

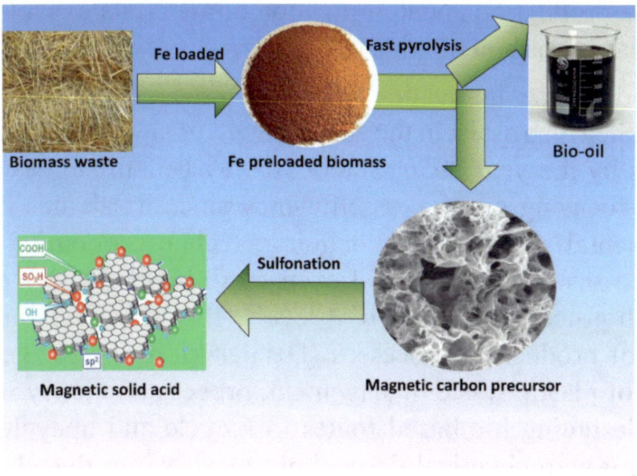

Figure 3.7. The trans-methylation reaction exemplifies how Brønsted-acid catalysts can transform conventional methods into more efficient processes.

Source: https://media.springernature.com/full/springer-static/image/art%3A10.1038%2Fsrep02419/MediaObjects/41598_2013_Article_BFsrep02419_Fig1_HTML.jpg.

3.3.1. Definition and Importance of Process Efficiency

In contemporary chemical manufacture and production, process efficiency is a determining parameter in economic viability as well as environmental sustainability. Process efficiency is the ratio of useful output to the total input sum of any process. Process efficiency comprises an array of parameters including material efficiency, energy efficiency, and waste generation. High process efficiency is particularly crucial for production industries looking to reduce expenditure as well as prevent wastage of resources. By concentrating on optimizing the outputs in terms of minimizing the inputs, businesses are not only able to maximize profitability but also minimize their impact on the environment. The relationship between atom economy and process efficiency stresses the fact that high efficiency is often at the expense of requiring innovative solutions to chemical reactions and materials design, highlighting the importance of these principles in the global context of sustainable chemical engineering practices. Growing focus on process efficiency is being fueled by the imperative for companies to adopt sustainable processes as a means to stem climate change. As led by developments in CO_2 adsorption technology, improvement of process efficiency can result in significant greenhouse gas emission reduction. Research on methods for CO_2 capture is indicative of priorities on the production and optimization of materials capable of efficiently adsorbing industrial carbon dioxide emissions.

Not only do these innovations enhance process efficiency in chemical industries, but they also assist in the achievement of aggregate goals such as net-zero emissions by the year 2050. Such two-way benefit suggests the strategic importance of focusing on process efficiency since it satisfies both economic and environmental needs. The ongoing techno-economic investigations into such processes within this field even more highlight key parameters for scaling up such processes to industrial levels, enhancing feasibility as well as sustainability of production processes (Dziejarski et al., 2023). Furthermore, in the context of plastic waste management, process efficiency is an important parameter in designing bio-based routes to recycle and upcycle plastics. The increasing use of petrochemical-derived plastics has introduced unprecedented environmental issues, and there is a requirement for innovative solutions that can support circular economy. Recent advances in enzyme-based biocatalysis highlight how optimization of the process can transform waste materials into valuable resources, thereby simplifying efficiency in waste treatment processes. By focusing on creating effective, enzyme-based techniques for breaking down commodified plastic, researchers are moving towards plastic governance solutions that are sustainable. This development not only reduces reliance on incineration and landfilling but also highlights the potential of a recyclable plastics lifecycle through effective recycling technologies (Tournier et al., 2023). Therefore, integration of process efficiency in the establishment of bio-catalytic approaches not only aligns with sustainability goals but also increases economic viability.

3.3.2. Metrics for Measuring Process Efficiency

Measurement of process efficiency in chemical manufacture is central to promoting sustainability and waste minimization. Atom economy and the E factor are two of the most widely accepted measures in these respects. Atom economy measures the selectivity of a chemical reaction in terms of the proportion of reactants that successfully undergo conversion to desired products, as given in the formula:

A high atom economy indicates that the reaction was more efficient, with an aim towards 100% in order to minimize waste production. This is consistent with the principles of sustainable processes, forcing chemists to optimize reactions and select procedures that promote green outcomes, such as the use of green catalysts, which are essential to improve atom efficiency as well as process efficacy (Prabhu, 2020). Other than atom economy, the E factor is an ancillary metric that quantifies the environmental expense of waste per unit of product formed. This number, as kilograms of waste produced per kilogram of product, provides a complete understanding of process efficiency. Some of the advancements over time have enabled distinguishing between E factors that are simple and those that are complete, enabling a clearer understanding of waste in the process.

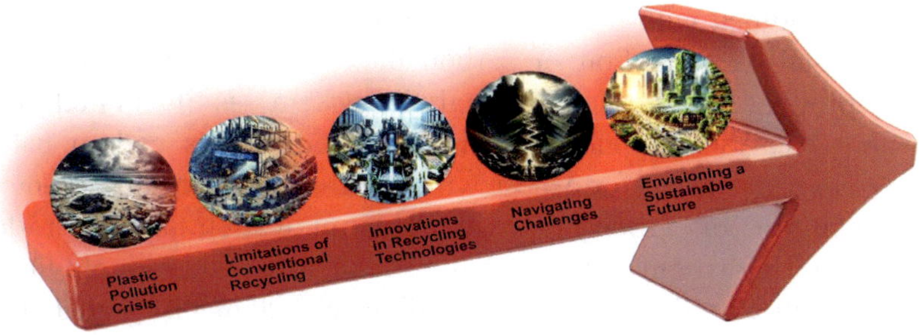

Figure 3.8. Over the years, various refinements have emerged, enabling distinctions between simple and complete E factors, fostering a more nuanced understanding of process waste.

Source: https://pub.mdpi-res.com/sustainability/sustainability-16-10401/article_deploy/html/images/sustainability-16-10401-g001.png?1732775327.

Additionally, emphasizing the mass-based measurements has brought us Process Mass Intensity (PMI) and Reaction Mass Efficiency (RME), thereby providing further proofs for considering chemical processes from the perspective of sustainability through a multi-faceted methodology. Such quantifications not only establish benchmarks for industrial operations but also trigger ongoing improvement toward greener approaches (Sheldon, 2018). While atom economy and the E factor are significant mass-based quantifications, they need to be complemented by assessments based on greater scales of environmental as well as financial impacts. Life Cycle Assessment (LCA) is a further mechanism that adds more complexity to the analysis through the quantification of the overall environmental impact throughout a products life cycle. Through the integration of life cycle perspectives with mass-based measurement, chemists can build a more integrated picture of sustainability with the ability to make well-informed decisions when designing and optimizing chemical processes. Last but not least, the unification of these metrics forms a culture of continuous improvement consistent with the wider goals of sustainable chemistry: waste minimization, efficiency maximization, and promotion of environmentally responsible practice in chemical manufacture (Prabhu, 2020) (Figure 3.8).

3.3.3. Case Studies of Efficient Chemical Processes

The research on efficient chemical processes has been of significant interest, especially in terms of atom economy and process efficiency. The most sophisticated methods, such as electrocatalysis, have the capability to bring about tremendous improvements in order to reduce greenhouse gas emissions combined with the optimal use of resources. A good case study is the electrocatalytic reduction of carbon dioxide to valuable chemicals and fuels. Employing a novel electrocatalytic system, as illustrated by nickel single-atom catalysts supported on sulfur-doped porous carbon nanofibers, has been observed to perform best with a Faradaic efficiency of more than 95% for synthesis of carbon monoxide. This innovation not only reflects the significance of high atom economy but also the potential of employing CO_2 to generate valuable products and hence aiding in sustainability goals (Varad et al., 2025). Moreover, multicomponent reactions (MCRs) are another efficient method of acquiring effective chemical synthesis. MCRs are applied with multiple reactants in a single-step reaction to produce complex products with high productivity and less waste. A demonstration example of the process is the spiroheterocycle synthesis with taurine as a green catalyst. The method efficiently brings isatins, 1,3-diketones, and 1-naphthylamine together in water as the solvent to provide high purity and good yields of 80–95% within 15–20 minutes. The solvent-free nature of using water and the recyclability of taurine promote a green process without the loss of efficiency. The inherent simplicity and operational convenience of this process are the hallmark of the versatility of MCRs in current synthetic chemistry, reinforcing the role of atom economy (Verma et al., 2024). Briefly, the case studies of successful chemical processes underscore the necessity of embracing frontier strategies for atom economy and environmental footprint minimization. Through application of state-of-the-art methods such as electrocatalysis and MCRs, chemists can address the pressing needs of sustainability in chemical manufacturing. Not only do these strategies lead to the generation of valuable products from wastes, but also to a circular economy in chemical manufacturing. And as the research continues to uncover new catalysts and more environmentally friendly reaction conditions, the potential for radical improvement of both process efficiency and environmental performance will continue to grow. Such studies serve as a catalyst for further research in the area, ultimately to spur the goals of sustainable chemical practice (Varad et al., 2025; Verma et al., 2024).

3.4. ENVIRONMENTAL IMPACT

The environmental impact of industrial processes is a serious problem in modern society, particularly for industries such as pharmaceuticals, where the production of chemicals may cause significant pollution and waste. Application of the principles of Green Chemistry is a crucial aspect in mitigating these adverse effects through a focus on atom economy and process efficiency. Optimizing the utilization of materials in final products reduces waste and the environmental footprint of chemical manufacture. Acting to use safer, renewable materials first and reduce energy demands, for example, saves money but also responds to increasing regulatory calls for drug manufacture to be more sustainable. Applications of atom economy enhance efficiency through the reduction of byproducts, thus countering the issue of environmental pollution brought about by traditional production processes, as is observed in recent literature on sustainable development of pharmaceuticals (Achieng, 2025). In electrochemistry, new approaches that include Green Chemical Principles offer new solutions for minimizing environmental harm. Green electrodeposition, for instance, is a process that has high respect for waste effluent management and the use of non-toxic materials. Material utilization is efficient through low-temperature and low-pressure reactions, minimizing waste production to very small levels against conventional processes. Using energy and atom efficiency measures, electrochemists are able to design environmentally friendly processes that are not only compliant with current environmental regulations but also result in successful management of resources. Encouraging the transition to green technology in electrochemical application is to lead the way into more open environmental practice in the manufacturing sector required for sustainable growth and smaller ecological footprints (Roy, 2022). Ultimately, the synergy of atom economy, process efficiency, and environmental footprint begins to be more universally seen as a corner stone of sustainable industrial practice. Practice of Green Chemistry not only proves to be more regulatory compliance but also fosters breakthrough innovation in various industries. As industries look towards increased efficiency and less wastage output, educational coursework and research on sustainable processes must be taken into account as a necessity. By fostering an environment that also focuses on environmental conservation as much as economic viability, future generations will be more able to address the ecological concerns related to manufacturing. The convergence of productive methods and sustainable practices offers an environment for natural rehabilitation and resilience with a focus on redirecting success in terms of ecological and social welfare rather than purely financial language (Achieng, 2025) (Figure 3.9).

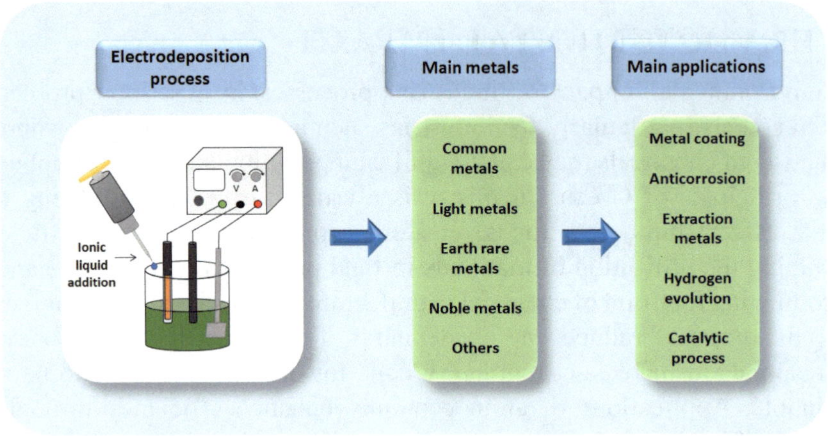

Figure 3.9. Green electrodeposition, for instance, exemplifies a sustainable approach that prioritizes the control of waste effluents and the use of non-toxic materials.

Source: https://pub.mdpi-res.com/metals/metals-12-02095/article_deploy/html/images/metals-12-02095-ag.png?1670323561.

3.4.1. Role of Atom Economy in Reducing Waste

Atom economy is the focus of the search for sustainable chemical processes, particularly waste minimization. Atom economy is the ability of reactants to be transformed into useful products without much wastage, with the focus that as many atoms be included in the end product as possible. Traditional synthetic procedures usually result in plenty of by-products, which form waste, and therefore cause both environmental and economic issues. By giving top priority to improving atom economy, it is possible for chemists to minimize waste generation to a large extent, thus endorsing green chemistry principles. For instance, new processes such as electrochemical CO_2 reduction show how improved atom economy, apart from reducing the accumulation of harmful by-products, allows waste to be converted into valuable chemicals in line with global efforts towards sustainability (Ha et al., 2024). In addition to enhancing efficiency, atom economy-driven strategies can lead to drastic reduction in the utilization of resources. One of the best examples of this practice is the use of room-temperature ionic liquids (RTILs) and deep eutectic solvents (DESs) in organic electrosynthesis. RTILs and DESs have the ability to dissolve organic compounds efficiently as well as serve as electrolytes, reducing the need for large amounts of supporting electrolyte salts. This diminution of additional components is to boost the reduction in waste directly, thus making the process more sustainable by and large. The recyclability of DESs also allows for more than one reaction cycle, which not only boosts atom economy but also encourages the closed-loop approach to utilizing resources in organic synthesis (Steffensen et al., 2024). The employment of these solvents is symptomatic of a broader shift towards cleaner processes that prioritize efficiency and environmental responsibility. Ultimately, the achievement of high atom economy will not merely be an issue of technical efficiency; it is also going to require a cultural shift in the chemical industry. Stakeholders including researchers, educators, and policymakers will need to adopt and encourage practices grounded in sustainability. This involves not only the development of new methodologies with atom economy included but the education of the next generation of chemists in minimizing waste in process design in chemistry. By instilling a culture with high respect for atom economy, the chemical industry can facilitate significant advancement toward reducing waste with potential to revolutionize chemical manufacturing. With rising global challenges such as climate change and consumption of resources, atom economy principles will come in handy in ensuring the chemical industry is not just sustainable but also promotes ecological balance and sustainability (Ha et al., 2024; Steffensen et al., 2024).

3.4.2. Process Efficiency and Resource Conservation

It is crucial to grasp how process efficiency and resource conservation work together in designing sustainable practices in various agricultural and industrial businesses. In particular, the use of resource-saving technologies has acted as a potent impetus to enhance productivity and avoid losses. For example, evidence confirms that innovative strategies such as zero tillage in wheat and direct seeded rice (DSR) are instrumental in India's Indo-Gangetic Plains' rice-wheat production system. These practices not only increase the yield but also reduce overall cost of production. To be precise, implementation of these resource conservation approaches has been found to result in the reduction of farmers' costs per hectare by 20–22%, thereby boosting net incomes by roughly 25–30%, thereby linking economic sustainability with sustainable agriculture (Kumar et al., 2025). In addition, making manufacturing and industry processes more effective usually involves adopting circular economy-based sustainability principles. By adopting design innovation and manufacturing process innovation on the basis of current equipment design, businesses can save vast resources and reduce waste creation. Setting an example using current machine tool components, such as stamping presses used in automobile production, has the beneficial feature of using materials anew. A quantitative study on such prevailing factors that was part of research proved that algorithm development can assist in thorough analysis of manufacturing cost and delivery time. These results not only impact decision-making but also justify incorporating resource conservation along with operational efficiency, thus making the environment sustainable for manufacturing (Centoamore et al., 2024). Incorporating practice-based training and monetary incentives for practicing such advanced resource conservation strategies is essential. The barriers to effective implementation, such as the high cost of machinery and shortage of skilled personnel, should be addressed through policy interventions and targeted support services. By prioritizing improved extension services and training in resource management at the top, players in the agricultural and industrial sectors can significantly increase the level of adoption of resource-conserving technologies. Lastly, equipping practitioners with the necessary tools and information will promote a culture of continuous improvement consistent with the general principles of atom economy and process efficiency, and serve as an impetus for environmentally friendly advancement across various disciplines. These activities, synergistically combined, can lead to worthwhile enhancements in environmental conservation and economic resilience (Kumar et al., 2025; Centoamore et al., 2024) (Figure 3.10).

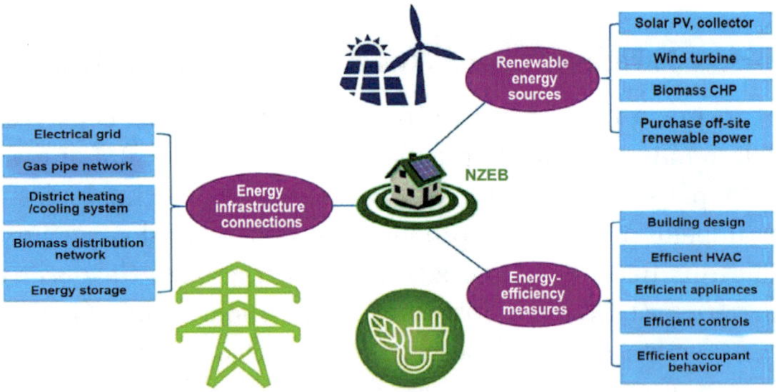

Figure 3.10. The application of resource-saving technologies has emerged as a significant factor in enhancing productivity while minimizing waste.

Source: https://pub.mdpi-res.com/buildings/buildings-13-02658/article_deploy/html/images/buildings-13-02658-g001.png?1697958305.

3.4.3. Long-Term Environmental Benefits of Improved Practices

The pursuit of improved practices in industrial processes is, in fact, directly related to enhancing atom economy and process efficiency. By employing greener methods, industries can reduce waste and utilize more raw material, thus aligning with sustainable development principles. This not only provides economic gains for companies, in terms of cost saving and rationalization of resources, but also ensures environmental responsibility. As governments and regulation authorities increasingly encourage sustainable methods, businesses using improved methods position themselves favorably regarding compliance as well as competitiveness in the market. The shift is not a reaction one but rather an outcome of an active strategy to global challenges, as has been seen with recent policies making sustainability a priority in innovation policy design. Through the integration of such improvements, firms provide long-term environmental benefits that resonate in society and ecosystems, demonstrating the relevance of innovation in addressing pressing environmental concerns and achieving strategic goals (Schot et al., 2018).

For such improved practices, the use of biocatalysis emerges as a driving force towards sustainability. Advancements in molecular biology and biotechnology have made it possible to design novel enzymatic reactions, which can be tailored to meet precise needs of environmental safety and efficiency. The inherent properties of biocatalysts, including the possibility to operate under mild conditions and limiting the utilization of toxic solvents, make them highly suitable for industrial applications. Immobilization of enzymes also facilitates their reuse, reducing waste generation and enhancing process sustainability. Thus, biocatalysis is strongly in sync with the objectives of green chemistry and provides a perfect example of how industries can innovate to achieve both economic and environmental efficiency. The widespread application of biocatalysis, especially within the pharmaceutical industry, reflects the transformative power that improved practices hold for the development of a biobased economy that is not only economically viable but environmentally aware (Roger et al., 2017). Moreover, long-term environmental gains of improved practice are present in the longer-term impacts of sustainability initiatives on global ecosystems. Improved efficiencies of process can minimize pollution and resource depletion and promote biodiversity by way of diminished environmental footprints. For example, the integration of circular economy principles makes reprocessing and recycling of materials possible, resulting in considerable reduction in waste creation and promoting an integrated perspective towards resource management. As companies adopt these practices, they help meet targets established by frameworks such as the Sustainable

Development Goals that emphasize the requirement for change at the socio-technical level. By prioritizing sustainable innovations, industries not only save natural resources for future generations but also create strong ecosystems that are capable of being resilient in the face of climate change impacts. Thus, the inclusion of better practices is an intrinsic method of ensuring a sustainable future where economic growth is obtained without compromising ecological integrity (Schot et al., 2018) (Figure 3.11).

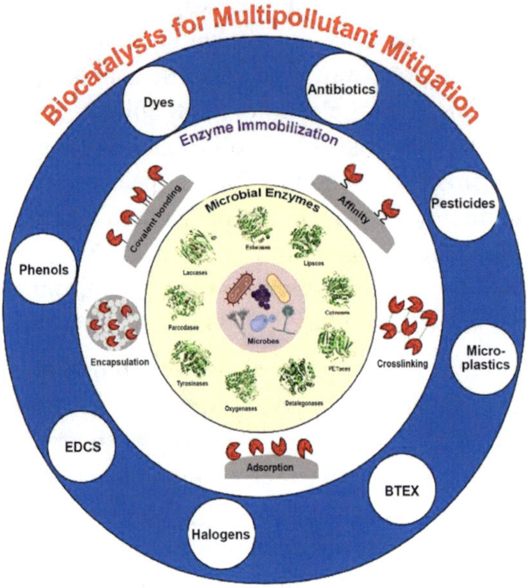

Figure 3.11. The intrinsic properties of biocatalysts, including their ability to operate under mild conditions and reduce reliance on hazardous solvents, greatly enhance their suitability for industrial applications.

Source: https://www.mdpi.com/ijms/ijms-25-08616/article_deploy/html/images/ijms-25-08616-g001.png.

3.5. ECONOMIC CONSIDERATIONS

In the field of 'Atom Economy and Process Efficiency,' there is a need to understand the economic considerations because it greatly contributes to fostering innovation and sustainability. A good atom economy is not just an environmental goal; it significantly affects the cost-effectiveness of chemical processes. When reactions yield more useful material than waste, operational costs associated with handling waste and supply of resources are lowered. This implies that less of the resources are being consumed, cutting down on the overall economic cost of production facilities. As such, industries that specialize in the efficient use of atoms can realize improved competitive edge in the marketplace. Synchronization of economic incentives and sustainable practices can precipitate a paradigm shift among the chemical industry, contributing to higher application of methods emphasizing efficiency and lower environmental impact, in harmony with existing demands for sustainability. Besides, the adoption of the principles of green chemistry directly relates to economic considerations. The intentional selection of reagents and solvents can optimize economic benefits at the expense of minimizing environmental effects. Proper solvent selection not only translates to higher sustainability but also enhances process productivity. For instance, traditional solvents typically have risky waste liabilities, which engage additional costs. Conversely, the creation of greener alternatives, such as ionic liquids or supercritical fluids, demonstrates the possibility of dramatically lowering ecological footprints. Nevertheless, it was observed that the application of these technologies will need to be coupled with enhanced commercial performance for it to be successful. This intersection of sustainability and profitability forms the basis of the general argument that efficiency in chemical processes can create profitable results while encouraging environmentally friendly practices. Finally, the transition to a circular economy presents a wealth of economic advantages that reinforce the argument for improved atom economy and process efficiency. As markets increasingly demand accountability for waste and resource use, businesses must adapt to this new world order. Through redesigning waste as a resource rather than an end product, businesses can design products for recycling and reuse right from the beginning. This is reflected in the ethos presented in green chemistry, where product design becomes the essence of sustainable design. More specifically, chemists today are compelled to develop materials not only to address market needs but to help deliver a wasteless future (Clark et al., 2016). The economic benefits derived from effective resource systems can therefore lead to lower production costs, higher customer loyalty, and a reputation for environmental care, all being essential in a competitive and green-minded business environment.

3.5.1. Cost-Benefit Analysis of High Atom Economy Processes

The atom economy principle is of crucial relevance to the assessment of chemical processes, particularly in determining the cost-benefit analysis of high atom economy processes. High atom economy means that a significant proportion of the initial materials consumed during a chemical reaction is used as desired products, minimizing waste. The infinitesimal creation of this waste not only decreases material expenses but also minimizes environmental effects arising from disposal. With industries considering more sustainable practices, it becomes significant to understand the economic effect of adopting high atom economy processes. The initial costs of such processes might be high; however, the long-term savings resulting from the reduced use of raw material and waste handling can enhance overall profitability, offering a strong justification for adoption. When considering the advantage of high atom economy processes, it is necessary to take into account not just individual profitability but also broader societal benefits.

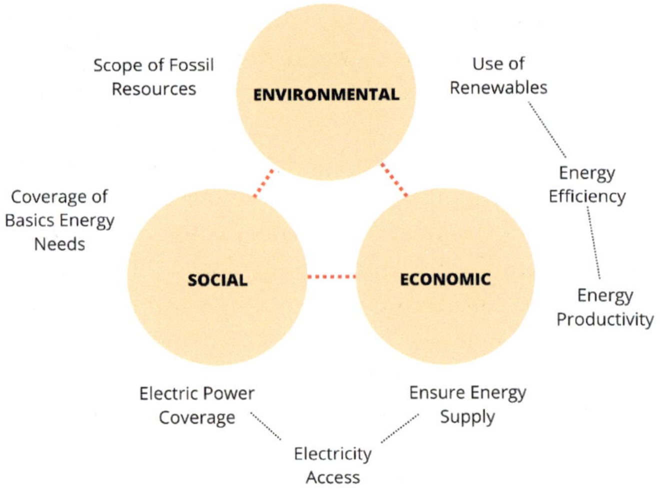

Figure 3.12. As industries strive for more sustainable practices, understanding the economic implications of adopting high atom economy processes becomes essential.

Source: https://pub.mdpi-res.com/sustainability/sustainability-15-14049/article_deploy/html/images/sustainability-15-14049-g001.png?1695356717.

The implementation of processes with an emphasis on atom economy can go a long way toward supporting a low-carbon economy by consuming less energy and producing fewer impurities. For instance, ammonia has been recognized as a potential energy carrier due to the fact that it contains high energy density and established transport infrastructure, making its potential as an energy carrier higher (Valera et al., 2018). In this case, the cost-benefit analysis must extend beyond monetary issues to include environmental benefits in the form of reduced greenhouse gas emissions and better utilization of resources. This holistic perception enables stakeholders to discover value in investing in technologies that emphasize atom economy, thereby driving the transition to more sustainable chemical practice. Of technologies in emergence, zinc-ion batteries portray the importance of atom economy in designing eco-efficient energy storage technology. As an alternative to conventional battery materials, zinc-ion batteries are attracting attention with reduced costs and improved safety features (Song et al., 2018). However, examining material utilization in zinc-ion batteries underscores the need for high atom economy on both an economic and environmental scale. The challenges involved in the creation of suitable cathode materials highlight the importance of optimization of atom economic processes towards the search for effective energy storage solutions. A complete cost-benefit analysis will not only aid in the identification of the most effective materials for the said batteries but also spur innovations aimed at sustainability goals. The strategic deployment of atom economy in energy storage systems eventually results in improved process efficiency and movement towards a greener tomorrow (Figure 3.12).

3.5.2. Impact on Industry Profitability

The profitability of industries is critically dependent on the efficiency of processes, because optimizing atom economy—a principle that focuses on reducing waste while maximizing the utility of the reactants—can minimize costs by a lot. When companies use practices that increase the value of chemical reactions, they are not only preventing harm to the environment but also maximizing the use of resources. This optimization translates directly to cost savings, with companies benefiting from reduced material costs and disposal fees. As companies focus on sustainability as much as on economic performance, investing in high atom economy processes is not only an ethical choice but a strategic one tied to enhancing product viability and market competitiveness. These changes, in addition to supporting profitability, also align with consumer expectations for environmentally friendly production processes and create a symbiotic relationship between financial health and environmental concerns. Continuing to look further at the implications of these changes, most process-efficient industries will tend to experience a favorable change in market positioning. Through the adoption of new process technologies and streamlining existing methodologies, companies can make processes easier to operate, ultimately leading to better throughput and reduced lead times. This improvement enables companies to respond more successfully to market changes and customer demands. As the market shifts towards more sustainable approaches, companies that adopt high atom economy principles not only boost their visibility but also create brand loyalty among environmentally conscious consumers. This dynamic world demands that businesses remain agile, embracing new technology and methodologies that guarantee efficiency in a bid to defend and sustain their profitability. Ultimately, process efficiency is an ant to competitive success in a world where economic and environmental challenges are inseparable. Additionally, regulatory circumstances play a role in dictating industry profitability in the context of atom economy and process efficiency. Stricter environmental laws and increasing calls for corporate transparency to sustainability can increase the cost of business for non-compliant firms. But proactive firms that make adjustments to maximize atom economy have a head start. Early movers in this regard stand to avoid fines and penalties while simultaneously gaining access to incentives that might be offered for sustainable business. For instance, businesses that comply with laws are set to benefit from incentives or tax subsidies for encouraging greener initiatives. Not only does this legal alteration encourage responsible production but also underscores the financial advantage of adopting high-efficiency strategies. With increasing standards in an industry, firms that invest in atom economy could realize greater profitability through conformity as well as differentiation in the market, ultimately transforming success in the competitive marketplace today.

3.5.3. Investment in Sustainable Practices

Investment in sustainable operations is increasingly regarded as a major factor to enhance atom economy and process efficiency across various industries. As businesses are increasingly under pressure to minimize their environmental impact, aligning the manufacturing process with sustainable goals is critical. With investment in newer technologies formulated from Industry 4.0, businesses can achieve higher production efficiencies while simultaneously addressing sustainability challenges. For instance, the integration of mobile technology and nanotechnology has the potential to contribute immensely to process efficiency, particularly in sectors such as automobiles and electronics, where the application of the two technologies can yield dramatic improvements on sustainability metrics. Moreover, companies ought to prioritize the screening of discrete technologies so that the implementation of technologies will be a significant factor in helping to meet sustainability goals while optimizing overall production performance (Bai et al., 2020). While industries look toward decarbonization and greenhouse gas emission decline, investments in green activities are imperative for attainment of climate goals on a global scale. Net-zero emission by 2050 requires scaling and multi-faceted strategies, and supply- and demand-side interventions are imperative. In the meantime, such key technologies as carbon capture, electrification, and zero-carbon hydrogen utilization are essential for facilitating process efficiency with a green production base. In addition, the adoption of material-efficient designs and circular economy principles will guarantee less material wastage and resource durability. Shaped by robust policies that incentivize sustainability, firms are able to create a conducive setting for innovation and technology uptake, guaranteeing that investment in sustainable practices indeed yields tangible ecological benefits while maintaining competitiveness in the marketplace (Rissman et al., 2020). The intersection of atom economy investment and green practice is characterized by efficiency and environmental concern. All-around techniques that capitalize on energy efficiency, waste reduction, and sustainable procurement can reorganize the traditional production paradigms into strong, sustainable systems. Utilizing improvements in chemical catalysis and separation technology, for instance, can make it feasible to develop cleaner processes for high-emitting industries such as chemicals and plastics. Moreover, the balance between investment in economic development and thriving sustainably is a reflection of the importance of investment in research and development. With the investment in green innovation by companies, they not only reduce their carbon footprint but also open up space for competitiveness in the more ethical consumer age today. Thus, the strategic emphasis on sustainable investment can drive consistent enhancement in atom economy and process efficiency to the advantage of the industry and society (Figure 3.13).

Figure 3.13. As businesses face growing pressure to minimize their environmental impacts, aligning production methods with sustainable objectives becomes paramount.

Source: https://www.datocms-assets.com/64859/1700478075-analyzing-the-impact-of-sustainable-logistics-practices-on-business-1700478074967.png.

3.6. CHALLENGES AND LIMITATIONS

In chemical reactions, atom economy and process inefficiency issues are complicated. A strong limitation is the inherent yield versus purity conflict that can constrain the overall value and efficiency of chemical reactions. High atom economy is usually desired to minimize waste and maximize the yield of reactant as product. However, achieving such efficiency tends to result in less selective reactions with the accompanying introduction of impurities reducing the purity of the product. This requires additional purification steps, which again reduce overall efficiency and are more expensive, negating the suggested advantages of high atom economy. Such complex interdependencies underscore the challenges of maximizing atom economy with good purity levels in industrial and academic settings. These issues require new approaches and a reconsideration of traditional production paradigms.

Beyond this, scaling processes with high atom economy is another monumental task. While laboratory-scale reactions will yield tremendous atom use efficiencies, scaling up the reactions to industrial scales usually reveals the performance mismatches. Factors such as heat transfer, mixing efficiency, and reaction time could be quite different for large-scale systems, leading to unanticipated inefficiencies. The variances usually manifest in lower yields or longer production spans than the positive numbers achieved in laboratory tests. Additionally, the search for higher atom economy in commercial applications may unwittingly lead companies to seek excessively stringent measures that overlook the actual and operational constraints. Thus, reconciling theoretical efficiency and practicality remains a crucial hurdle in process design and optimization, with the need for ongoing research and innovation to bridge this gap (Canela et al., 2017). Beyond such technical challenges, regulatory and economic considerations are still determinative in the adoption of principles of atom economy. Industries have strict regulations regarding waste management and environmental impact that might dictate allowable processes and materials. In aiming for high atom economy, companies must sacrifice their production processes to meet the regulations, which are not always in favor of the most efficient and sustainable processes. Economic constraints contribute to the complexity of such an environment, as productive investment in the technologies required to maximize atom economy may not fall proportionally into economic benefit, particularly within highly competitive economies. The interplay between regulatory frameworks, profitability, and process procedure optimization requires an integrated strategy wherein decision-makers consider not only the technical feasibility but also the long-range implications of their production choice on the external environment and economy.

3.6.1. Technical Challenges in Achieving High Atom Economy

Achieving high atom economy in chemical reactions comes with some technical challenges that must be tackled to realize optimum efficiency and minimize waste. One of the primary challenges is the development of strategies that enable efficient building of complex molecules such that the bulk of the reactants are not left behind as unwanted by-products, but as wanted by-products. Conventional synthetic paths are not atom-economical since they consist of multiple reaction steps which all contribute to material loss. Chemists thus need to design new methods that not only maximize yield but also keep paths simple to avoid unwanted by-products. The integration of advanced reaction conditions, i.e., greener solvents or solvent-free systems, can be employed to maximize atom economy, but these are steps which have to be reflected on deeply as to their appropriateness and effective functionality in different chemical contexts.

Figure 3.14. Technological innovations—such as those utilized in the development of programmable neutral atom arrays—have demonstrated potential for enhancing atom economy by enabling precise control over chemical reactions.

Source: https://www.science.org/do/10.1126/science.zrpwizg/abs/_20250710_nf_atom_computing_optical_trap.jpg.

Another key factor towards the realization of high atom economy is the optimization of reaction conditions such that they favor desired yields and suppress the formation of waste. For instance, in the research of novel materials and new chemical pathways, scientists must employ rigorous analysis techniques to measure reaction efficiency quantitatively. Advances in technology—like those in the development of programmable neutral atom arrays—proved to have potential to advance atom economy by controlling chemical reactions with high accuracy (Fangde et al., 2025). The problem is the efficiency of operation of such technologies in diverse reaction conditions and scales. As researchers work to upscale laboratory-scale successes into industrial applications, they find themselves being confronted by scalability challenges, reproducibility, and the capacity of such innovations to be flexible. In addition, being flexible in developing commercial strategies and accommodating market demands is critical in the advancement of high atom economy practices across various industries.

A strategy that considers six points of atom economy as well as economic considerations can introduce commercialization and realization of efficient chemical processes. As it is evident from the satellite service commercialization case study, one should understand customer requirements and trace market changes (Makedon et al., 2025). The interaction of innovative concepts and economic viability plays a pivotal role in determining the practicability of new methods for optimizing atom economy. Finally, technical challenges are overcome based on inter-disciplinary practices entailing inter-collaboration among chemists, engineers, and marketing strategists to facilitate sustainable chemical manufacturing practices (Figure 3.14).

3.6.2. Economic Barriers to Implementing Efficient Processes

The use of effective processes across many different industries is greatly inhibited by economic factors, which may present themselves in many different forms, including large initial capital outlays and regular maintenance expenses. Several companies struggle with the daunting prospect of having to use significant financial capital for converting current processes to more efficiency-based methods. This economic expense could deter firms, especially small- and medium-sized enterprises, from adopting advanced technologies that optimize atom economy. Ultimately, the reluctance to invest in advanced techniques, such as those that accompany nanotechnology, signifies a broader hesitation inherent in economic thought. As highlighted by Malik et al. (2023), industries are more conscious of the possible benefits of nanotechnology but continue to face the issue of economic feasibility in such applications, necessitating reconsideration of funding models to finance innovation and enhance productivity without sacrificing financial health.

Economic barriers are typically supplemented with poor governmental and institutional support models that can hinder the achievement of effective process adoption. Policy measures and regulatory environments facilitate or hinder technological advancement. For instance, companies can be deterred from making investments in atom-economy-promoting energy-saving technologies without robust incentives like carbon pricing or subsidies for sustainable operations. Rissman et al. (2020) assert that effective policies can spur innovation and enhance low-carbon process uptake. However, when there is no policy intervention, companies may not be able to see the need to invest in transformational processes even if transformation makes long-term economic sense. The interplay between profitability and policy intervention therefore underscores the need for a resource realignment on a strategic basis to encourage companies to embrace new technologies that enhance efficiency in operations.

In addition, market volatility and unstable raw material prices introduce an added dimension of economic challenges in implementing effective processes. Industries with high reliance on specific materials are the most vulnerable to price shocks, thus creating uncertainty regarding the process improvement viability. This is most applicable in the cement and steel industries where the economic landscape is profoundly reliant on global supply chain dynamics. Companies will not invest in productive processes that are worthwhile if they do not wish to depreciate the monetary investment return by volatile market forces. Rissman et al. (2020) are of the opinion that demand-side interventions like material-efficient design and the circular economy promise promising directions for evading such risks. But short of a reliable economic system underpinning predictability, the inherent challenges of implementing useful processes, on economic grounds, would likely persist, hindering atom economy and process efficiency improvements in general (Figure 3.15).

Figure 3.15. Industries that rely heavily on specific materials are particularly susceptible to price shocks, which can create uncertainty regarding the feasibility of process improvements.

Source: https://web-assets.bcg.com/e4/14/f1a459fe4cc6b3221192b17b08af/boosting-resilience-with-production-as-a-service-rectangle.jpg.

3.6.3. Regulatory and Compliance Issues

In process efficiency and atom economy contexts, compliance and regulation issues become seminal factors that influence the dynamics of business operations of modern enterprises. The digital economy has transformed business operations dramatically by the way it handles labor practices and thus creates new sources of compliance complexity. Modern businesses must confront the legal ambiguities of new labor relations with flexible and tele-commuting work arrangements. These regulatory issues can not only threaten operational efficiency but also stifle innovations toward improved atom economy. As the regulatory framework trails behind the changing technology, companies need to prioritize regulatory compliance both to stay out of legal troubles and to establish a long-term business strategy that aligns with the new employment dynamic (Liu, 2025). This demands an active policy-making approach considering the nuances of digital transformation in workforce management. Also, the introduction of advanced technologies presents with it the challenges and opportunities in regulatory compliance. While the organization is engaging AI, blockchain, and data analytics to automate and optimize atom economy, there are equally challenging compliance issues in protecting data as well as issues of ethics.

Their integration necessitates a cooperative partnership between regulators and companies in such a manner that compliance systems not only remain relevant but are also responsive to the rapid rate of innovation. For instance, the application of data analytics can significantly improve the oversight of compliance, thereby reducing the risk of violations that can result from misunderstanding complex regulations. However, this technological adoption is constantly hindered by inadequate infrastructure and resistance to change, particularly in emerging markets (Alex-Omiogbemi et al., 2024). Thus, business organizations must value spending in robust compliance programs that can leverage technology to enhance efficiency and transparency in processes. To navigate the intricate landscape of compliance and regulatory issues, it is essential that business organizations adopt a strategic model which considers both business objectives as well as the law. This involves implementing an electronic compliance management system in order to ensure transparency in employment relationships and terms, thereby minimizing disagreements arising from algorithmic management practices. Further, organizations have to emphasize legal literacy in their employees to allow them to understand compliance requirements and how these influence atom economy and process efficiency. By the culture of compliance, companies can improve operational resilience and flexibility in processes. Besides, when regulatory environments shift, businesses must be attentive and responsive to the extent of altering their compliance strategies in order to cater to evolving demands, building sustainable practices that ultimately translate into improved overall performance (Alex-Omiogbemi et al., 2024; Liu, 2025).

3.7. CONCLUSION

In summary of atom economy treatment and process efficiency, it is important to recognize their central roles in enhancing sustainable chemistry practice. Atom economy provides a method for quantifying the efficiency with which raw materials are consumed during chemical reactions. Good atom economy implies greater ratios of reactant material converted to products of interest while saving waste and resource usage. Chemical manufacturing sustainability serves not only the needs of the environment but is also placed alongside economic sustainability. With growing art in green chemistry, its practitioners are now more implementing these concepts to come up with effective processes that meet stringent regulation while, at the same time, maintaining efficiency. This strategic interfacing is called for because global focus has now shifted towards increased industrial sustainability and wastage curbing has now become an emergency (Roger, 2016). Moreover, process efficiency improvements supplement atom economy by optimizing reaction conditions and reducing overall productivity. Efficient processes strive to minimize energy consumption, maximize product yield, and minimize chemical use, thereby decreasing the environmental footprint of chemical manufacture. Developments such as the development of new catalysts and new technologies are evidences that efforts are aimed at increasing process efficiency in many chemical industries. These advancements not only facilitate sustainable operations but also underpin competitive advantages in an increasingly environmentally conscious market. With the adoption of an integrated strategy that harmonizes atom economy with process efficiency, industries can most prominently minimize their negative impacts on the environment while promoting a circular economy strategy (Nitta et al., 2014). Finally, the marriage of atom economy and process efficiency ultimately leads to a revolution in the way chemical industries operate. The coming together of scientists, policymakers, and industrial leaders is very important in driving the application of these principles on a bigger level. The transition to cleaner processes also relies not just on advances in technology but on the development of a culture of sustainability within research and development communities. As we advance into the future, issues resulting from climate change and insufficiency of resources will necessitate an unyielding emphasis on achieving the maximum atom economy and process efficiency. This convergence of principles not only serves to continue developing scientific understanding but also promises long-term accomplishment for chemical manufacture as it combines with international demands for sustainability, promising maximum potential for providing grand benefit to society (Roger, 2016).

3.7.1. Summary of Key Points

In considering the atom economy and efficiency of process principles, it is necessary to appreciate their central positions in sustainable chemistry. Atom economy describes the extent to which atoms from reactants are efficiently integrated into the end products, with the aim of reducing waste in chemical reactions. An atom economy process is indicative of an efficient use of resources by the end process, essentially reducing the environmental impact and enhancing the overall sustainability of chemical synthesis. It means that this principle requires techniques to prefer reactions that yield fewer by-products, thereby becoming more resource effective. These methods not only enhance environmental responsibility but also enhance economic sustainability, both of which make them attractive to industries coming under increasing regulatory and public pressure to operate sustainably. Process efficiency is more than stoichiometric efficiency in atom economy; it encompasses all features of a chemical process, including energy consumption, reaction time, and material utilization. A systems approach to process efficiency considers the complete lifecycle of a chemical product from production to use and disposal. An illustration is that issues concerning recycling plastics demand in-depth understanding of atom economy employed in their manufacture, as well as their end-of-life management. This implies examination of various recycling strategies and how effectively they reduce environmental burdens while securing maximum resource recovery. Data derived from such analyses are also significant for the encouragement of innovation in chemical processes since they provide a context for increased efficiency as well as sustainability along the product life cycle that is increasingly significant on the background of global environmental goals (Li et al., 2022). Also, the catalytic process complexity, in particular hydrogen production through proton exchange membrane (PEM) water electrolysis, is the intersection of atom economy and process efficiency. The achievement of stable and efficient electrocatalysts toward the oxygen evolution reaction (OER) remains an area of high-priority research in hydrogen energy technology. The relationship between stability and catalyst activity is very significant, as knowledge of the mechanisms can result in optimizations of both the performance and the longevity of the systems. Improved catalysts are important in order to achieve high efficiencies in the production of hydrogen and, consequently, towards a green energy future. With increasingly more recent studies in this area continually revealing new insights, the implications for atom economy and process efficiency in chemical engineering as a whole are enormous, confirming the necessity of holistic solutions to the issues involved with discovering more eco-friendly solutions (An et al., 2021).

3.7.2. Future Directions for Research and Development

The need for improved atom economy and process efficiency is of the utmost importance for sustainable industrial processes, and therefore future research and development pathways are paramount. Among the most promising directions is the development of low-temperature plasma (LTP) technologies, which have been well considered for their future potential in many applications, including plasma-assisted additive manufacturing and electrochemical conversion electrification. The most recent 2022 Roadmap on LTP shows significant advancement and comes with emerging challenges in the area, particularly in data-driven plasma science. Through deliberate application of these technologies, scientists are able to make reactions optimize the use of atoms at the cost of generating minimal waste, thus squarely aligning the principles of green chemistry. Consequently, LTP technology research can contribute to the tools of chemists and engineers to pave the way for sustainable production processes that satisfy demanding sustainability standards (Adamovich et al., 2022). Secondly, advances in catalytic technology, i.e., proton exchange membrane (PEM) water electrolysis, are another key direction towards process efficiency enhancement. Water electrolysis presents a viable method for hydrogen production, particularly through the oxygen evolution reaction (OER) at the anode, which is a persistent challenge due to stability issues with the current electrocatalysts. Recent research has emphasized the development of innovative and long-lived catalysts that can function under harsh acidic conditions and be efficient. By focusing on the activity and stability of noble and non-noble metal-based catalysts, future research can provide insights into the physicochemical architectures that will enable more efficient OER processes (An et al., 2021). Focusing on the design of sustainable catalytic solutions can be profoundly vital in unlocking the potential of hydrogen as a clean energy source, which can have profound implications for the overall sustainability of chemical processes. Furthermore, interdisciplinary approaches will be crucial to the future of atom economy and process efficiency research. The merging of computational chemistry, materials science, and data analytics has the capability to synergistically drive innovations that enable the discovery of better synthetic routes and methodologies. For example, machine learning algorithms applied to predict reaction outcomes and improve reaction conditions can significantly decrease trial-and-error practices typically employed in the lab. Lastly, integrating the environmental impact of new technologies through a life-cycle assessment method will guarantee that atom economy gains place not only output efficiency but also ecological sustainability at the forefront. As these various lines of research converge, the innovations that follow are on the cusp of revolutionizing industrial processes, solidifying the commitment to sustainable practices within the chemistry community.

3.7.3. Final Thoughts on the Importance of Atom Economy and Process Efficiency

The value of atom economy and process efficiency in contemporary chemical processes cannot be exaggerated. With escalating demands for sustainable practices, chemists and engineers are increasingly putting value on these concepts in their strategies. Atom economy focuses on the ratio of reactants consumed to form useful products, giving great weight to minimizing waste and maximizing the utilization of resources. This approach not only conforms to green chemistry philosophy but also suffices in economic requirements through material cost saving and overall efficiency. With growing environmental consciousness, there is a shift towards prioritizing high atom economy practice, hence promoting more environmentally friendly manufacturing techniques in order to reduce reliance on finite materials and drastically reduce carbon footprint. At the center of this argument lies the use of life cycle analysis (LCA), which presents a holistic framework for measuring the sustainability of chemical processes. Through the examination of all phases of product life, from the generation of raw materials to the end-of-life disposal, LCA shows the environmental and financial implications of different modes of production. The method is based on both the merits and limitations of current processes so that scientists can identify where adjustments are necessary. For example, focusing on high value-added products could yield crucial information about atom economy and process efficiency. According to the relevant literature, feedstock, reaction conditions, and catalysts could significantly influence these aspects, and hence research in such processes would be pivotal for sustainable chemistry innovation and efficient production (Bos and Meesters, 2025). Along with enhancing the economy of atom is the creation of catalytic reactions for maximizing the yield of important intermediates, for instance, hydroxymethylfurfural (HMF) from biomass. Catalytic innovation enabling effective carbohydrate conversion not only maximizes the sustainability of chemical reactions but also produces valuable intermediates for biofuels and other applications. Recent accounts have covered a series of catalytic routes, with emphasis placed on concerted catalysis, which presents higher selectivity and efficiency in such processes. This development shows that the application of highly advanced catalytic processes under the precepts of atom economy has the potential to yield high-impact outcomes, demonstrating that such processes are able to facilitate the synthesis of sustainable fuels from renewable energy (Amarasekara et al., 2010). Collectively, this underscores the supreme importance of atom economy and process efficiency in defining a more sustainable future for chemistry.

REFERENCES

1. Achieng, N. M. (2025). Green chemistry in pharmaceuticals: Reducing environmental impact. *Newport International Journal of Public Health and Pharmacy.* https://www.semanticscholar.org/paper/fcf7c9c0751460e055a99f3481f52db214e8d25d (accessed on 24th October 2025).
2. Adamovich, I., Agarwal, S., Ahedo, E., Alves, L. L., Baalrud, S., Babaeva, N. Y., Bogaerts, A., et al. (2022). The 2022 plasma roadmap: Low temperature plasma science and technology. *Journal of Physics D: Applied Physics, 55*, 373001. https://doi.org/10.1088/1361-6463/ac5e1c (accessed on 24th October 2025).
3. Alex-Omiogbemi, A. A., Sule, A. K., Omowole, B. M., & Owoade, S. J. (2024). Conceptual framework for advancing regulatory compliance and risk management in emerging markets through digital innovation. *World Journal of Advanced Research and Reviews.* https://www.semanticscholar.org/paper/22247129363367cb1d983b1cf09dde5959b4ac13 (accessed on 24th October 2025).
4. Amarasekara, S., Antal, M. J., Balaban, A., Bao, J., Bicker, M., Bilgen, S., Binder, J. B., et al. (2010). Towards the conversion of carbohydrate biomass feedstocks to biofuels via hydroxylmethylfurfural. https://core.ac.uk/download/12356021.pdf (accessed on 24th October 2025).
5. An, L., Wei, C., Lu, M., Liu, H., Chen, Y., Scherer, G. G., Fisher, A. C., et al. (2021). Recent development of oxygen evolution electrocatalysts in acidic environment. *Advanced Materials, 33*. https://doi.org/10.1002/adma.202006328 (accessed on 24th October 2025).
6. Arndt, T., Erkens, M., Gressner, A. M., Holtkamp, M., et al. (2007). Synergism of microwaves and ultrasound for advanced biorefineries. https://core.ac.uk/download/pdf/82265972.pdf (accessed on 24th October 2025).
7. Bai, C., Dallasega, P., Orzes, G., & Sarkis, J. (2020). Industry 4.0 technologies assessment: A sustainability perspective. *International Journal of Production Economics, 229*, 107776. https://doi.org/10.1016/j.ijpe.2020.107776 (accessed on 24th October 2025).
8. Bogaerts, A., Tu, X., Whitehead, J. C., Centi, G., Lefferts, L., Guaitella, O., Azzolina-Jury, F., et al. (2020). The 2020 plasma catalysis roadmap. *Journal of Physics D: Applied Physics, 53*, 443001. https://doi.org/10.1088/1361-6463/ab9048 (accessed on 24th October 2025).
9. Bos, H. L., & Meesters, K. P. H. (2025). Sustainability evaluation of high value-added products. https://core.ac.uk/download/pdf/29256560.pdf

(accessed on 24th October 2025).

10. Castelletto, S., Peruzzo, A., Bonato, C., Johnson, B. C., Radulaski, M., Ou, H., Kaiser, F., et al. (2022). Silicon carbide photonics bridging quantum technology. *ACS Photonics, 9*, 1434–1457. https://doi.org/10.1021/acsphotonics.1c01775 (accessed on 24th October 2025).
11. Centoamore, P., & Pinto, L. (2024). Remanufacturing assessment of machine tools under a circular economy perspective: A resource conservation initiative. *Sustainability*. https://www.semanticscholar.org/paper/ad4b8ec49b0c6a489b7d767c5ffc6ac1a8204f7a (accessed on 24th October 2025).
12. Charruault, L., Genêt, J. P., Gladiali, S. G., Michelet, V., et al. (2006). Alkoxy- and hydroxy cyclization of enynes catalyzed by Pd(II) and Pt(II) catalysts. https://core.ac.uk/download/11687004.pdf (accessed on 24th October 2025).
13. Cherkasov, N., Hessel, V., Ibhadon, A. O., Lang, J., et al. (2016). Low temperature plasma-catalytic NOx synthesis in a packed DBD reactor: Effect of support materials and supported active metal oxides. https://core.ac.uk/download/151161090.pdf (accessed on 24th October 2025).
14. Clark, J. H., Farmer, T. J., Herrero-Davila, L., Sherwood, J., et al. (2016). Circular economy design considerations for research and process development in the chemical sciences. https://core.ac.uk/download/42627202.pdf (accessed on 24th October 2025).
15. Dziejarski, B., Serafin, J., Andersson, K., & Krzyżyńska, R. (2023). CO_2 capture materials: A review of current trends and future challenges. *Materials Today Sustainability, 24*, 100483. https://doi.org/10.1016/j.mtsust.2023.100483 (accessed on 24th October 2025).
16. Guerrero Zúñiga, L. A., Ramirez-Gallardo, A. C., López Sánchez, M. F., & Garcia-Cruz, I. (2025). Hydrodeoxygenation of anisole to produce value-added products: ortho-, meta-, para-cresol. *Renewable Energy, Biomass & Sustainability*. https://www.semanticscholar.org/paper/6292c728f66c9f1be4ff8972c31c506bc30ac7f9 (accessed on 24th October 2025).
17. Ha, T. H., & Oh, J. (2024). (Invited) Selective and economical CO_2 reduction in acid. *ECS Meeting Abstracts*. https://www.semanticscholar.org/paper/87933b2e8c924800b4b2d631ad1adb6ac0a50b5d (accessed on 24th October 2025).
18. Hourcade, J. C., Shukla, P. R., & Mathy, S. (2025). Cutting the climate-development Gordian knot – Economic options in a politically constrained world. https://core.ac.uk/download/pdf/7308561.pdf (accessed on 24th October 2025).

19. Kabeyi, M. J. B., & Olanrewaju, O. A. (2022). Sustainable energy transition for renewable and low carbon grid electricity generation and supply. *Frontiers in Energy Research, 9.* https://doi.org/10.3389/fenrg.2021.743114 (accessed on 24th October 2025).
20. Kallitsakis, M. G., Nikopoulos, K. D., & Lykakis, I. N. (2025). Hydrazine as a reducing agent in catalytic transfer hydrogenation processes: Up-to-date overview approaches. *ChemCatChem, 17.* https://www.semanticscholar.org/paper/caac24d745f490776f72eca8e9af8df6abec74ff (accessed on 24th October 2025).
21. Kumar, S., Prajapati, A., & Yadav, J. (2025). Economic efficiency and profitability of resource conservation technologies in the rice-wheat cropping system in India's Indo-Gangetic Plains: A review. *Agricultural Reviews.* https://www.semanticscholar.org/paper/72e83f79847abcee77cb176f7e1c3d9756401bc6 (accessed on 24th October 2025).
22. Li, H., Aguirre-Villegas, H. A., Allen, R. D., Bai, X., Benson, C. H., Beckham, G. T., Bradshaw, S. L., et al. (2022). Expanding plastics recycling technologies: Chemical aspects, technology status and challenges. *Green Chemistry, 24*, 8899–9002. https://doi.org/10.1039/d2gc02588d (accessed on 24th October 2025).
23. Liu, F., Zhao, F., Li, Y., Yang, W., Yao, X., Han, W., Meng, Z., et al. (2025). Experimental study of TeO_2 anomalous acousto-optic deflector for manipulating neutral atom arrays. *Acta Physica Sinica.* https://www.semanticscholar.org/paper/320db1af70248b64fc662f45d710e443f4fbba3f (accessed on 24th October 2025).
24. Liu, J. (2025). Challenges and countermeasures for the compliance of enterprise employment management in the context of the digital economy. *Scientific and Social Research.* https://www.semanticscholar.org/paper/5ce99d0523b4077c44028d71ac9e2f2ac6136ab1 (accessed on 24th October 2025).
25. Luo, H., Barrio, J., Sunny, N., Li, A. Y., Steier, L., Shah, N., Stephens, I. E. L., et al. (2021). Progress and perspectives in photo- and electrochemical-oxidation of biomass for sustainable chemicals and hydrogen production. *Advanced Energy Materials, 11.* https://doi.org/10.1002/aenm.202101180 (accessed on 24th October 2025).
26. Makedon, V., Yarmolenko, L. I., Chumak, T. V., & Ye Zaporozhchenko, O. (2025). Ensuring the implementation of commercialization strategies for satellite services in the digital economy. *Academy Review.* https://www.semanticscholar.org/paper/5f022b0b09ed09d581a1b44641bd5d1f647d6879 (accessed on 24th October 2025).

27. Malik, S., Muhammad, K., & Waheed, Y. (2023). Nanotechnology: A revolution in modern industry. *Molecules, 28*, 661–661. https://doi.org/10.3390/molecules28020661 (accessed on 24th October 2025).

28. Mariscal, G., & Araume, C. (2017). Fintech and entrepreneurial finance: What's coming next? https://core.ac.uk/download/141440904.pdf (accessed on 24th October 2025).

29. Martínez-Huitle, C. A., Rodrigo, M. A., Sirés, I., & Scialdone, O. (2023). A critical review on latest innovations and future challenges of electrochemical technology for the abatement of organics in water. *Applied Catalysis B: Environment and Energy, 328*, 122430. https://doi.org/10.1016/j.apcatb.2023.122430 (accessed on 24th October 2025).

30. Minguella Canela, J., & Muguruza Blanco, A. (2017). Digital manufacturing: What are we able to print? https://core.ac.uk/download/157811997.pdf (accessed on 24th October 2025).

31. Modak, V. A., Chen, K. H., Nasr, O., Ashebir, M. E., Qorbani, M., & Chen, L. C. (2025). Enhancement of CO_2 electroreduction by Ni SACs embedded in chalcogenide-doped carbon nanofibers: An electrochemical study. *ECS Meeting Abstracts*. https://www.semanticscholar.org/paper/a88e321b0670322d30f67d616b29ecbdbaa0f109 (accessed on 24th October 2025).

32. Naik, S., Bharti, R., & Sharma, R. (2025). Eco-friendly heterocyclic synthesis via multicomponent reactions using solid base catalysts: An overview. *Current Organic Chemistry*. https://www.semanticscholar.org/paper/916176e1baddfdbb50bfb987c841c898fab76947 (accessed on 24th October 2025).

33. Nitta, N., Wu, F., Lee, J. T., & Yushin, G. (2014). Li-ion battery materials: Present and future. *Materials Today, 18*, 252–264. https://doi.org/10.1016/j.mattod.2014.10.040 (accessed on 24th October 2025).

34. Płotka-Wasylka, J., & Wojnowski, W. (2021). Complementary green analytical procedure index (ComplexGAPI) and software. *Green Chemistry, 23*, 8657–8665. https://doi.org/10.1039/d1gc02318g (accessed on 24th October 2025).

35. Prabhu, D. (2020). Role of catalysis in sustainable chemistry. *Catalysis Research, 6*. https://www.semanticscholar.org/paper/cb58536c8ec2fa9095aeca104ec5b58f9f284292 (accessed on 24th October 2025).

36. Rissman, J., Bataille, C., Masanet, E., Aden, N., Morrow, W. R., Zhou, N., Elliott, N., et al. (2020). Technologies and policies to decarbonize global industry: Review and assessment of mitigation drivers through

2070. *Applied Energy, 266*, 114848–114848. https://doi.org/10.1016/j.apenergy.2020.114848 (accessed on 24th October 2025).

37. Rossi, L. I., & Velasco, M. I. (2012). Alternatives to free molecular halogens as chemo selective reactants: Catalysis of organic reactions with reusable complexes of halogen metal salts. https://core.ac.uk/download/224998706.pdf (accessed on 24th October 2025).
38. Roy, S. (2022). Perspectives and impact of green electrodeposition. *ECS Meeting Abstracts*. https://www.semanticscholar.org/paper/0d950eb839d50778e5be69e2abc9df20596a5905 (accessed on 24th October 2025).
39. Saffar, A. (2025). Tetrahydrodipyrazolopyridine covalently immobilized on silica-coated magnetic nanoparticles as a basic heterogeneous catalyst for multicomponent synthesis of dihydropyrano[2,3-c]pyrazole derivatives and Knoevenagel condensation. *Scientific Reports, 15*. https://www.semanticscholar.org/paper/2872439e08cd799f198c0df40d23be18f6a8fba0 (accessed on 24th October 2025).
40. Saint Jean, M., & Oltra, V. (2025). Incrementalism of environmental innovations versus paradigmatic change: A comparative study of the automotive and chemical industries. https://core.ac.uk/download/pdf/6297038.pdf (accessed on 24th October 2025).
41. Sakhahari, K. V. (2025). Green chemistry and catalysis: Innovations toward a cleaner and sustainable world. *International Journal of Pharmaceutical Research and Applications*. https://www.semanticscholar.org/paper/4a3292276b1c806584dc1e557279ec5e81c0f3e8 (accessed on 24th October 2025).
42. Samir, A., Ashour, F. H., Abdel Hakim, A. A., & Bassyouni, M. (2022). Recent advances in biodegradable polymers for sustainable applications. *npj Materials Degradation, 6*. https://doi.org/10.1038/s41529-022-00277-7 (accessed on 24th October 2025).
43. Schot, J., & Steinmueller, W. E. (2018). Three frames for innovation policy: R&D, systems of innovation and transformative change. *Research Policy, 47*, 1554–1567. https://doi.org/10.1016/j.respol.2018.08.011 (accessed on 24th October 2025).
44. Sheldon, R. (2018). Metrics of green chemistry and sustainability: Past, present, and future. *ACS Sustainable Chemistry & Engineering, 6*, 32–48. https://www.semanticscholar.org/paper/0633fef03ad3fb4fbb2e63eba4db98d643a59ea7 (accessed on 24th October 2025).
45. Sheldon, R. A. (2016). The E factor 25 years on: The rise of green chemistry and sustainability. *Green Chemistry, 19*, 18–43. https://doi.org/10.1039/c6gc02157c (accessed on 24th October 2025).

46. Song, M., Tan, H., Chao, D., & Fan, H. J. (2018). Recent advances in Zn-ion batteries. *Advanced Functional Materials, 28*. https://doi.org/10.1002/adfm.201802564 (accessed on 24th October 2025).
47. Steffensen, A., Lundberg, H., & Riisager, A. (2024). Deep eutectic solvents as novel solvent-electrolyte system for electro reductive transformations. *ECS Meeting Abstracts.* https://www.semanticscholar.org/paper/ec026c9206510961f464cd28873a0aaf5e14c0b8 (accessed on 24th October 2025).
48. Tournier, V., Duquesne, S., Guillamot, F., Cramail, H., Taton, D., Marty, A., & André, I. (2023). Enzymes' power for plastics degradation. *Chemical Reviews, 123*, 5612–5701. https://doi.org/10.1021/acs.chemrev.2c00644 (accessed on 24th October 2025).
49. Valera-Medina, A., Xiao, H., Owen-Jones, M., David, W. I. F., & Bowen, P. (2018). Ammonia for power. *Progress in Energy and Combustion Science, 69*, 63–102. https://doi.org/10.1016/j.pecs.2018.07.001 (accessed on 24th October 2025).
50. Verma, A., Pathak, G., Kumar, S., Khatri, V., Chhatwal, R. J., & Arya, D. K. (2024). Multicomponent synthesis of structurally diverse Spiro heterocycles using bio-organic catalyst in aqueous medium. *Current Organocatalysis.* https://www.semanticscholar.org/paper/eda401e453990eb6ce69bb777e17dd0ff8ed7b4b (accessed on 24th October 2025).
51. Zhu, J., Hu, L., Zhao, P., Lee, L. Y. S., & Wong, K. Y. (2019). Recent advances in electrocatalytic hydrogen evolution using nanoparticles. *Chemical Reviews, 120*, 851–918. https://doi.org/10.1021/acs.chemrev.9b00248 (accessed on 24th October 2025).
52. Zhu, S., Yu, T., Xu, T., Chen, H., Dustdar, S., Gigan, S., Gündüz, D., et al. (2023). Intelligent computing: The latest advances, challenges, and future. *Intelligent Computing, 2.* https://doi.org/10.34133/icomputing.0006 (accessed on 24th October 2025).

CHAPTER 4

SAFER SOLVENTS AND REACTION MEDIA

Contents

INTRODUCTION

Green chemistry represents a new science that operates as an essential technique for mitigating the adverse environmental consequences of traditional chemical operations. Fundamentally, green chemistry is centered on reducing hazards in addition to increasing sustainability, particularly through the usage of safer solvents and reaction media. Conventional solvents are liable to pose immense risk to human health and the ecosystem due to their volatility and toxicity, and hence an effort should be made to seek less hazardous alternatives. By emphasizing the use of environmentally compatible solvents, scientists are targeting minimizing not only the chemical imprint of artificial processes but also the adverse effects on the ecosystem and community health. This commitment to sustainability is focusing on the use of innovation in deploying safer solvents that maintain the efficacy of chemical reactions and reduce hazard profiles. In recent years, the use of materials like nanocellulose has increasingly been recognized as green solvents and reaction medium alternatives. Nanocellulose materials are especially famous for their unique characteristics, such as surface area, renewability, and biocompatibility, making them further relevant to the field of sustainable chemistry application. Functionalization of nanocellulose widens its potential for application across a broad range of sectors, from nanocomposites to waste water treatment, with focus placed on limiting the generation of harmful waste. As scientists examine the uses of nanocellulose, the materials not only offer an alternative to traditional solvents but also embody the tenets of green chemistry in terms of creating renewable resources and minimizing environmental impacts. Recent advancements in preparing and functionalizing nanocellulose show the potential for adding the materials to cleaner solvent systems for use in chemical processes (Trache et al., 2020). Furthermore, metal-based nanoparticle research offers another avenue towards enhancing reaction media safety and efficiency in green chemistry. The unique antibacterial properties of metal nanoparticles like silver, gold, and copper have positioned them as prime target candidates for green application in medicine and biotechnology. Such nanoparticles exhibit broad-spectrum antimicrobial activity due to their non-specific toxicity mechanisms, which reduce the risk of bacterial resistance. As developers seek to enter these materials into green chemistry systems, the emphasis on non-toxic but effective alternatives serves to reinforce the overall goal of establishing sustainable chemical processes that are safer for the environment and human health (Elena et al., 2020). Thus, the inclusion of more environmentally friendly solvents and reaction media using state-of-the-art materials is a critical step toward establishing green chemistry programs.

Importance of Solvents in Chemical Reactions

Choice of solvent plays a crucial role in the direction of chemical reactions and impacts not just the efficacy of the reactants and toxic nature of the process but also has impact on solubility and reaction rate. Solvents are the phase in which reactants meet, and thus preferred in this case are non-toxic and bio-renewed derived solvents, which minimize environmental loads. Traditional solvents are often harmful to human health and the environment, and this drives the innovation and employment of safer alternatives. If chemists were able to prioritize the utilization of safer solvents, they would be able to raise reaction conditions without decreasing product yields. As such, priority cannot be given highly enough to the selection of appropriate solvents, particularly in the broader context of sustainable chemistry to reduce destructive waste and strive towards environmentally friendly practices. Recent advances bring to the fore the use of solvent systems in designing biocatalytic reactions, an integral part of green chemical technologies.

Figure 4.1. By focusing on the use of safer solvents, chemists can enhance reaction conditions while still maintaining product yields.

Source: https://greenchemuoft.wordpress.com/wp-content/uploads/2015/07/columns-edit.jpg.

Biocatalysis presents the chance to carry out reactions under milder conditions, typically in water, without the need for harsh organic solvents. This shift relies on enhancements in molecular biology and protein design, which allow enzyme functions to be adapted for these more secure solvents. In this case, the use of biocatalysts not only enhances reaction efficiency but also conforms to green chemistry principles promoting processes as safe and sustainable as possible. In pursuit of such, there is a requirement for ongoing research into the impact of solvents on enzyme activity, which can rethink active pharmaceutical ingredient manufacture in alignment with environmentally responsible practice (Roger et al., 2017). The employment of safer solvents in chemical reactions is also important for increasing the sustainability of nanomedicine therapy. With the evolution of the field, the employment of nanocarriers and drug delivery systems demands solvents that should not undermine the biocompatibility of materials employed at the nanolevel. Recent literature offers evidence that carefully chosen solvent systems can impact the efficacy and safety profile of nanomedicines significantly while delivering the anticipated therapeutic effect. With the focus on green solvents in the design of such future systems, researchers not only face the problems of toxicity but are also paving the way for new means of treatment minimizing the detrimental side effects. Overall, the greater emphasis on the importance of solvents in chemical reactions is indicating a paradigm shift towards improvements and greener means in the area of green chemistry (Patra et al., 2018) (Figure 4.1).

Overview of Safer Solvents and Reaction Media

In recent years, the quest for greener solvents and reaction media has been a foremost area of interest in the field of green chemistry. Traditional solvents generally introduce gigantic health and environmental risks in terms of toxicity and volatility, whose releases result in perilous waste. In response to this, the quest for safer alternatives has gathered momentum, spanning bio-based solvents from renewable resources. These solvents are less harmful to human health not just in terms of being less toxic but also in reducing the carbon footprint associated with chemical processes. Furthermore, advancements in solvent-free processes and microwave-assisted processes are a perfect case in point for sophisticated methodologies reducing the need to use conventional solvents, reflecting a tremendous shift towards sustainability in chemical manufacturing processes. Further, reaction media choice is critical in optimizing the efficiency and sustainability of chemical processes. Solvent selection has immediate repercussions on rate of reaction, yield, and selectivity and thereby on overall process efficacy. Ionic liquids and supercritical fluids have, for example, been identified as possible reaction media due to their tunable properties that can be optimized for specific chemical transformations. Such alternatives are less toxic and recyclable or reusable, and in addition, they are green chemistry. With the future of synthesis, collaboration between scientists and industrialists in synthesizing green solvents and reaction media can help make greener processes common in various industries of industry, thus helping create a greener tomorrow. Despite the improvement, there are still issues with the widescale utilization of safer solvents. Formulation challenges must not only deal with the environmental impacts, but also compatibility with production processes already in place. A switch to these greener formulations involves careful evaluation procedures, such as life cycle analyses, that critically identify their environmental strengths and weaknesses. Second, the economic impact of replacing risky solvents must be weighed, especially for industries strongly reliant on traditional solvent systems. By meeting these challenges head-on, the transition to safer solvents can be most effectively realized, leading to improved chemical manufacturing practices. Future research will need to focus on the work of designing and regulating new solvents in line with tighter environmental conditions but guaranteeing functional qualities required by various applications (Neumann et al., 2022; Mrozik et al., 2021).

4.1. THE ROLE OF SOLVENTS IN CHEMICAL REACTIONS

The role of solvents in chemical processes is of pivotal significance, enabling a wide variety of processes ranging from solvation to reaction kinetics. The primary role of solvents is the dissolution of reactants, thus increasing molecular interaction and reaction rate. The ancient classical chemistry utilizes a variety of organic solvents such as benzene, toluene, and dichloromethane due to the occurrence of a broad variety of compounds that they can dissolve. However, the application of such toxic substances presents significant health and environmental concerns, and hence the necessity to switch to safer alternatives. Green chemistry values embrace approaches that minimize the utilization of toxic solvents and promote the utilization of safer solvents that can execute the same functions without compromising efficiency in reaction products. This shift in paradigm is needed in environmentally friendly chemical processes and reflects the evolving mindset regarding solvent functions in green chemistry applications. Teasing further discussion on safer solvents, it is clear that water and other innocuous media are coming to the forefront as alternatives to traditional organic solvents. Water, also known as the universal solvent, has numerous benefits, including low toxicity, sustainability, and economic affordability. Its use in chemical reactions not only minimizes ecological footprints but also reflects the green chemistry concept of minimizing waste and using renewable resources. Supercritical carbon dioxide and ionic liquids are examples of superchemicals that further illustrate the diversity of application of safer solvents, which are capable of conducting a variety of reactions while keeping hazard generation to a minimum. These benign solvents open up new possibilities for chemists to perform reactions in mild conditions with more efficiency and with more safeguards for the environment in chemical processes. The transformation toward these alternatives is a turning point toward accepting fully the philosophies of green chemistry. Not only is the employment of safer solvents an environmental morality, it also radically transforms the business landscape of chemical synthesis. By looking back at solvent selection criteria, scientists are compelled to consider the biodegradability and toxicity of reaction media at an initial stage of chemical design. This foresight initiates the design of safer chemical processes that are protecting human health as well as the environment. At the center is the adherence to established rules of green chemistry, which seek to avoid hazardous chemicals in all phases of chemical production. As the science progresses and advances, exploration of new environmentally safer solvents is leading to a pivotal shift towards sustainable operations, eventually shaping the future of chemical reactions and instilling a culture of valuing public and ecological health (Chen, 2025; Sharma et al., 2025) (Figure 4.2).

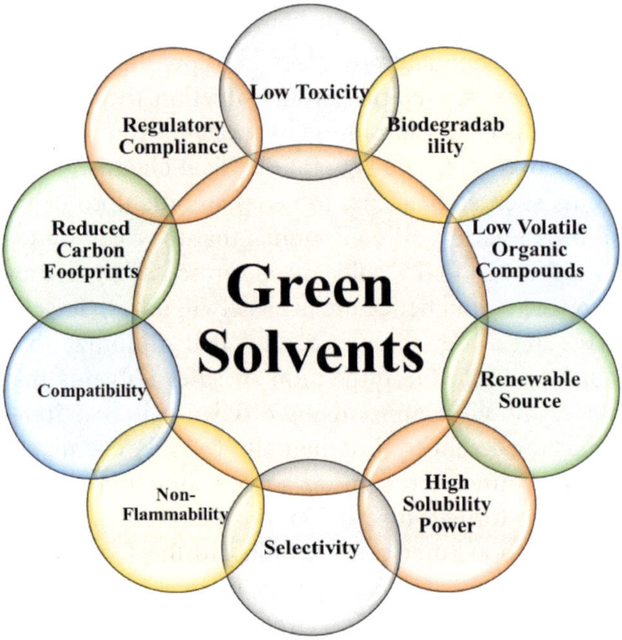

Figure 4.2. In traditional chemistry, a variety of organic solvents, such as benzene, toluene, and dichloromethane, have been widely utilized due to their ability to dissolve a broad spectrum of compounds.

Source: https://pub.mdpi-res.com/energies/energies-16-05852/article_deploy/html/images/energies-16-05852-g006.png?1691473392.

4.1.1. Functions of Solvents in Reaction Mechanisms

Solvent choice is a key control of chemical reaction efficiency and selectivity as well as a dominant influence on reaction mechanisms. Solvents carry out different functions, from dissolving reactants and facilitating molecular interactions to influencing reaction rates and pathways. In green chemistry, emphasis is on the selection of safer solvents with reduced environmental impact while maintaining or enhancing reaction efficiency. For instance, polar solvents can stabilize charged transition states, typically increasing reaction rates, while nonpolar solvents would be best for reactions involving nonpolar substrates. By optimizing solvent systems through design, chemists are able to achieve increased yields and decreased by-products, which are aligned with sustainable practices of waste reduction and toxicity in chemical reactions. Besides, the solvent-reactant interaction is capable of initiating some reaction pathways, such as micellar catalysis or solvation-induced enhanced reaction selectivity. The solvents also influence the thermodynamic stability of transition states and intermediates, thereby directing the reaction towards desired products. For all these instances, the use of biocatalysis is an important aspect within green chemistry procedures, where water and other nonharmful solvents are competing reaction media. The modification of reaction conditions, including the solvent environment, allows unimaginable catalytic efficiencies and selectivities to preserve chemical processes as sustainable and green but meet industry standards (Roger et al., 2017). The incorporation of safer solvents into reaction mechanisms could revolutionize a vast array of chemical processes, from synthesis to industrial applications. Above all, nanomedicine, which entails the application of nanostructured materials for targeted drug delivery, showcases the practical relevance of solvent functionality. Use of safer and more environmentally friendly solvents not only enhances the therapeutic agent's efficiency but also minimizes the risk of hazards posed by toxic solvents. With advancements in the science of green chemistry, the emphasis on new solvents and their roles within reaction mechanisms remains at the forefront of developing sustainable approaches in all industries, ultimately bridging the gap between environmental stewardship and industrial competitiveness (Patra et al., 2018).

4.1.2. Impact of Solvents on Reaction Efficiency

The choice of the solvent plays a central role in determining the effectiveness of chemical reactions, particularly in the context of green chemistry. Traditional organic solvents, though powerful, are rather enormous threats to the environment and human health due to being toxic and persistent substances. Scientists hence now seek safer alternatives that are in congruence with green practices. Advances in solvents, including the creation of biodegradable and non-toxic solvents, offer promising avenues to enhance the effectiveness of reactions at the expense of adverse environmental impacts. The application of surfactants like Plantacare 2000 (PL-2000) is one such case, illustrating how renewable biosurfactants enhance the effectiveness of palladium-catalyzed reactions without sacrificing green chemistry principles. Environmental compatibility improvement projects the option to maximize synthetic processes through the selection of proper solvent systems. Another key advance in solvent technology is the development of deep eutectic solvents (DES), where a number of components are combined to create a stable system that might function as both catalyst and solvent. The paper which reveals the use of a DES of Choline chloride (ChCl) and ammonium thiocyanate demonstrates how these systems are capable of catalyzing reactions while providing high yield and efficiency.

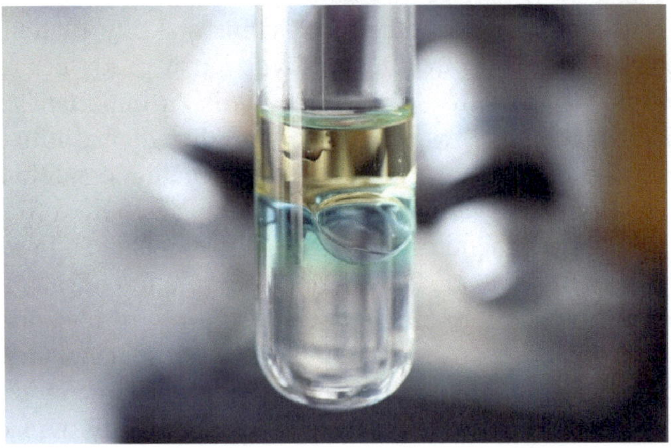

Figure 4.3. Traditional organic solvents, while effective, often pose significant hazards to the environment and human health due to their toxicity and persistence.

Source: https://www.quimidroga.com/wp-content/uploads/2024/01/disolventes-alternativos.jpg.

This new approach simplifies reaction setups, guarantees sustainability, and minimizes waste compared to the use of traditional solvents. As a multifunctional agent, DES can bypass or mitigate the use of toxic reagents and organic solvents, thereby underscoring the merit of these as greener options in organic synthesis. Their ability to recycle and reuse them makes them more desirable, as they facilitate the practice of a circular economy in the production of chemicals that leads to lower environmental impact on chemical operations. Solvent effects on reaction efficiency are not merely a matter of yield and rate of reaction; they are also about the long-term implications of sustainability and environmental stewardship. Safer reaction media and solvents are also a key factor in the green metrics of entire chemical processes. Based on the performance comparison of PL-2000 systems versus DES systems, one can see that both promote efficient reactions as well as fulfill green chemistry requirements via reduction of toxicity and environmental hazard. These breakthroughs in solvent technology reveal growing awareness on the part of scientists of the need for selecting solvents to be used with sustainable methods. Lastly, the transition to safer solvents can redefine chemical synthesis to reconcile efficiency of operations with regard to the environment and set a healthy precedent for future research and industrial practice towards green chemistry (Figure 4.3).

4.1.3. Environmental Implications of Traditional Solvents

The ecological effects of traditional solvents in chemical processes are relevant and widespread. Traditional solvents such as benzene, toluene, and chloroform have been utilized for decades due to their effectiveness in catalyzing reactions and product extraction. Their volatile organic compounds (VOCs) are, however, the cause of air contamination and smog generation as well as adverse human and wildlife health effects. The use of these solvents also results in toxic spills and inappropriate disposal practices, contaminating the ground and water bodies. Further, production of traditional solvents is usually very energy-consuming and with release of greenhouse gases, contributing to global warming. In order to mitigate their impacts, there has to be shift towards more safe solvent alternatives under green chemistry in order to minimize ecological footprints without compromising reaction efficiency and effectiveness in processes in industries. As a response to the environmental hazards of traditional solvents, green chemistry scientists and researchers are shifting toward greener solvents. The greener solvents are typically derived from renewable feedstocks and are engineered to have reduced toxicity, volatility, and increased biodegradability. For instance, ethyl lactate and glycerol based bio-based solvents decrease petrochemical-based feedstock reliance and decrease toxic emissions during their operation in chemical processes.

Recent work emphasizes the development of solvent-free reaction media and aqueous solutions even more reducing the environmental burden of traditional solvents. New methods, such as supercritical fluid technology and ionic liquids, promise potential in revolutionary solvent use and adding significant efficiency with an equal focus on sustainability. In this way, the promotion of a shift to safer alternatives is a notable achievement in green chemistry. Despite major leaps, there remain problems in the widespread adoption of cleaner solvents. The largest problem is the financial feasibility and performance parity of these alternatives with traditional solvents. Incumbent process companies may be less inclined to adopt unknown, but more environmentally friendly, alternatives because of perceived risk in product quality or reaction yield. In addition, more research must be done to verify the long-term effects on the environment of newer solvents to make sure that they do not inadvertently establish new ecological problems. For instance, while certain bio-based solvents are less dangerous, their production may still include resource intensive processes or land cover alteration that will offset their sustainability benefits. Therefore, there must be an overall plan that weaves together technological innovation and rigorous life cycle assessments in order to fully understand and regulate the environmental impact of solvent use in chemical manufacture (Figure 4.4).

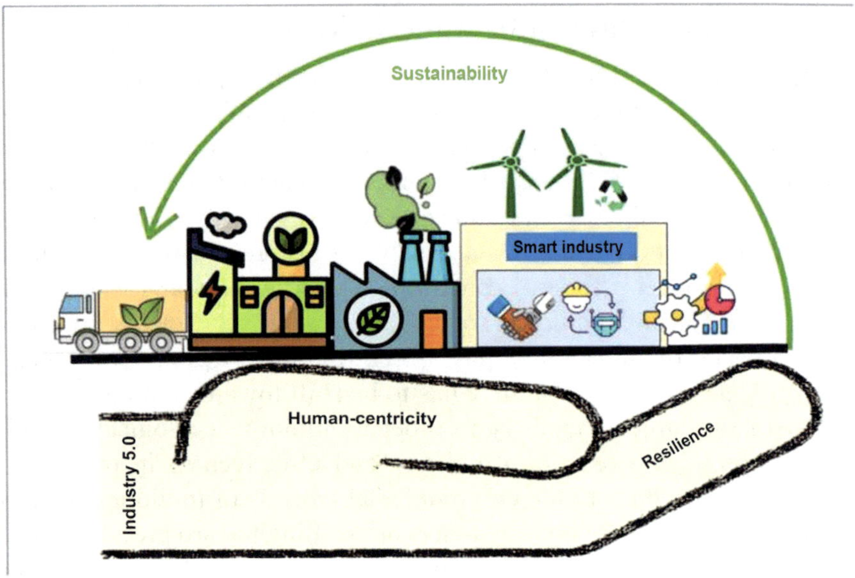

Figure 4.4. Industries reliant on established processes may resist transitioning to less familiar, albeit environmentally friendly, options due to perceived risks in product quality or reaction efficiency.

Source: https://media.springernature.com/lw685/springer-static/image/art%3A10.1007%2Fs42824-025-00176-7/MediaObjects/42824_2025_176_Figa_HTML.png.

4.2. CHARACTERISTICS OF SAFER SOLVENTS

In green chemistry, safer solvent properties are key to green redesign of chemical processes to make them more environmentally friendly. Safer solvents must have low toxicity, meaning that they do very little harm to human health and the environment. This characteristic significantly reduces the risks of solvent exposure, making them a preferable alternative to traditional solvents, which often possess harmful qualities. In addition, they should also be highly biodegradable to minimize long-term impacts on the environment. Being in line with the values of sustainability permits not only short-term safety benefit but also environmental harmony as a central concern of contemporary chemical practice. Such characteristics facilitate transition to green practice, where safer solvents can avoid toxic effects typically caused by conventional substances, transforming academic and industrial chemical operations. Another important characteristic of safer solvents is that they reduce waste and maximize efficiency in chemical reactions. Safer solvents are intended to boost the rate and yield of reactions, which is very significant for economic as well as environmental reasons. Being able to utilize fewer solvents or a reduced quantity can lead to a reduced demand for purification procedures and lower the overall environmental footprint. In developing new chemical processes, the properties of solvents need to be examined closely to determine if they encourage not only the desired chemical conversions but also align with sustainable development. Furthermore, application of natural extracts as solvents, highlighted in green synthesis approaches, provides a clear example of how such materials can be doubly useful as reactants as well as reaction mediums and thus decrease their footprint on the environment while maintaining efficiency in applications such as nanoparticle synthesis (Singh et al., 2018). Moreover, more evolved applications of safer solvents include the fields of nanomedicine and targeted drug delivery systems, where their safety can also play a pivotal role in therapeutic advantages. Safer solvents in these cases must be biocompatible and non-toxic to ensure patient safety as well as facilitate effective drug delivery mechanisms. For instance, solvents in the support of stability of nanomedicine formulation ensure its enhanced efficacy and safety profile. The trapping of therapeutic compounds in biodegradable material has provided new methods of treating long-term conditions, therefore presenting an opportunity to apply safer solvents in new applications (Patra et al., 2018). Generally, characteristics of safer solvents are included in being not just responsible for reducing harmful waste but also enhancing opportunities in biotherapeutic practice, a massive step into the practice of green chemistry principles.

4.2.1. Low Toxicity and Biodegradability

The value of low toxicity and biodegradability of solvents and reaction media cannot be overstated, particularly under the context of green chemistry. Because the discipline aims to minimize unwanted chemicals and environmental impact, the utilization of solvents with such properties becomes imperative. Low-toxicity solvents reduce the possibility of deleterious health effects for researchers and consumers alike, leading to healthier working environments. Moreover, use of biodegradable solvents is also a progression towards reducing environmental footprints as they naturally decompose and do not pose lasting pollution. By prioritizing solvents that possess these properties, chemists can integrate their research in harmony with the aspects of sustainability, directly responding to the call for more environmentally friendly alternatives within industrial processes. Biodegradable solvents not only benefit the environment but also streamline chemical processes. For instance, plant-based solvent feedstocks like biodiesel or ethanol can be used in many chemical processes without impacting performance. The reality that these solvents will decompose into harmless byproducts when they are being discarded means that the possibility of their harmful environmental impact is precluded.

Figure 4.5. Solvents with low toxicity reduce the risk of adverse health effects on both researchers and consumers, creating safer working environments.

Source: https://paintingbestpractices.com/wp-content/uploads/2024/04/solvents_cans.jpg.

Furthermore, the use of such materials may simplify compliance with tougher environmental regulations. With a move towards more sustainable industries in the world, the application of low toxicity and biodegradable solvents is a fundamental step towards reducing waste and chemical production that is accountable to both human and environmental well-being, aligning to new standards that benefit both. The pursuit of low toxicity and biodegradability does not come without challenges, particularly in the context of complex blends of materials used in current applications, as with the example of lithium-ion batteries. They encompass a range of materials with recycling and disposal issues, highlighting the necessity of continued improvement in recycling technology, as suggested by research on battery materials and their environmental impact. In addition to innovative industry, safer solvent and reaction media design must also take into account future developments in material science. Addressing these challenges with emphasis on low toxicity and biodegradability not only contributes to the sustainability agenda but also sets the ground for further designing of more ethical management practices for numerous chemical substances, highlighting the inseparable connection of green chemistry and broader environmental objectives (Neumann et al., 2022; Mrozik et al., 2021) (Figure 4.5).

4.2.2. Renewable and Sustainable Sources

The search for green and renewable sources in green chemistry has gained momentum due to the need to reduce environmental impacts and enable circular economy. With their prime consideration on materials drawn from biogenic sources, researchers are increasingly focusing on solvents with lower toxicity and waste generation during processes while ensuring greatest efficiency in chemical reactions. This is not merely a trend but a necessary paradigm shift that will reshape the traditional processes that are petroleum solvent-based. With the utilization of renewable resources, the resulting chemical processes can even lower their carbon footprint and contribute towards sustainability efforts, thereby keeping pace with science advancement and societal requirements for green practices. Among the solvents available that are taking an increasing lead is 2-methyltetrahydrofuran (2-MeTHF), a biogenic solvent that is derived from renewable feedstocks such as furfural and levulinic acid. 2-MeTHF possesses some desirable features making it a valuable product for use in organic synthesis, particularly in biocatalysis. It possesses a low water solubility, which allows efficient recovery such that it can be reused and waste products reduced in chemical reactions. As production continues to consider the suitability of 2-MeTHF as a replacement for traditional solvents, this solvent higher qualitie not only make the process cleaner but also serve as a model for innovation in the use of sustainable resources in real-world applications. Its beneficial characteristics demonstrate that 2-MeTHF may have far-reaching uses in both industrial and educational applications, becoming a pioneer among the move towards safer solvents of green chemistry (Andrés et al., 2018). On top of that is the use of renewable materials in chemical reactions, keeping pace with a broader trend of development for nanomedicine. With scientists seeking the application of nanotechnology in drug delivery, the role of sustainable materials has become more significant. Nanomedicines often rely on the application of nanomaterials based on biodegradable or green materials, which indicate the essence of sustainability in green chemistry. Through developing methods that leverage the capability of clean sources, scientists can create more effective and targeted treatments that have a lower environmental impact on drug manufacturing. This integration of nanotechnology and green practice illustrates continued commitment to methodology enhancement that enhances scientific returns with a focus on ecological stewardship (Patra et al., 2018) (Figure 4.6).

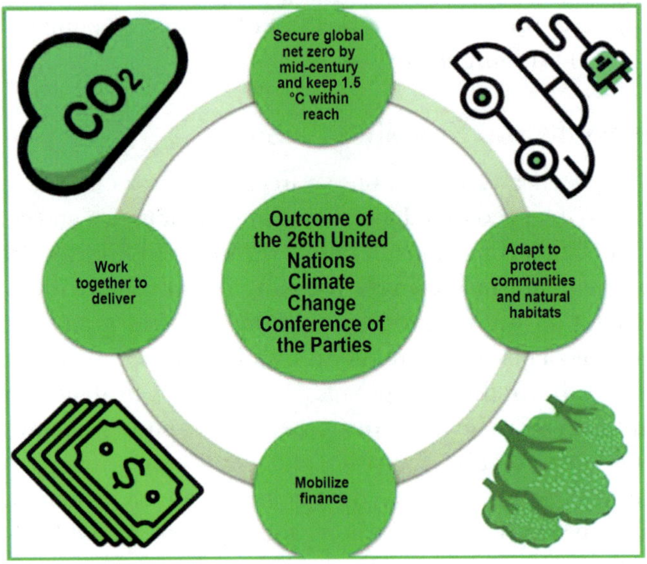

Figure 4.6. By utilizing renewable resources, the resulting chemical processes can potentially lower their carbon footprint and contribute to sustainability efforts, thereby aligning with both scientific advancements and societal expectations for environmentally responsible practices.

Source: https://media.springernature.com/lw685/springer-static/image/art%3A10.1007%2Fs10311-022-01435-8/MediaObjects/10311_2022_1435_Fig1_HTML.png.

4.2.3. Non-Flammability and Stability

For green chemistry, stability and non-flammability of solvents are crucial in ensuring safer reaction media for a broad spectrum of chemical operations. Traditional organic solvents are normally flammable and contribute to volatile organic compound (VOC) emissions that are toxic to humans and the environment. On the other hand, green solvents are continuously being developed with inherent properties that reduce flammability and hence reduce risks associated with handling and storage. The use of non-flammable solvents adds more safety measures in laboratories and encourages a greener approach to chemical experimentation. This shift in solvent choice not only places people first but also conforms with regulatory frameworks aimed at reducing harmful material in laboratories. Stability is yet another important parameter that dictates the selection of solvents in green chemistry. A stable solvent contributes to consistent reaction conditions, thereby reproducibility and reliability of experimental outcomes. This matters in the context of long storage shelf life and demands of chemical reagents. Non-flammable solvents, such as deep eutectic solvents (DESs), are gaining more popularity due to their favorable thermal and chemical stability.

Figure 4.7. Non-flammable solvents, such as deep eutectic solvents (DESs), have emerged as promising alternatives due to their favorable thermal and chemical stability.

Source: https://pub.mdpi-res.com/molecules/molecules-29-04767/article_deploy/html/images/molecules-29-04767-ag.png?1728463968.

DESs possess low volatility, lowering the environmental impact even more and rendering them safer to handle (Wu et al., 2021). Progress toward such flame-resistant and stable solvents holds high potential for use in different chemical processes like battery materials and catalysis, contributing to the enhanced progress of green and efficient chemical methods. The pairing of stability and non-flammability in reaction media not only makes lab activities safer but also creates opportunities for revolutionary applications in industries relying on green chemistry. According to existing research, the ability to leverage the utilization of non-flammable solvents effectively can greatly enhance the performance of electrochemical energy storage devices, such as lithium-ion batteries (Neumann et al., 2022). This two-way focus of safety and functionality promises the development of both academic study and industrial applications. With ongoing progress of green chemistry, striking a balance between optimizing these attributes as much as is feasible and maximizing reaction efficiency will be important. Emphasizing non-flammability and stability will not only render chemical procedures safer but also contribute to worldwide initiatives of sustainable processes for chemical production and utilization of resources (Figure 4.7).

4.3. TYPES OF SAFER SOLVENTS

This change to green chemistry has promoted the development and application of safer solvents, which are central to reducing the impact of chemical processes on the environment. Bio-based solvents have been among the safer solvent classes that have received great attention due to their renewability and lesser toxicity. These solvents, often derived from organic sources such as plants and agricultural waste, are environmentally friendly and reduce toxic waste. Ethyl lactate, for instance, is an ideal bio-based solvent that makes an excellent substitute for traditional petroleum-based solvents. Its low level of toxicity and biodegradability make it suitable for various applications in pharmaceutical products and also in paint formulations. As the study progresses, the continuous assessment of bio-based solvents holds promise in redefining industrial practice towards the utilization of solvents and hints at the possibility of bypassing harmful environmental effects without compromise on performance. Another recent safer solvent is ionic liquids. Ionic liquids are liquid at room temperature salts characterized by having very low vapor pressure and good thermal stability. Ionic liquids have been lauded for the broad range of materials with which they can dissolve and therefore have applications in a wide range of chemical processes, including separation, catalysis, and organic synthesis. Their unique properties enable them to replace more dangerous and explosive solvents that have so far been employed in the industry. Further, ionic liquids may be tailored to specific applications, which results in their enhanced performance while reducing the environmental impact. However, the integration of ionic liquids into mainstream chemistry requires a composite assessment of their life cycle, as the energy-intensive synthesis steps of the ionic liquids may pose sustainability challenges. Yet their application in green chemistry is a milestone in the development of safer and more environment-friendly alternatives to solvents. Apart from bio-based solvents and ionic liquids, the use of supercritical fluids, particularly supercritical carbon dioxide ($scCO_2$), as a game-changing technique in green chemistry has been discussed. Supercritical fluids possess unique characteristics that allow them to act both as a liquid and a gas and, as a result, are effective solvents for a wide variety of compounds with none of the harmful attributes of conventional solvents. The use of $scCO_2$ in extraction processes demonstrates its capability for selectively dissolving target molecules, enhancing efficiency while considerably reducing solvent waste and toxicity. This process gains increasing applications in the recovery of natural sources of materials for flavors, aromas, and drugs, confirming its versatility and efficacy. Constant advancement in the field of supercritical fluid technology confirms its promise for safer, more environmentally friendly chemical processes as well as to promote environmental responsibility through industrial use.

4.3.1. Water as a Universal Solvent

Waters unique molecular structure and polarity render it the unmatched solvent, essential to hundreds of biological and chemical processes. Its polar nature allows water to excel in dealing with many ionic as well as polar substances, dissolving them very efficiently. The property enables basic biochemical reactions, such as metabolic processes in living things, where water acts as the vehicle to transport solutes and reaction. In green chemistry, water's application as a solvent place itself in alignment with the sustainability paradigm, reducing the necessity of toxic organic solvents that are environmentally and health threatening. Referral to water as the universal solvent emphasizes its central role in supporting the practice of safe and sustainable chemical operations, hence being a main constituent in developing greener practice. Although it has myriad advantages, use of water as a solvent does have problems with certain chemical reactions. Solubility behavior of water occasionally decreases the effectiveness of reactions that require non-polar or low-solubility materials. Further, water's high heat capacity and high dielectric constant can influence the reaction kinetics differently than conventional organic solvents. However, advances in green chemistry aimed to fill these gaps with new approaches, such as employing co-solvents or surfactants to enhance the solubility of non-polar substances in water-based mediums. By probing how to optimize water-based reactions, researchers can tap into waters environmentally friendly characteristics and maximize its applications in various industrial uses, holding out a possible path to the future for safer substitutes for conventional solvents. Apart from acting as a catalyst for reactions, waters biodegradable nature is significantly to blame for environmental sustainability because it lowers the ecological footprints of chemical manufacture. For the majority of scenarios, the by-products of reactions in aqueous systems are less harmful and more manageable than those created in the use of standard organic solvents. For example, in biocatalysis, lipases are very efficient when used in aqueous systems, which generally offer excellent selectivity and functionality for a variety of industrial processes, including the production of biodiesel and food processing. By harnessing water-based systems and enzymes as catalysts, industries are even capable of reducing their consumption of toxic solvents, promoting a circular economy in which waste is minimized and resources are maximally reused. These activities collectively pinpoint the significance of water as a general solvent in the broader discourse on greener reaction media and safer solvents in green chemistry (Neumann et al., 2022; Chandra et al., 2020) (Figure 4.8).

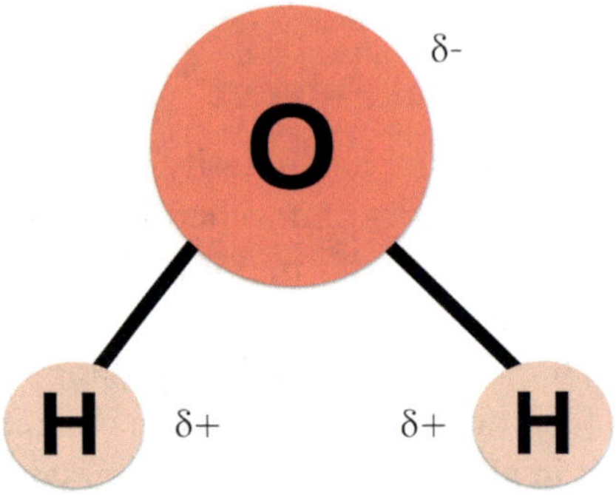

Figure 4.8. Waters unique molecular structure and polar characteristics render it an unparalleled solvent, crucial for countless biological and chemical processes.

Source: https://cdn.kastatic.org/ka-perseus-images/c8f5a3e5a16f2fc0a4fdd54eab-7204ca39f836be.png.

4.3.2. Bio-Based Solvents (E.g., Ethanol, Glycerol)

The search for greener chemistry has placed the limelight on bio-based solvents, which are a more eco-friendly alternative to traditional organic solvents. Amongst these, ethanol and glycerol are especially favored due to their renewability and lower environmental impact. Ethanol, which is obtained through the fermentation of biomass, is not only biodegradable but also less toxic to human beings and the environment compared to common solvents like toluene or acetone. Furthermore, glycerol, which is a byproduct of biodiesel, has fascinating physicochemical properties like high viscosity and low vapor pressure, making it a good medium for various chemical reactions. With the development of research in green chemistry steadily picking up, bio-based solvents like ethanol and glycerol are increasingly gaining interest as potential alternatives to be able to effectively eliminate the formation of toxic waste without hindering synthetic efficiency and yield of product. The use of bio-based solvents in organic synthesis has been demonstrated to bring numerous advantages in line with green chemistry principles. Solvents minimize health risks associated with volatile organic compounds, which are highly recognized with toxicological implications. Studies suggest that the application of bio-based solvents leads not only to a reduction in environmental emissions but also optimizes reaction outcomes with better yields without the need for toxic chemicals. For example, recent studies indicate that the use of ethanol as a solvent can effectively minimize the energy requirements of certain reactions, thus making synthetic processes more sustainable. Moreover, glycerol has been found to enable reactions under mild conditions, counteracting the high-energy requirement of classical solvent systems while enabling effective product recovery and recycling along chemical processes, showing promising future directions for sustainable practices in chemistry. While the advantages of bio-based solvents are evident, there are issues with their adoption en masse. Possible increased expenses and limited access over conventional solvents present significant barriers to their adoption in the entire chemical industry. For example, while ethanol is readily accessible in most regions of the globe, fluctuations in farm produce yield can affect pricing and availability. In addition, the changeover will see an upgrading of current infrastructure and practices that are quite dependent on conventional solvents. Still, the design of innovative catalytic systems that utilize bio-based solvents can also somehow reverse some of these issues. Research has proven that biobased solvents not only reduce toxicity but enhance the performance of catalysts, such as in research involving glycerol to conduct multicomponent reactions with very high product yields without the use of hazardous organic solvents (Zheng, 2025; Lashaki et al., 2022). Thus, continued research and development funding is needed to unlock the full potential of bio-based solvents in making cleaner chemical processes possible.

4.3.3. Ionic Liquids and Supercritical Fluids

The application of ionic liquids and supercritical fluids as solvents and reaction media is a significant advancement in greener chemical processes. Ionic liquids are molten salts at room temperature, and they exhibit certain unique properties such as low vapor pressure, high thermal stability, and adjustable solvation. These attributes make them effective in dissolving high quantities of organic and inorganic compounds with high efficiency. Their ability to act as catalysts and solvents cuts down on the use of traditional organic solvents, thus conserving the environment. Ionic liquids have proven to be effective media for production processes of sustainable ingredients across a variety of industries such as pharmaceutical and cosmetic that are squarely aligned with green chemistry principles (Chemat et al., 2020). On the other hand, supercritical fluids, particularly supercritical carbon dioxide, have attracted attention because of their effective solvating abilities without the toxic side products of most traditional solvents. Supercritical fluids operate at elevated pressure and temperature, where they exhibit properties intermediate between gases and liquids. They play a role as solvents to allow efficient extraction and reaction operations with minimal use of toxic chemicals, thereby enhancing the yield of the products of concern.

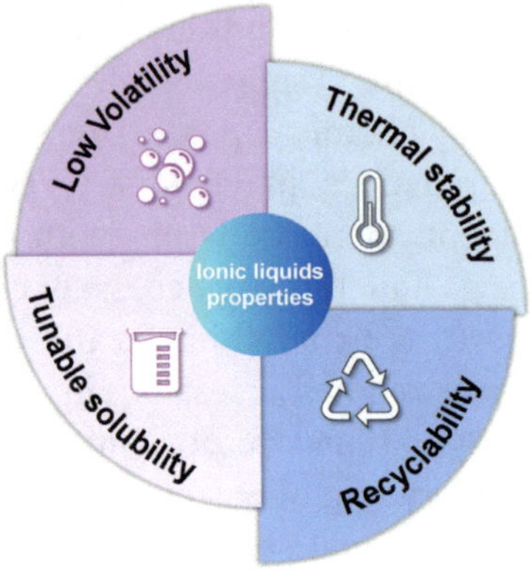

Figure 4.9. Ionic liquids, which are salts in a liquid state at ambient temperatures, offer unique properties such as low vapor pressure, high thermal stability, and tunable solvation characteristics.

Source: https://www.mdpi.com/molecules/molecules-30-00353/article_deploy/html/images/molecules-30-00353-g003.png.

The use of supercritical fluids in green chemistry, particularly in the synthesis of nanomaterials and their derivatives, optimizes the sustainability of nanoparticle production. This approach not only minimizes the environmental footprint but also minimizes the danger associated with the toxic solvents traditionally utilized in such operations (Singh et al., 2018). Generally, the utilization of ionic liquids and supercritical fluids is revolutionizing the landscape of solvent application in chemical processes with safer and more efficient solutions. Supercritical fluids and ionic liquids both assist in reducing environmental and health risks associated with conventional solvents, a crucial aspect of green chemistry. With these novel solvents currently in the process of being developed for many applications, their ability to catalyze sustainable processes in industries will keep increasing. The switch to such safer alternatives promotes not only environmental friendliness but also versatility of chemical reaction media, thereby promoting commercial and environmental sustainability goals. The new paradigm rests on the imperative to develop and implement safer solvents in ongoing endeavors towards greener chemistry solutions that meet the demands of today's society and nature (Figure 4.9).

4.4. ADVANTAGES OF USING SAFER SOLVENTS

The use of cleaner solvents in chemical processing is a big step in the practice of green chemistry, which aims to reduce the environmental impact associated with the use of conventional solvents. Cleaner solvents, such as those derived from renewable resources or designed to have little or no harmful properties, have several advantages over conventional organic solvents. One of the major benefits is the reduced generation of harmful waste during chemical reactions that normally take place with VOCs. Replacing such harmful solvents with safer substitutes allows scientists to minimize the generation of hazardous waste and make chemical reactions more environmentally friendly. The application of safer solvents thus addresses global initiatives towards environmental preservation and employment of sustainable practices in the chemical process. In addition to minimizing environmental challenges, safer solvents are also beneficial for the safety and health of laboratory personnel.

The majority of conventional solvents not only possess toxic properties but also pose severe health risks to scientists via inhalation, skin contact, or accidental ingestion. Safer solvents, which are often less toxic and less volatile, render the workplace safer. The shift has the function of protecting the health of scientists and employees, thereby promoting a more sustainable and responsible method of conducting chemical research and industry. The practice using safer solvents may also mitigate regulatory pressures associated with hazardous substances, enabling an environment-friendly setting for research and innovation without the ongoing threat of being out of compliance. Besides, new solvents that are safer have opened the door for greater efficiency in chemical reactions. For instance, the utilization of green solvents has the capacity to yield improved reaction yields, such as higher yield and shorter reaction times. Materials such as nanocellulose have been identified as being appropriate as green solvents since they are readily available, have good mechanical properties, and are suitable with various functionalization's (Trache et al., 2020). Additionally, metal nanoparticles were found to exhibit significant antibacterial activity when used in safer solvent media, exhibiting their effectiveness against many types of pathogens (Elena et al., 2020). By promoting such developments, the implementation of safer solvents not only improves environmental and health conditions but also enhances the overall efficiency and applicability of chemical processes, further supporting their significance in the realm of green chemistry.

4.4.1. Reduced Environmental Impact

The emphasis on reducing environmental impact in Green Chemistry is a novel approach to the inherent unsustainability of traditional chemical process. Optimizing less toxic solvents and reaction media, Green Chemistry significantly minimizes the generation of toxic waste and restricts the release of toxic substances into the environment. One of the most attractive aspects of this strategy is that renewable feedstocks are used to replace non-renewable raw materials that cause pollution and also resource depletion. Such substitution not only follows nature's law but also renders manufacturing processes more sensitive to stringent environmental regulations. Therefore, this commitment to sustainability ignites a shift in paradigm towards improved sustainable manufacturing and utilization of chemicals in most industries, which eventually leads to minimized environmental footprints and improved human health outcomes (Achieng, 2025).

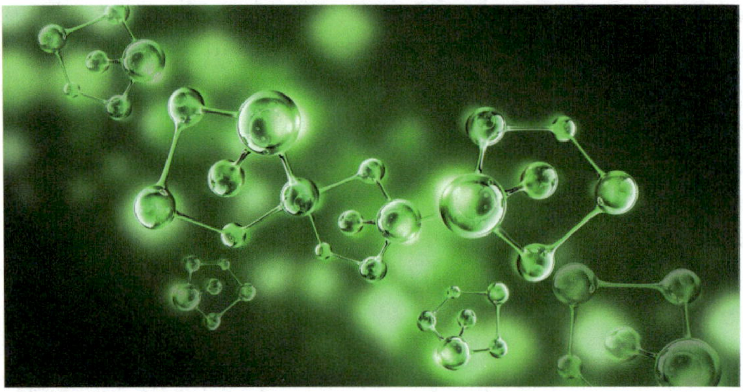

Figure 4.10. By prioritizing safer solvents and reaction media, green chemistry significantly diminishes the generation of hazardous waste and curtails the release of toxic substances into the environment.

Source: https://www.silicycle.com/media/images/mea-green-chemistry.png.

Another extremely crucial aspect of achieving a minimized environmental footprint is chemical reaction efficiency. Green Chemistry promotes atom economy principles, which enhance maximum utilization of all the materials used in the process into the finished product. The principle is significant in minimizing waste and optimizing the use of resources, and as such, fewer by-products are generated in the synthesis process. In addition, the application of catalytic processes optimizes the efficiency of chemical reactions through lower energy requirements as well as eliminating harmful waste. Such innovations not only make the production facility more sustainable but also save companies money. By reengineering traditional practices and integrating more sustainable methods, the chemical industry can minimize its environmental impact significantly without sacrificing productivity and profitability (Chen, 2025). Also, a key component of minimizing environmental impacts is selecting safer solvents. The majority of conventional solvents are volatile organic compounds

(VOCs) that contribute to the pollution of the air and can harm workers' and consumers' health. Green Chemistry, in contrast, encourages the utilization of less toxic solvents, such as water or bio-based solvents, that minimize negative effects on ecosystems and human health. This tactical move toward safer solvents not only complies with evolving regulatory needs but also responds to market pressures for greener products. Through innovation in solvent selection and reaction medium, the chemical industry has the potential to lead broader changes in manufacturing practice toward a greener future. Lastly, the commitment towards reducing environmental impact through these measures puts the crucial relevance of Green Chemistry under limelight to address the pressing global concerns (Achieng, 2025) (Figure 4.10).

4.4.2. Improved Safety for Laboratory Personnel

The shift towards greener reaction media and solvents in green chemistry is a significant advancement in lab protocols that directly affects the safety of laboratory personnel. The majority of traditional solvents present an array of hazards from toxicity to flammability and environmental problems. By employing more gentle substitutes, laboratories not only reduce the possibility of chemical exposure, but also reduce the possibility of spills, fire, or accidents due to chemical reactions. This reevaluation of solvent usage is based on the tenets of green chemistry, which value human health and safety as well as environmental sustainability. Enhanced safety procedures can reduce the physical and psychological burden of laboratory activity, providing a more favorable working environment for staff and promoting a safety culture that pervades laboratory practice. Beyond direct physical safety, the consequences of employing safer solvents also include the end goal of diminishing risks to laboratory staff over a longer term. The use of safer materials can improve overall health and well-being among lab workers, reducing stress levels in the workplace and promoting an active safety culture. Improved training and education are responsible for this change; as researchers get more exposure to safer materials, they become better equipped to recognize dangers of hazardous materials. This emphasis on safety also creates a sense of collective responsibility among lab workers, inducing a proactive attitude towards potential risk. Enhanced safety data sheet communication and safe handling procedures continue to empower staff by ensuring they remain informed about chemicals' hazards within their working environment. This multi-faceted approach to safety supplements the increasing volume of research towards safer approaches to chemistry, reiterating once again that enhanced safety procedures are key to laboratory performance. In addition to the immediate benefit to personnel, greener solvents and reaction media affect laboratory sustainability and regulatory compliance in basic ways.

As regulatory pressure on the environment mounts, laboratories that prioritize the use of green chemistry principles become leaders in responsible science practice. Through minimizing toxic waste and reducing chemical footprints, laboratories not only enhance the safety element of their operations but also promote public health and environmental goals at large. Moreover, the use of safer solvents is in line with emerging trends in the chemical industry, particularly the demand for sustainable development. As is handled in battery recycling and application of microbial enzymes within the literature, innovative solutions to processes and materials are indicative of a broader movement towards sustainability in a business that inherently values the safety of everyone involved (Neumann et al., 2022; Chandra et al., 2020). Such innovation is met with public trust and creates the environment in which scientists in laboratories can excel at their job, ultimately propelling science and safety forward in tandem.

4.4.3. Enhanced Reaction Selectivity and Yield

Improved selectivity and yield in the field of green chemistry are the real targets, particularly with regards to deployment of safer reaction media and solvents. Selectivity in product formation is not just crucial in terms of obtaining maximum output but also in avoiding wastage formation beyond limits and by-product formation, thereby encompassing the sustainability philosophy. By adjusting solvents' characteristics such as polarity, viscosity, and chemical compatibility, one can exert considerable control over the thermodynamics and kinetics of reactions. For instance, utilizing safer, lower-toxicity solvents in catalytic reactions can induce improved reaction rates and selectivity, resulting in improved overall yields of desired products without enhancing hazards from volatile organic compounds. The new strategy of electrochemical approaches has shown considerable advancement in enhancing reaction selectivity and yield in green chemistry. The technique incorporates the direct utilization of electrical energy to drive chemical transformations, typically allowing milder conditions and increased precision in the generation of the desired product. For example, electrochemical dehalogenative deuteration of the aromatic halides is a compelling case for efficiency with the use of D2O as an economical deuterium donor without the use of hazardous reagents or metal catalysts. Such kinds of methodologies are indicative of the manner in which ideal reaction conditions can offer much cleaner purification profiles and reduced environmental impact, hence facilitating the development and application of safer solvents in green chemistry applications (Lu et al., 2020). Besides, advances in reaction medium design, such as ionic liquids and supercritical fluids, have transformed the chemistry of synthesis. These new mediums provide unique properties that enable higher solubility and reactivity, leading to better reaction yields and selectivity. For instance, ionic liquids stabilize reaction intermediates and optimize catalyst efficiency, properties that are crucial to achieving desired outcomes in complex synthetic pathways. Experiments have revealed that these novel media not only increase the effectiveness of the chemical reactions but also promote cleaner chemistry with minimized solvent waste. The strategic use of safer media for reactions, thus, has a central role to play in resolving the problems of selectivity and yield of chemicals, paving the way for green chemistry in modern chemistry (Jiang et al., 2020) (Figure 4.11).

Figure 4.11. The selective formation of products is not only crucial for maximizing output but also for minimizing waste and by-product formation, thereby aligning with the principles of sustainability.

Source: https://images.squarespace-cdn.com/content/v1/5b2d5eb9ee17596af71cb0f6/1628695062088-6JSFHUPO1Q3TI1AHYVNN/Plastic-recycling-800x553.jpg.

4.5. CHALLENGES IN IMPLEMENTING SAFER SOLVENTS

The transition to greener solvents in green chemistry is faced with challenges that have several facets and limit longer use across industries. Among the principal challenges is the inherent cost of developing and applying new solvent substitutes. Traditional solvents are usually inexpensive due to several factors, such as available supply chains and lower manufacturing costs. In contrast, green reaction solvents and media, although environmental friendly and healthy safe, usually require enormous initial research, development, and production scale-up costs. Consequently, such an economic barrier discourages most organizations, particularly small and medium-sized enterprises, from embracing greener substitutes. As companies aim for short-term cost reduction, product life-long green consequences of safer solvents can be sacrificed, freezing dependence on harmful chemicals in the chemical process (Popescu et al., 2025). The second main challenge is the regulatory scenario for chemical manufacturing and application. Usage of safer solvents must meet existing regulations, which perhaps fail to recognize the intricacies and potential benefits of such alternatives. Regulatory bodies prefer conventional chemicals with known use history, in the process, creating a disincentive towards innovative, clean alternatives. Furthermore, the lack of standardized testing and assessment of new solvents can lead to problems related to their performance, safety, and environmental impact compared to conventional solvents. Producers therefore avoid discontinuing established practice and adopting cleaner technologies lest they put product quality at risk or have concerns about noncompliance. This policy lag necessitates re-examination of policy systems for integrating and advancing the utilization of safer, sustainable solvents (Achieng, 2025). Finally, the lack of shared awareness and training for safer solvents presents a significant implementation barrier. Successful integration includes not just adequate technical awareness but also organizational cultural transformation emphasizing sustainability. Many chemical professionals may not know about the principles of green chemistry or the potential applications of safer solvents. Thus, it is essential to develop an educational platform that emphasizes the importance of solvent selection and its environmental implications. Both education providers and industry stakeholders must collaborate to develop courses and training programs grounded on greener chemical methodologies. By increasing learning and skill acquisition in solvent safe practices, the industry can move towards a future of more sustainable chemical production gradually, overcoming implementation challenges in the process (Popescu et al., 2025).

4.5.1. Cost Considerations and Economic Feasibility

In seeking greener chemistry, economic profitability and cost are essential considerations in determining the viability of safer solvents and reaction media. Traditional solvents, in most instances derived from petrochemical feedstocks, bring with them environmental and economic costs that become ever less sustainable in today's regulatory context. Conversion to greener alternatives can involve more up-front raw material or processing technology cost. Such costs must be weighed against the long-term benefits of reduced health risk, environmental protection, and compliance with stringent regulatory demands. Ultimately, effective economic analysis involving direct and indirect costs may discover that it is worthwhile to invest in safer solvents through reduced liabilities and enhanced public reputation for companies that hold sustainability in high regard. Moreover, immobilized enzymes present a clear example of economics in green chemistry. The utilization of enzymes, particularly in pharmaceutical and chemical industries, can radically change the economics of production processes. Immobilized enzymes, for instance, enable the use of equimolar substrate concentrations and permit the reactions to be performed at lower temperatures, thus saving energy and resources when producing. The efficiency of such processes says a lot regarding the importance of careful choice of reaction media that would enhance enzyme performance at the expense of cost-effectiveness. Basso et al. (2019) demonstrates the economic benefits accrued from the use of immobilized enzymes, which can produce better product purity and ease downstream processing, thereby leading to lower costs of production than in traditional chemical processes. Lastly, economic feasibility of employing safer solvents must not only consider material price but also consider the impact of process efficiency. For example, emerging solvents and reaction media can influence reaction speeds and selectivity, which can lead to greater throughput and less waste generation. Such increased application of such solvents may subsequently provide a competitive advantage to firms spending money on R&D. The addition of safer solvents can also help to require less stringent disposal requirements, further saving costs on operations. As the market continues to be more driven by sustainability, companies that take action to address this cost implications will be positioned for sustainable long-term success in a constantly evolving economic climate, making it all the more important to take into account both upfront costs and ongoing savings.

4.5.2. Regulatory Hurdles and Compliance

The transition to cleaner solvents and reaction media in green chemistry is commonly impeded by a multitude of regulatory obstacles that influence development and implementation. These regulations generally arise from the need to safeguard public health and the environment, demanding extensive documentation and compliance before new materials are approved for use. As an illustration, green solvents must meet some minimum standards of safety, which could be significantly varied between different jurisdictions. This disparity is a hindrance to researchers and producers who want to innovate and offer new solvent alternatives to the global market. Regulatory frameworks inadvertently discourage the utilization of safer alternatives by burdening them with stringent requirements during their examination, thus favoring traditional solvents despite their environmental drawbacks. The inconsistency between regulatory systems may be an obstacle to green chemistry development, which decelerates the development towards cleaner and more sustainable chemical practice. Also, the complexity in authorizing new solvents has a tendency to create significant hold-ups that deter investment in coming up with safer chemical substitutes. Companies that engage in research and development can be intimidated by venturing into new solvents in anticipation of lengthy and uncertain regulatory timelines. The compliance landscape is characterized by an exhaustive array of testing protocols and evidence requirements, which can burden resources and result in increased costs.

Figure 4.12. Companies engaged in research and development may find themselves hesitating to pursue innovative solvents due to the fear of lengthy and uncertain regulatory timelines.

Source: https://www.ft.com/__origami/service/image/v2/images/raw/https%3A%2F%2Fd1e00ek4ebabms.cloudfront.net%2Fproduction%2Ffbbda67c-fd2b-4536-a9d9-ff8425e3afe1.png?source=next-article&fit=scale-down&quality=highest&width=700&dpr=1.

This economic strain is particularly challenging for smaller firms and start-ups that lack the capital to navigate intricate regulatory pathways. Furthermore, as the push for sustainable practices intensifies, regulatory agencies and policymakers face increasing pressure to adapt existing frameworks. This demands a delicate equilibrium between ensuring safety and encouraging innovation in developing environmentally friendly solvent solutions. Regulatory coordination with the green chemistry industry is needed to enhance compliance streamlining and enhance the adoption of safer solvents. Interdisciplinary solutions can lead to more flexible and efficient regulatory pathways taking into account the distinctive chemical properties and safety profiles of novel substances. For example, applying advanced technological techniques such as predictive modeling and artificial intelligence can assist in facilitating a better assessment of new materials, streamlining their approval procedure. In addition, research into alternative testing approaches that reduce the application of animal-based testing may be a means through which regulatory standards can be met while promoting ethical practice. By confronting the regulatory issues head-on, stakeholders can help find a balance between the safety, sustainability, and innovation goals, ultimately bringing about a friendly regulatory situation for green chemistry research in safer solvents and reaction media (Ampomah et al., 2024; Maria-Titirici et al., 2024) (Figure 4.12).

4.5.3. Limited Availability of Certain Safer Solvents

The transition to greener solvents under green chemistry is pivotal in enhancing the sustainability of chemical processes. The transition, however, is highly constrained by the reduced availability of safer solvents that meet the stringent demands of both performance and environmental hazard. While traditional solvents provide high performance and efficiency in a range of chemical reactions, they are normally accompanied by health and environmental risks that do not meet the aims of green chemistry. The lack of alternatives capable of delivering a comparable function to toxic solvents reinforces reliance on such toxic substances and stifles innovation in the design of safer solvents. There is thus a need for a combined effort by researchers and industry stakeholders to increase the number of safer solvents available, thereby facilitating cleaner reaction media and processes to enable sustainable operations. The constraints relative to the limited availability of safer solvents are compounded further in specialized applications, whose performance standards are highly specific. All but the very most solvent alternatives identified to be safer or more sustainable might lack the appropriate properties ideal for challenging chemical reactions. For instance, in industries such as pharmaceuticals and biochemistry, precision is paramount, and a solvent's ability to dissolve certain compounds or facilitate certain reactions can be the difference between success and failure in a project. This need for high-performance solvents means that the green chemistry revolution must not only seek to design new safer solvents but also enhance the properties of existing substitutes. These technologies could open up new routes of chemical synthesis that are in harmony with green chemistry practices, and thus reduce dependence on traditional solvents with environmental or health risks. Secondly, green chemistry research must involve a thorough analysis of the lifecycle impacts of green solvents. Limited availability is most often made up for by the requirement for thorough studies to determine the ecological and toxicological profiles of such potential alternatives. For example, while substances such as nanocellulose are recognized as green candidates for alternatives since they can be replenished and are biocompatible, their real application in various chemical processes remains under research (Trache et al., 2020). Second, the development of recycling technology for substances like lithium-ion batteries shows how vital it is to close the solvent supply gap with broader sustainability initiatives (Neumann et al., 2022). Lastly, instilling a thorough understanding of safer solvents (their accessibility, role, and environmental impact) will be the solution to overcoming limitations and advancing the field of green chemistry.

4.6. CASE STUDIES IN GREEN CHEMISTRY

Green chemistry seeks to transform the way chemical reactions occur by lowering the ecological cost of traditional solvent utilization. Sample studies are presented to demonstrate innovative adoption of safer solvents and reaction media in accordance with green chemistry, showing actual instances where traditional solvents have been replaced by less harmful ones. For instance, a bioactive compounds extraction case study has identified supercritical carbon dioxide to be a green solvent that greatly reduces hazardous waste while ensuring efficiency in extraction. Such conversions not only ensure that exposure to hazardous materials is minimized but also result in economic benefits due to the simplified processes of waste management. This shift towards safer solvents aligns with the larger objective of promoting sustainable processes within the chemical industry, with a view to mitigating the negative ecological effects. The application of nanocellulose as a safer alternative within chemical reactions represents a remarkable feature of current case studies, whose evidence indicates that it possesses the ability to be utilized as a green reaction medium. Recent studies have further identified nanocellulose's positive attributes, including high aspect ratio, renewability, and biocompatibility, that favor it as a solvent and stabilizer in various uses. The presence of hydroxyl functional groups on nanocellulose also favors easy modification with modification avenues that enhance its functionality as well as providing for environmental stewardship. Case studies have further illustrated how cellulose nanocrystals can easily replace toxic solvents in polymer composites so as to offer a safer medium for industrial use. As researchers continue to develop better processes for the processing and use of nanocellulose, the substance increasingly becomes a backbone within the search for greener chemical processes, underscoring its importance within contemporary green chemistry debates (Trache et al., 2020). Further studies into the environmental impact of conventional solvents and alternatives also underscore the necessity for an industrial chemistry paradigm shift. Recent evaluations have identified the pollution routes and environmental implications of conventional solvents in a systematic approach, with the conclusion that the shift towards safer alternatives is not just desirable, but essential for protecting the environment. The recycling and upcycling of chemical waste through case studies have identified potential solutions to the problem, with special emphasis on employing alternative solvents like ionic liquids and bio-based solvents. These have enlightened pathways of effective recovery of resources, showing that modifications in the use of solvents can, to a significant degree, reduce dangerous waste. By maintaining its concentration on ecologically friendly behavior, green chemistry courses not only advance science boundaries but also encourage joint initiatives among stakeholders that are keen on furthering sustainable development (Mrozik et al., 2021).

4.6.1. Successful Applications of Safer Solvents in Industry

Transition towards cleaner chemical processes is greatly facilitated by the utilization of safer solvents in various industrial processes. Traditional organic solvents tend to be environmentally and health risks, necessitating the exploration of other possibilities that are safer. Industries are now gravitating towards solvents produced from renewable feedstocks, such as bio-based solvents, that are less toxic and more biodegradable. For instance, in drug manufacturing, solvents like ethanol and methanol have been replaced by more environmentally friendly substitutes without affecting reaction efficacy or solubility. This shift not only portrays the ideality of green chemistry but also raises the industrys commitment to sustainability as it reduces landfill waste and toxic emissions, while demonstrating that safety and efficacy are compatible with chemical manufacturing processes (Arshiya et al., 2025). In polymer science, safer solvents have led to the development of new materials with minimized environmental effects. Supercritical carbon dioxide ($scCO_2$) is a revolutionary approach that can be used as a solvent without the health risk of conventional organic solvents. It has been applied in polymer synthesis with very promising outcomes, particularly in production techniques aimed at producing high-performance materials. Employing $scCO_2$ facilitates efficient reaction and product recovery with ease, reducing waste generated in polymerization processes. Moreover, ionic liquids being alternative media have gained popularity due to their volatility and the ability to dissolve a wide range of compounds, making them optimum for highly complex polymer systems. These advancements underscore the role of safer solvents in driving industry to innovation while maintaining concern for the environment (Babbar, 2024). Employing new solvent technologies is changing the landscape of chemical manufacturing, as evidenced by the visibility of solvent-free reactions and biocatalytic reactions. These strategies have been welcomed in various industries, like agrochemicals and fine chemicals, due to their efficacy and reduced environmental impact. Biocatalysis, employing enzymes in the absence of solvent, not only streamlines manufacture but also enables milder conditions of reaction that can be achieved at lower energy cost. Additionally, employment of safer solvents such as water or biodegradable solvents instead of conventional solvents also enhances green chemistry principles compliance. This whole-of-approach helps to further the development of sustainable practices across industries, that aligns the manufacturing of chemicals to global environmental frameworks and generates a circular economy and sets a precedent for continuous innovation (Arshiya et al., 2025).

4.6.2. Research Advancements in Solvent Alternatives

Recent research has become increasingly willing to engage in the design of greener solvent replacements, recognizing the need to shift away from traditional organic solvents that are toxic to the environment and to human health. An example of such development is the research into the use of water as a sustainable solvent, particularly for uses such as dye-sensitized solar cells (DSSCs). WBDSSCs are in themselves difficult, having low photovoltaic efficiency as the main issue in response to issues of dye solubility and adhesion on semiconductor surfaces. Resolving these issues entails new strategies in material science, such as maximizing the photoanode/electrolyte interface and dye selection for enhanced stability and performance. The use of green materials not only advances green chemistry but also fuels the advancement of renewable energy technologies, consolidating the roots of sustainability in the synthesis and application of chemicals (Spadaro et al., 2023). Apart from aqueous solvents, there has been considerable progress in exploring other green substitutes such as ionic liquids (ILs) and supercritical fluids (SCFs). Ionic liquids, which are liquid salts at room temperature, are characterized by zero vapor pressure and adjustable properties that can be engineered to suit the needs of desired chemical reactions. This versatility enables scientists to create ionic liquids with minimized environmental effects and optimized reaction efficiency. Similarly, supercritical fluids, and particularly supercritical carbon dioxide, have become a solvent in reaction and extraction processes due to their unique properties that can lead to more energy-efficient utilization and reduced use of hazardous materials. Such emerging solute technologies are part of an overarching effort to move beyond the limitations of traditional solvents and to foster research on greener chemical strategies (Adepu et al., 2021). Besides, controlled drug delivery system development is another convincing instance of safer solvent alternatives' potential. In drug delivery in pharmaceutical application, conventional drug delivery mechanisms are saddled with poor bioavailability and ineffectiveness in targeting sites within the body. Advances in controlled drug delivery systems were aimed at designing biocompatible solvents and carriers for enhancing drug solubility as well as release rates. Such a shift in drug delivery technology is a concerted effort to increase therapeutic efficacy with diminished side effects associated with traditional formulations. With increased research geared towards addressing the challenges of integrating safer solvents into such systems, increased potential for more efficient and sustainable health outcomes emerges, underscoring the pressing need for advancement within the scope of solvent alternatives (Spadaro et al., 2023).

4.6.3. Comparative Analysis of Traditional vs. Safer Solvents

Use of solvents in chemical reactions is of significant environmental and health significance, particularly when traditional solvents such as benzene and dichloromethane are employed. The traditional solvents are generally associated with toxicological risks, including carcinogenesis and environmental damage. Conversely, more benign solvents, including bio-based and solventless solvents, provide significant advantages from a green chemistry perspective. By employing safer solvents, the generation of risky waste is reduced, exposure to harmful chemicals is minimized, and the sustainability of the lab processes is promoted. In a comparative study, even if classical solvents may have certain desirable qualities such as polarity or boiling point, the emergence of safer solvents points towards the need to prioritize human and environmental health as well as chemical efficiency only. Shifting to greener solvents not only aligns with green chemistry principles but also signifies a shift in scientific ethos towards sustainability. Greener solvents are usually synthesized using renewable feedstocks, and their application can lead to reduction of toxic emissions and energy consumption. Techniques such as recycling of solvents and using supercritical carbon dioxide as a solvent are some of the most important innovative methods of reducing the environmental disadvantage of traditional solvents. Furthermore, research demonstrates that safer solvents are as efficient for dissolving many reactants as their traditional equivalents and reduce the environmental expenses of chemical processes. This positive outcome is not merely in accordance with the principles presented within White Analytical Chemistry but also reflects the harmony of performance and sustainability in contemporary chemical practice (Paweł et al., 2021). In a comparison of the effectiveness of traditional solvents and safer solvents, it is clear that new approaches are able to produce comparable, if not superior, outcomes in most reactions. For instance, greener solvent systems are used to prepare metal-based nanoparticles with broad antimicrobial activity as opposed to more toxic and conventional counterparts (Elena et al., 2020). This enhances not only the safety profile of the scientific process but also pushes the boundaries of interest in green chemistry to a wider community, which in turn translates to a responsibility culture among scientists. As the science evolves, future research must continue to refine these more secure methods more sharply, reassuring that scientific excellence and environmental sustainability are compatible but not conflicting goals towards achieving chemical innovation (Figure 4.13).

Figure 4.13. Safer solvents are often derived from renewable resources, and their implementation can lead to a reduction in harmful emissions and energy consumption.

Source: https://www.moserbaersolar.com/wp-content/uploads/2024/10/solar-panel-eco-impact-scale-810x500.jpeg.

4.7. CONCLUSION

Overall, greener chemistry demands the utilization of safer reaction media and solvents that not only reduce the environmental footprint but also make chemical processes more sustainable. Application of green solvents, such as supercritical CO_2, is an indicator of a significant shift in the approach to chemical reactions. The outcomes of modern studies confirm that such ecologically friendly reaction media can surpass traditional organic solvents in yield and efficiency. So, combination of heterogeneous catalysts with supercritical CO_2 results in serious advantages, which is a promising trend for the chemical industry. Combating the constraints of catalyst reusability and performance remains significant, as evidenced by the results that show heterogeneous catalysts are capable of performing well over a sequence of repetition with low waste and energy use, in compliance with green chemistry principles (Sonia et al., 2022). Furthermore, the versatility of materials such as nanocellulose (NC) is an indicator of innovative methods' ability to perform in the development of sustainable reaction media. Its unique properties render NC not only a fine green material but also an avenue for improvement of catalysis and functionality of solvents. The structure advantage of NC, owing to its affordability and biocompatibility, offers an avenue for enormous research and development. Recent studies have established the fact that NC could be effectively modified to adjust its chemical properties to accommodate a range of applications, hence expanding its scope in green catalytic reactions. The study of nanocellulose as a renewable resource provides new routes to the development of new and non-toxic reaction media, pushing the green chemistry agenda forward while providing enormous prospects for environmental benefit (Trache et al., 2020). Lastly, the general advancement in safer solvents and reaction media reflects an increasing emphasis on sustainability in the chemical community. As catalysts and solvents move towards greener alternatives, revisiting and refining existing protocols to make them more useful and effective is necessary. The integration of green chemistry principles not only encourages innovative research but also encourages a culture of accountability for safeguarding the environment. This commitment to the use of safer solvents, as exemplified by promising developments in both heterogeneous catalysis and nanocellulose utilization, forms the basis for future research. Such work represents important strides toward rendering the ecologically harmonious manufacture of chemicals achievable, where advancements in this area redound to the benefit of industry and the environment (Sonia et al., 2022; Trache et al., 2020).

4.7.1. Summary of Key Points

In the evolution of green chemistry, the selection of safer solvents and reaction media plays a key role in the minimization of environmental and health impacts. Current advancements focus on phasing out traditional solvents that pose enormous risk to human life and the environment. The creation of bio-based solvents, which are usually derived from renewable resources, is now an effective solution to the phaseout of volatile organic solvents that contribute to air pollution and greenhouse gas emissions. With the utilization of non-toxic, biodegradable solvents derived from renewable materials, chemists can perform safer and more efficient chemical reactions. This shift not only fosters green chemistry philosophy but also enhances the overall sustainability of the practice of chemical production. Moreover, emphasis on the optimization of reaction media can significantly influence the outcome of chemical reactions. Employing water as a solvent, for example, is in line with the principles of green chemistry through the promotion of safer operations that minimize the creation of harmful waste. The effect of solvent choice goes beyond safety; it has a significant impact on reaction rates, yield, and selectivity. The use of environmentally friendly solvents can make chemical reactions more efficient by having less resource consumption and fewer overall costs. Having a clear understanding of the intricate interaction between solvent properties, reaction conditions, and the environment will enable researchers to make informed choices when designing new synthetic protocols and enhance the sustainability of their processes. Lastly, in recycling and reusing solvent materials, there is an important component in the implementation of a circular economy in green chemistry. The materials recycling technology of lithium-ion batteries emphasizes the complexity in resource management sustainability. As recent reports outline, the environmental impact of spent lithium-ion battery disposal presents a pressing need for effective recycling strategies. Improving the recycling process is essential given the different chemistries employed during battery assembly, thus requiring difficulties in the formation of standard recycling practices. In the coming future, an interconnected concept of regulatory frameworks, technological advances, and innovative approaches will be vital for making the chemical sector not only reduce its footprint on the environment but also thrive in a sustainable world (Neumann et al., 2022; Mrozik et al., 2021).

4.7.2. Future Directions for Research in Green Chemistry

As the green chemistry area has picked up momentum, future studies will more and more center on the development of innovative solvent systems that minimize environmental impact while enhancing process efficiency. One of the most promising areas of research is in deep eutectic solvents (DESs), which have been recognized as promising alternatives to traditional solvents due to their lower toxicity and biodegradability. By logical questioning of the structure–composition–property–performance relationships of DESs, researchers envision to optimize their application as reaction media in a wide variety of chemical processes. This would facilitate their utilization not only in organic reactions but also in energy storage and conversion systems that require advanced electrolytes. The finding of DESs as charge transfer agents could be of crucial value to cleaner production modes in various industries, opening a new era for sustainable chemical practice (Wu et al., 2021). Incorporating hydrogels into green chemistry is another promising direction for future research, particularly their potential as cleaner media for reactions and electrolytes. Hydrogels can be synthesized from both natural and synthetic polymers, which renders them equally versatile in design and functioning. Their biodegradability and biocompatibility contribute further to their applicability based on the principles of green chemistry. As hydrogels continue to find more research avenues into their mechanism in many aspects, future applications for energy purposes are more likely to involve substitutes like flexible, self-healable hydrogels. These hydrogels can potentially reduce the possibility of device failure as well as enhance the life cycle of devices like batteries and supercapacitors, crucial towards the realization of sustainable energy technologies. Applications based on these materials will open up innovations that significantly lower the environmental footprint of electronic devices and energy storage devices (Bashir et al., 2020). The future of green chemistry research also needs interdisciplinarity among fields such as chemistry, materials science, and engineering to facilitate the innovation required for safer solvents. An interdisciplinary research environment can give the appropriate environment for unorthodox thinking and idea exchange, thus giving rise to new materials prepared as per the high standards that green chemistry promotes. By fostering partnerships across sectors, researchers can better address challenges related to the scalability and commercialization of novel solvents, ensuring their practical application in industrial settings. Such collaborative work has the potential to make discoveries that overcome the frontiers of present knowledge, triggering the implementation of safer solvents that are in harmony with environmental goals, thereby paving the way towards advancements in cleaner, greener processes of chemical production (Wu et al., 2021; Bashir et al., 2020).

4.7.3. The Importance of Continued Innovation in Solvent Use

The evolution of solvent use in chemical processes indicates the necessity for continued innovation in this area, particularly under the green chemistry umbrella. Traditional solvents are often environmentally and health risk-disposed in their vapor phase VOC emissions and potential toxicity. While science grapples with the problem of such solvents, science has gradually transitioned towards developing safer alternatives minimizing adverse effects while maintaining efficiency in chemical reactions. This is not merely a trend but a necessity because the effect of such toxic solvents on the environment is not confined to the laboratory alone but extends much more into broader ecosystems and public health. A convergence on innovation, apart from aiding in the design of safer solvents, also aligns with sustainability goals, driving the need for environmentally friendly methods in various industries that are engaged in the manufacture of chemicals. Furthermore, the drive for innovation in solvents is hand-in-hand with recycling and resource recovery principles, particularly for industries with high reliance on lithium-ion battery technologies. The human constituents in such batteries are the call for better solvent systems, especially with the mounting e-waste from spent batteries. Recycling spent lithium-ion batteries is more critical since their environmental and safety concerns must be addressed to prevent toxic waste. New solvent systems can enhance the effectiveness of recycling operations by recovering precious metals while eliminating toxic byproducts. Current recycling processes, such as pyrometallurgy and hydrometallurgy, pose significant environmental concerns; it is therefore crucial to come up with new sustainable green approaches. Optimization of solvent use can therefore not only assist in sustainable recycling but also align with circular economy programs according to established research studies (Neumann et al., 2022). The discussion of the application of new solvents also must consider the social implication of such advancement, particularly the promotion of public education and green chemistry practices acceptance. With increasingly secure solvents and reaction media on the market, their application can contribute in a lasting manner to industrial and consumer practices. Public collaboration and legislation support of green chemistry technology are essential to facilitating the transition towards sustainable chemical manufacturing practices. Furthermore, the creation of learning frameworks that educate stakeholders on the benefits of safer solvents can enable their integration into existing processes. By embracing the culture of innovation and commitment to sustainability, stakeholders can take the front seat in revolutionizing solvent usage standards, and ultimately realize significant enhancement in chemical process safety and efficiency. The focus on new solvent systems is thus not merely a question of technical innovation but a vital step towards a greener tomorrow (Roy et al., 2021).

REFERENCES

1. Achieng, N. M. (2025). Green chemistry in pharmaceuticals: Reducing environmental impact. *Newport International Journal of Public Health and Pharmacy*. https://www.semanticscholar.org/paper/fcf7c9c0751460e055a99f3481f52db214e8d25d (accessed on 24th October 2025).
2. Adepu, S., & Ramakrishna, S. (2021). Controlled drug delivery systems: Current status and future directions. *Molecules, 26*, 5905–5905. https://doi.org/10.3390/molecules26195905 (accessed on 24th October 2025).
3. Alcántara, A. R., & Domínguez de María, P. (2018). Recent advances on the use of 2-methyltetrahydrofuran (2-MeTHF) in biotransformation's. *Current Green Chemistry, 5*, 86–103. https://doi.org/10.2174/2213346105666180727100924 (accessed on 24th October 2025).
4. Ampomah, W., Morgan, A., Koranteng, D. O., & Nyamekye, W. I. (2024). CCUS perspectives: Assessing historical contexts, current realities, and future prospects. *Energies, 17*, 4248–4248. https://doi.org/10.3390/en17174248 (accessed on 24th October 2025).
5. Arshiya, A. S., Challa, M. S., Majji, M. K., Umme, K., Suneetha, A., & Patibandla, J. (2025). ECO-Logic reactions: Shaping the future of sustainable molecular science. *World Journal of Current Medical and Pharmaceutical Research*. https://www.semanticscholar.org/paper/47de7be2a84c33aa2ba7ebf2a55bcaffe902d9b5 (accessed on 24th October 2025).
6. Asenbauer, J., Eisenmann, T., Kuenzel, M., Kazzazi, A., Chen, Z., & Bresser, D. (2020). The success story of graphite as a lithium-ion anode material–Fundamentals, remaining challenges, and recent developments including silicon (oxide) composites. *Sustainable Energy & Fuels, 4*, 5387–5416. https://doi.org/10.1039/d0se00175a (accessed on 24th October 2025).
7. Babbar, R. (2024). Green chemistry and its applications. *Journal of Pharmaceutical Technology, Research and Management*. https://www.semanticscholar.org/paper/22a2ff31f002a034fa93718f2f5d1593eb55e08f (accessed on 24th October 2025).
8. Bashir, S., Hina, M., Iqbal, J., Rajpar, A. H., Mujtaba, M. A., Alghamdi, N., Wageh, S., et al. (2020). Fundamental concepts of hydrogels: Synthesis, properties, and their applications. *Polymers, 12*, 2702–2702. https://doi.org/10.3390/polym12112702 (accessed on 24th October 2025).
9. Basso, A., & Şerban, S. (2019). Industrial applications of immobilized enzymes—A review. *Molecular Catalysis, 479*, 110607–110607. https://

doi.org/10.1016/j.mcat.2019.110607 (accessed on 24th October 2025).

10. Chandra, P., Enespa, Singh, R., & Arora, P. K. (2020). Microbial lipases and their industrial applications: A comprehensive review. *Microbial Cell Factories, 19*. https://doi.org/10.1186/s12934-020-01428-8 (accessed on 24th October 2025).

11. Chemat, F., Abert Vian, M., Fabiano-Tixier, A.-S., Nutrizio, M., Režek Jambrak, A., Munekata, P. E. S., Lorenzo, J. M., et al. (2020). A review of sustainable and intensified techniques for extraction of food and natural products. *Green Chemistry, 22*, 2325–2353. https://doi.org/10.1039/c9gc03878g (accessed on 24th October 2025).

12. Chen, E. (2025). Exploring green chemistry: Sustainable approaches to chemical synthesis. *Revita Latinoamericana de Química*. https://www.semanticscholar.org/paper/37ef2209f3997399352c45ee568c965c16604a1a (accessed on 24th October 2025).

13. Jiang,D.,Xin,K.,Yang,B.,Chen,Y.,Zhang,Q.,He,H.,&Gao,S.(2020).Total synthesis of three families of natural antibiotics: Anthrabenzoxocinones, fasamycins/naphthacemycins, and benastatins. *CCS Chemistry, 2*, 800–812. https://doi.org/10.31635/ccschem.020.202000151 (accessed on 24th October 2025).

14. Khandebharad, A., Arde, S. M., Pardeshi, S. C., Dhotre, B. K., & Sarda, S. (2024). Multifunctional deep eutectic solvent-catalyzed synthesis of dihydropyrimidinethiones: A sustainable approach for green and efficient reactions. *Current Organocatalysis*. https://www.semanticscholar.org/paper/58722bcf14e6d6be92e86d8e6fd8c0c35ffa09e0 (accessed on 24th October 2025).

15. Lashaki, M. J., Hajinasiri, R., Hossaini, Z., & Nami, N. (2022). Ag/KF/CP@MWCNTs catalyzed green synthetic pathway for bio-active new thiazolopyrimidines and oxazolopyrimidines. *Polycyclic Aromatic Compounds, 43*, 6088–6106. https://www.semanticscholar.org/paper/d5c32bdcbe9c77b56ee496571f75f2382b2d495a (accessed on 24th October 2025).

16. López, S., García-Vargas, J. M., García, M. T., Rodríguez, J. F., Gracia, I., & Ramos, M. J. (2022). Copper-containing catalysts for azide–alkyne cycloaddition in supercritical CO_2. *Catalysts, 12*, 194–194. https://doi.org/10.3390/catal12020194 (accessed on 24th October 2025).

17. Lu, L., Li, H., Zheng, Y., Bu, F., & Lei, A. (2020). Facile and economical electrochemical dehalogenative deuteration of (hetero) aryl halides. *CCS Chemistry, 3*, 2669–2675. https://doi.org/10.31635/ccschem.020.202000512 (accessed on 24th October 2025).

18. Mrozik, W., Rajaeifar, M. A., Heidrich, O., & Christensen, P. A. (2021). Environmental impacts, pollution sources and pathways of spent lithium-ion batteries. *Energy & Environmental Science, 14*, 6099–6121. https://doi.org/10.1039/d1ee00691f (accessed on 24th October 2025).
19. Muratov, E., Bajorath, J., Sheridan, R. P., Tetko, I. V., Filimonov, D., Poroikov, V., Oprea, T. I., et al. (2020). QSAR without borders. *Chemical Society Reviews, 49*, 3525–3564. https://doi.org/10.1039/d0cs00098a (accessed on 24th October 2025).
20. Neumann, J., Petrániková, M., Meeus, M., Gamarra, J. D., Younesi, R., Winter, M., & Nowak, S. (2022). Recycling of lithium-ion batteries—Current state of the art, circular economy, and next generation recycling. *Advanced Energy Materials, 12*. https://doi.org/10.1002/aenm.202102917 (accessed on 24th October 2025).
21. Nowak, P. M., Wietecha-Posłuszny, R., & Pawliszyn, J. (2021). White analytical chemistry: An approach to reconcile the principles of green analytical chemistry and functionality. *TrAC Trends in Analytical Chemistry, 138*, 116223–116223. https://doi.org/10.1016/j.trac.2021.116223 (accessed on 24th October 2025).
22. Patra, J. K., Das, G., Fraceto, L. F., Campos, E. V. R., Rodriguez-Torres, M. del P., Acosta-Torres, L. S., Díaz-Torres, L. A., et al. (2018). Nano based drug delivery systems: Recent developments and future prospects. *Journal of Nanobiotechnology, 16*. https://doi.org/10.1186/s12951-018-0392-8 (accessed on 24th October 2025).
23. Poizot, P., Gaubicher, J., Renault, S., Dubois, L., Liang, Y., & Yao, Y. (2020). Opportunities and challenges for organic electrodes in electrochemical energy storage. *Chemical Reviews, 120*, 6490–6557. https://doi.org/10.1021/acs.chemrev.9b00482 (accessed on 24th October 2025).
24. Popescu, A., & Ionescu, E. (2025). Green chemistry: Pioneering sustainable solutions for a safer planet. *International Journal of Emerging Trends in Chemistry (IJETC)*. https://www.semanticscholar.org/paper/31f5ae82fc1f3d33117ee3b223eee5059152b322 (accessed on 24th October 2025).
25. Roy, J. J., Rarotra, S., Krikstolaityte, V., Wu, Z., Cindy, Y. D. I., Tan, X. Y., Carboni, M., et al. (2021). Green recycling methods to treat lithium-ion batteries e-waste: A circular approach to sustainability. *Advanced Materials, 34*. https://doi.org/10.1002/adma.202103346 (accessed on 24th October 2025).
26. Sánchez-López, E., Gomes, D., Esteruelas, G., Bonilla, L., López-Machado, A., Galindo, R., Cano, A., et al. (2020). Metal-based nanoparticles as

antimicrobial agents: An overview. *Nanomaterials, 10*, 292–292. https://doi.org/10.3390/nano10020292 (accessed on 24th October 2025).

27. Sharma, A., Sharma, R., & Singh, A. (2025). Deciphering the meticulous role of safer auxiliaries and solvents: A perspective. *Journal of Biochemical and Molecular Toxicology, 39*(8), e70411. https://www.semanticscholar.org/paper/506c118d62eeeb2c48a78731f90930dc85180a66 (accessed on 24th October 2025).
28. Sheldon, R. A., & Woodley, J. M. (2017). Role of biocatalysis in sustainable chemistry. *Chemical Reviews, 118*, 801–838. https://doi.org/10.1021/acs.chemrev.7b00203 (accessed on 24th October 2025).
29. Singh, J., Dutta, T., Kim, K.-H., Rawat, M., Samddar, P., & Kumar, P. (2018). 'Green' synthesis of metals and their oxide nanoparticles: Applications for environmental remediation. *Journal of Nanobiotechnology, 16*. https://doi.org/10.1186/s12951-018-0408-4 (accessed on 24th October 2025).
30. Spadaro, D., Barichello, J., Citro, I., & Calogero, G. (2023). Environmentally friendly water-based electrolyte for dye-sensitized solar cells: Future prospective and outlook. *Solar, 3*, 229–252. https://doi.org/10.3390/solar3020015 (accessed on 24th October 2025).
31. Titirici, M. M., Johansson, P., Crespo Ribadeneyra, M., Au, H., Innocenti, A., Passerini, S., Petavratzi, E., et al. (2024). 2024 roadmap for sustainable batteries. *Journal of Physics Energy, 6*, 041502–041502. https://doi.org/10.1088/2515-7655/ad6bc0 (accessed on 24th October 2025).
32. Trache, D., Tarchoun, A. F., Derradji, M., Hamidon, T. S., Masruchin, N., Brosse, N., & Hussin, M. H. (2020). Nanocellulose: From fundamentals to advanced applications. *Frontiers in Chemistry, 8*. https://doi.org/10.3389/fchem.2020.00392 (accessed on 24th October 2025).
33. Wu, J., Liang, Q., Yu, X., Lü, Q. F., Ma, L., Qin, X., Chen, G., et al. (2021). Deep eutectic solvents for boosting electrochemical energy storage and conversion: A review and perspective. *Advanced Functional Materials, 31*. https://doi.org/10.1002/adfm.202011102 (accessed on 24th October 2025).
34. Zheng, B. (2025). A review of the application of green solvents in organic synthesis. *Theoretical and Natural Science*. https://www.semanticscholar.org/paper/422a65c3c03c0ea4a3c21c9bb52fb98c0b0f716b (accessed on 24th October 2025).

CHAPTER 5

CATALYSIS IN GREEN CHEMISTRY

Contents

INTRODUCTION

The increased global interest in environmental degradation has provoked a revolution in the chemistry science toward sustainable processes. Catalysis in green chemistry leads the revolution as it aims at attaining maximum efficiency without waste and detrimental byproducts. Green chemistry makes possible the production of processes that are both environmentally sound and economical through the employment of catalysis. The relevance of this field is borne out by its ability to facilitate a spectrum of chemical transformations under milder conditions, where previously difficult or impossible conversions are made facile. This has opened new windows for innovation, where chemists can design novel modes of synthesizing crucial compounds on the basis of green chemistry. Catalytic innovations have also made it possible to increase the exploration of materials and their properties, bolstering the role of green chemistry even more. One excellent instance of such research is the synthesis of metal-organic frameworks (MOFs) and covalent organic frameworks (COFs), which immensely expanded material science frontiers. The unique characteristics of these reticular materials render them candidate materials for the tuning of their properties, thereby enhancing their application in catalytic reactions. Reticular design principles are being used more and more to design catalysts specific to specific reactions, thereby enhancing selectivity and efficiency. Through these advancements, the role of catalysis in green chemistry has been found to possess profound importance in both fundamental research and industrial operations as a critical bridge between advances in theory and practical applications (Freund et al., 2021). Under these dynamic circumstances, organic photocatalysts have been of critical concern for their potential in enabling sustainable conversions in organic synthesis. The catalytic fabrication of acridinium photocatalysts represents an excellent instance of the shift towards the utilization of environment-friendly materials for catalysis. Recent studies have unveiled new strategies towards the modular and scalable fabrication of the photocatalysts, which increase their functionality while simplifying synthetic complexity. This represents a giant stride towards sustainable chemistry as the organic photocatalysts are capable of facilitating reactions under mild conditions with characteristic reactivities that match metal-based catalysts. Organic photocatalysis application is a paradigm for broader green chemistry trends that highlight low environmental impact and energy efficiency, showing the way catalysis integration within this field is creating real avenues for the realization of a more sustainable chemical future (Yan et al., 2021).

Importance of Catalysis in Chemical Reactions

Catalysis plays a crucial role in the pursuit of green chemistry through effective and sustainable catalysis of chemical processes. Essentially, catalysts decrease the activation energy necessary for reactions to induce elevated rates of reaction without being consumed in the process. This is absolutely essential in industry where time and resource efficiency are of the highest concern. Catalysis in green chemistry not only minimizes waste but also best curtails energy consumption and harmonizes production processes with eco-friendly sustainability goals. Catalysts allow for more discriminating reactions, which result in increased yields and fewer by-products produced, which are a prime concern in traditional chemical processings. Thus, catalysis adoption is important to establish methodologies that correspond to the principles of green chemistry, and thereby advancing a greener means of chemical production. Biocatalysis has been an important focus in catalysis where chemical conversions utilize natural biological catalysts such as enzymes. Enzyme optimization through advances in molecular biology and protein engineering has paved new ways toward greener chemical processes.

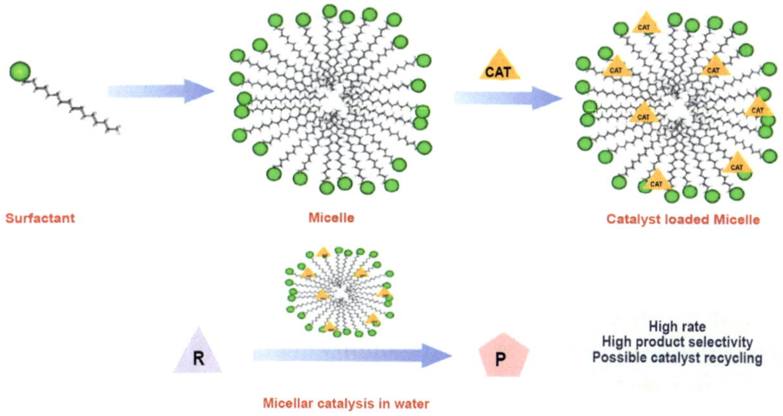

Figure 5.1. Catalysis plays an indispensable role in advancing green chemistry by facilitating chemical reactions efficiently and sustainably.

Source: https://pub.mdpi-res.com/molecules/molecules-28-04809/article_deploy/html/images/molecules-28-04809-g001.png?1686911182.

Among the most significant advantages of biocatalysis is its ability to operate under milder reaction conditions that further reduce energy usage and environment damage. Second, biocatalytic processes can be designed to follow specific sustainability guidelines, in harmony with the principles of green chemistry. For instance, biocatalysis finds successful applications in industrial production of active pharmaceutical ingredients, attesting to its suitability in the production of required compounds while being environmentally friendly. Thus, the implementation of biocatalysis is changing industries' approach to designing the synthesis of chemicals, testifying to the importance of catalysis in the provision of chemical reaction sustainability (Roger et al., 2017). Moreover, research of photoelectrochemical and photocatalytic phenomena has been of significant interest, in particular for their application in renewable energy devices. The phenomena harness sunlight to drive chemical reactions, such as water splitting for hydrogen production, which are key to the integration of transparent energy sources. However, there are constraints in maximizing the use of these devices in terms of efficiency, namely, the capacity to utilize longer-wavelength photons as well as the stability of semiconductor materials used in the reactions. Technological advancements in the properties of photocatalytic materials, surface engineering, and modification can play a vital role in significantly improving the energy conversion efficiencies, hence opening avenues for their massive applications. As these technologies are advancing and becoming more efficient, they will come a long way towards fulfilling the requirement of fossil fuel reduction and ushering in a sustainable energy era as per green chemistry principles (Hisatomi et al., 2014). Consequently, the contribution of catalysis to chemical reaction far surpasses mere efficiency; it forms a fundamental component of the transition towards a green chemical industry (Figure 5.1).

Overview of the Role of Catalysis in Green Chemistry

Catalysis is central to the advancement of green chemistry principles, characterized by an emphasis on the design of processes minimizing waste, utilizing renewable materials, and being highly energy efficient. The integration of catalysis into green chemistry not only facilitates the development of sustainable synthetic pathways but also enhances the efficiency of existing chemical processes. By lowering activation energy barriers, catalysts accelerate chemical reactions, allowing for milder conditions of reaction and thus less energy usage. This turn towards more sustainable operation is crucial in meeting international challenges such as resource consumption and environmental pollution. Moreover, the application of catalysts during organic conversions leads to higher selectivity, which indirectly explains lower byproduct formation, further continuing the green chemistry principle of waste reduction. Metal-catalyzed reactions in the synthesis of complex organic compounds are a classic example of the new methodology adopted by green chemistry. Current studies name the strength of employing metal complexes such as Rh, Ru, and Pd to exhibit how the catalysts can be employed to induce regio- and stereoselective processes while also being compatible with substrates of a wide variety. Important breakthroughs in the area of polysubstituted quaternary ammonium salts (PQAS) note their cost-effectiveness and versatility in drug design and material science (Parthasarathy et al., 2025). These catalytic systems not only streamline the synthesis but also meet the sustainable demands through their foundation on green solvents and the reduction of hazardous waste. Such advancements are a significant step towards a greener chemical universe, endorsing the call for further research and development into catalytic methods.

In addition to traditional homogeneous catalysts, heterogeneous catalysis has introduced advanced materials such as magnetically recoverable nanoparticles to green chemistry. Modification and functionalization of Fe_3O_4 nanoparticles is a highly developed interface between heterogeneous and homogeneous catalysis that ensures effective separation and recycling of catalytic systems (Mondal et al., 2025). Such advances establish the potential of such nanomaterials to catalyze reactions rapidly with low environmental impact, particularly in alcohol oxidation reactions in which they can efficiently catalyze reactions and be easily recovered from the reaction mixture. The ongoing efforts in refining these catalyst systems identify selectivity and yield as key aspects of green chemistry goals. As the demand for green processes in chemical manufacture continues to increase, these novel catalytic technologies are a hopeful course of action for achieving environmentally friendly conversions and process efficiencies.

5.1. TYPES OF CATALYSTS

Green chemistry's catalyst landscape is wide with heterogeneous catalysts with different mechanisms. Homogeneous catalysts, which are soluble in the same phase as reactants, have high selectivity and catalytic activity. Homogeneous catalysts support reactions by bringing about a uniform medium in which reactants can efficiently communicate. They have found application in organic synthesis, particularly carbon-carbon bond formation, where selectivity control is highly critical. For instance, the application of homogeneous catalysts in reactions demonstrates fewer wastes generated and can support green chemistry principles through minimizing dangerous by-products since sustainable chemical processes aim for a goal (Sakhahari, 2025). Thus, the application of homogeneous catalysts is an important achievement towards efficient and environmentally friendly synthetic processes. In contrast, heterogeneous catalysts, which are in a phase other than that of the reagents, have certain benefits in the ease of recovery and recycling. These are usually employed in gas-phase reactions or where solid catalysts are suitable, i.e., those that are used in petrochemical processes. The surface properties of heterogeneous catalysts dominate their performance; e.g., variation in particle size and surface area can control catalytic activity. As such catalysts are not consumed in the reaction, they can be used several times, facilitating sustainability in the case of green chemistry (Sakhahari, 2025). Their efficiency in a broad range of industrial processes serves to highlight their importance in catalyzing the development of environmentally friendly catalysis, further proving their usefulness in creating a greener chemical industry. Another important category under the term catalysts is biocatalysts, which are derived from natural sources of biological origin, such as enzymes and whole cells. Biocatalysts facilitate chemical conversion under mild conditions, often with a high level of substrate specificity. This specificity constitutes an avenue to efficient and sustainable synthesis, with fewer by-products than is the case with traditional chemical catalysts. Moreover, biocatalysts are also green chemistry-compatible in the sense that they prefer to operate at ambient pressure and temperature, thus lowering the energy consumption. Their use in processes ranging from pharmaceutical production to food production helps to underscore their utility and sustainability. With increased demand for sustainable operations, biocatalysis is front and center with the potential for a revolutionary product that addresses the need for a greener, more sustainable future (Figure 5.2).

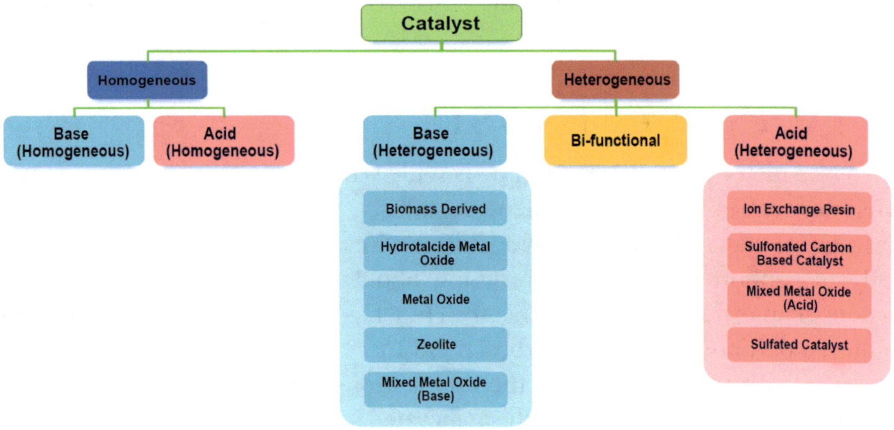

Figure 5.2. Types of catalysts.

Source: https://www.mdpi.com/catalysts/catalysts-15-00239/article_deploy/html/images/catalysts-15-00239-g001.png.

5.1.1. Homogeneous Catalysts

Homogeneous catalysts play a crucial role in the search for green chemistry through maximized reaction efficiency and selective conversion. These catalysts, usually dissolved in the same phase as reactants, facilitate reactions through their capability to reduce activation energy as well as control molecular pathways. In recent years, the synthesis of homogeneous catalysis with electrochemistry has been of interest, and thus new catalytic processes were formed that are more sustainable and more selective. By leveraging the unique features of transition metals in homogeneous systems, researchers have made significant progress in a range of organic synthesis reactions, thus contributing to greener approaches.

This is particularly important in regard to reduced waste generation and improved chemical process efficiency that align with green chemistry protocols to reduce harm to the environment (Cheng et al., 2021). Homogeneous catalyst performance is enhanced in conjunction with electrochemical processes, wherein oxidative and reducing conversions are enabled that would otherwise require harsh conditions. Notably, the incorporation of renewable energy resources into catalytic cycles has evolved into a revolutionary green chemistry approach. For instance, electrochemical homogeneous catalysis allows reactions based on controlled electron transfer, which can significantly minimize the use of traditional and sometimes harmful reagents. As demonstrated by recent studies, the ability to regulate catalyst reactivity through electrochemical means allows for a wider set of chemical conversions while mitigating environmental risks.

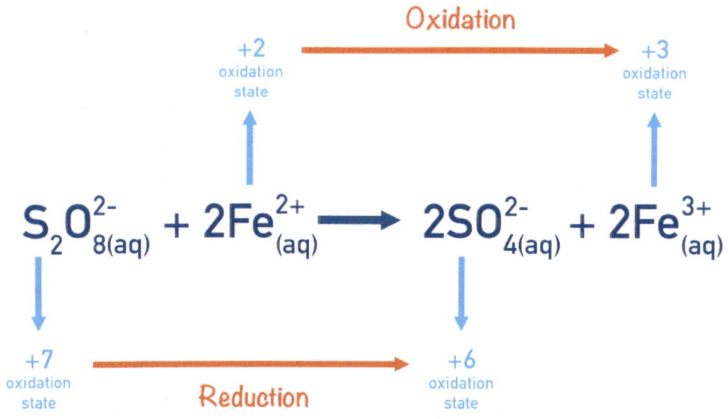

Figure 5.3. Homogeneous catalysts play a crucial role in advancing green chemistry by improving reaction efficiencies and enabling selective transformations.

Source: https://www.chemistrystudent.com/images/A2Inorganic/transitionelements/transitionelementshomogeneouscatalysis3.png.

The reality that these catalysts can catalyze multiple electron-transfer processes in the absence of excess reductants and oxidants indicates their potential to facilitate a more sustainable practice in molecular synthesis (Zhu et al., 2019). Moreover, research on new homogeneous catalysts of earth-abundant metals has drawn considerable attention to their potential for use in actual processes in green chemistry strategies. Aside from making equivalent or even better efficiency than their precious metal counterparts, the catalysts have no socioeconomic issues related to rare resource extraction and use. The key to their successful implementation is developing systems that optimize catalyst activity and durability without reducing energy input and producing waste. As research continues, it will be indispensable to their future use in sustainable catalysis to know the intrinsic character of these metals and how they behave when they bond with ligands. Through less reliance on expensive noble metals and better catalytic activity, homogeneous catalysts are ready to lead the way in environmentally friendly chemical processes (Cheng et al., 2021) (Figure 5.3).

5.1.2. Heterogeneous Catalysts

Heterogeneous catalysts are very crucial in the area of green chemistry, as they facilitate reactions with minimal impact on the environment. Catalysts, which exist in a phase different from that of the reactants, typically accelerate the reaction rate without being permanently altered. Their contribution to reactions like the xanthone synthesis underscores their key role in green chemistry. By enhancing the selectivity of the reaction and reducing byproduct generation, heterogeneous catalysts are harmonious with the principles of green chemistry and are therefore an essential field of research. Their elevated reusability is not just economically beneficial but also leads to less waste generation. This efficiency is reflected in several indices such as Reaction Mass Efficiency (RME) and Process Mass Intensity (PMI), which illustrate their superiority to traditional stoichiometric methods, therefore encouraging greener methods in industrial synthesis (Mendioroz et al., 2024). One of the best aspects of heterogeneous catalysis is its versatility and material utilized as catalyst supports. Silica supports doped with metals or their salts have gained popularity due to their advantageous characteristics such as low cost, recyclability, and environmental friendliness. The catalysts are not only high-yielding in organic synthesis but are also easy to process and apply in diverse uses ranging from pharmaceuticals to industry. They are good platforms for the building of biological scaffolds and natural products and have been of great value in making progress in medicinal chemistry. This flexibility emphasizes the potential for widely acceptable catalyst systems in addressing growing demands for sustainable chemical processes, highlighting the fact that green practices can exist synergistically with industrial efficiency (Mukherjee, 2024). The uses of heterogeneous catalysts are more than mere efficiency; they are amenable to the wider goal of sustainability as part of the practice of green chemistry. The development of such catalysts suggests increased effort towards reducing the environmental footprint of chemical reactions. As industries strive to meet strict environment regulations, the use of heterogeneous catalysts can lead to substantial reductions in energy usage and waste production. Moreover, the constant optimization of catalyst design and use places such systems in the forefront of catalytic science, underscoring their importance in the quest for sustainable technology. Their ability to propel high-volume production without being at odds with the tenets of green chemistry puts heterogeneous catalysts at a crossroads in the evolution of environmentally responsible manufacturing practices ultimately setting the course for more sustainable futures (Mendioroz et al., 2024; Mukherjee, 2024) (Figure 5.4).

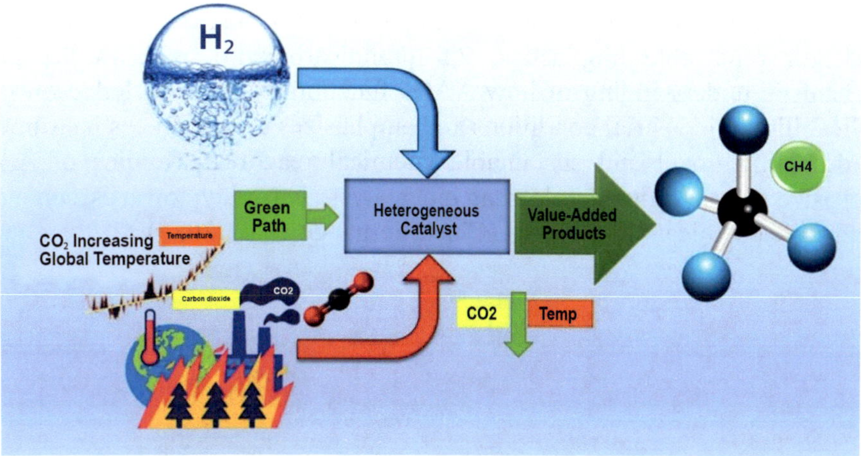

Figure 5.4. Heterogeneous catalysts are critical components in the framework of green chemistry, facilitating reactions while minimizing environmental impact.

Source: https://pub.mdpi-res.com/catalysts/catalysts-13-01514/article_deploy/html/images/catalysts-13-01514-ag.png?1702972830.

5.1.3. Biocatalysts

Biocatalysts are discovered in green chemistry to be principal characters in recasting chemical reactions with greener means. By tapping the natural catalytic ability of enzymes, biocatalysts accelerate reactions under gentle conditions, resulting in enhanced selectivity and less by-product creation. Not just does such efficiency enhance reaction yields but also severely mitigates the environmental burden otherwise inherent to traditional catalytic methodologies. With sectors turning increasingly to sustainable alternatives for substituting chemical procedures, biocatalysts are a promising choice, in accordance with the green chemistry guidelines of reducing hazardous materials and energy consumption in manufacturing processes. Furthermore, with the improved enzyme engineering and immobilization techniques, they have become increasingly versatile, with proofs of their capacity to function in an extremely broad range of applications and on complex substrates (Fatemah, 2025). One of the most fascinating biocatalyst families is the aryl-alcohol oxidases (AAOs), with the main alcohol oxidations to aldehydes. These enzymes are characterized by the excellent substrate promiscuity and have been identified in a wide range of organisms, from fungi, bacteria, and arthropods. Structural promiscuity of the AAOs opens the door to an unusually wide range of applications in biocatalytic processes that are pivotal to green chemistry. Notably, they can also be utilized to catalyze the formation of valuable bio-based chemicals, such as the building block for renewable bioplastics, 2,5-furandicarboxylic acid. Furthermore, mechanistic understanding of how AAOs function offers knowledge on their applicability to industrial conditions and emphasizes their value as instruments for designing novel and sustainable chemical reactions. Studies of AAOs emphasize the significant role that biocatalysts can play towards achieving sustainable chemical conversions (Cinca-Fernando et al., 2025).

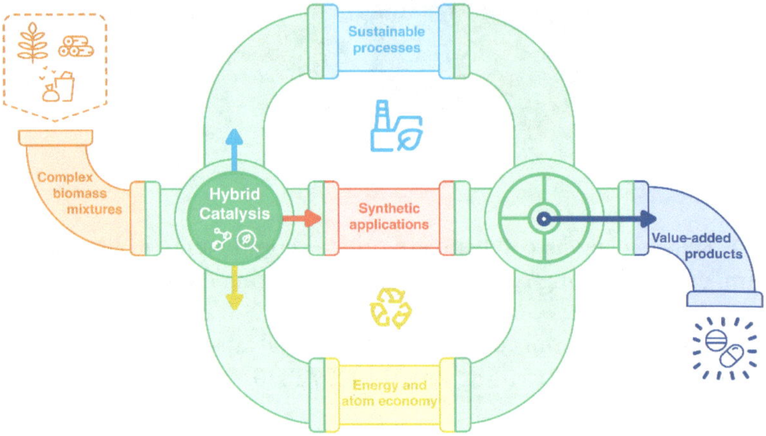

Figure 5.5. As industries increasingly seek eco-friendly alternatives to chemical processes, biocatalysts represent a promising avenue, aligning with the principles of green chemistry, which emphasize reducing hazardous substances and energy consumption in manufacturing processes.

Source: https://comptes-rendus.academie-sciences.fr/chimie/media/images/10_5802_crchim_398/graphical_abstract/EchevarriaGraphicalAbstract.png.

The utilization of biocatalysts, in this case enzymes like AAOs, in chemical manufacturing processes is a paradigm shift towards more eco-friendly processes. That they can operate under mild conditions and be specific to substrates improves process efficiencies as well as aligns with the overall goal of minimizing ecological footprints. While more and more studies continue to uncover the vast potential of biocatalysts, there is little doubt that their application can contribute significantly to waste minimization and energy use, meeting regulatory demands for sustainable actions. Future directions for biocatalysis are most probably to depend upon making such reactions scalable and resolving challenges regarding enzyme stability and cost. Thus, long-term research and application of biocatalysts are an important path towards expanding the goals of green chemistry in many industries (Fatemah, 2025) (Figure 5.5).

5.2. CATALYSIS AND SUSTAINABILITY

The intersection of catalysis and sustainability is increasingly at the forefront in pushing green chemistry forward. With the world grappling with environmental concerns, catalysts are indispensable tools that have the ability to minimize waste and maximize the efficiency of chemical processes. By facilitating reactions to happen under more subdued conditions and lowering the reliance on poisonous reagents, catalytic methods contribute to making sustainable techniques. For instance, current themed issues highlight catalysis innovation targeted at reducing waste and resources conservation in facilitating development towards a circular economy where resources can be reused and recycled efficiently (Wattanakit et al., 2024). Not only does this stress the capability of catalysis for promoting sustainability, but it also reflects the general commitment of the chemical industry toward environmental stewardship. Recent developments in sustainable catalysts demonstrate the possibility to transition away from conventional methods to greener alternatives. Synthesis of critical organic materials, e.g., hole transport layers for solar cells, demonstrates this shift. Applying alternative techniques, such as solventless processes and micellar catalysis, makes the synthesis of essential materials like Spiro-OMeTAD possible with dramatically lower environmental footprint. Interestingly, these processes have exhibited a remarkable reduction in the overall E-factor—a prime green chemistry indicator—exhibiting a significant diminishment in waste generated during synthesis. The efficiency of these environmentally friendly catalysts not only matches that of their conventional analogs but also acts as a precedent for reducing the environmental impact of chemical production (Mattiello et al., 2022). Lastly, this breakthrough is an indication of moving towards integrating sustainability more deeply into industrial processes. The catalytic role in promoting sustainability in green chemistry goes beyond efficiency; it also includes broader application of sustainable processes in chemistry. Waste valorization projects highlight the importance of converting waste and biomass into a useful product, thereby developing a circular economy based on minimizing resource depletion (Wattanakit et al., 2024). Catalysts are employed in the realization of successful conversions so industries can recycle waste materials into commercial chemicals. Through the promotion of sustainable practices and improving catalyst design, the aim is not only to minimize the environmental impact of chemical production but also to be economically advantageous through the utilization of once-discarded resources. With the expansion of the field of catalysis, demands for generating synergies among innovation and sustainability still remain extremely crucial in highlighting the revolutionizing potential of green chemistry in addressing global challenges.

5.2.1. Reducing Waste in Chemical Processes

The issue of waste generation in chemical processes is a prime issue in the field of green chemistry because it comes under direct opposition in the name of sustainability and environmental safeguarding. Moreover, traditional production of chemicals provides significant amounts of noxious byproducts, which can create detrimental effects on environments and human health. Green Chemistry principles emphasize eliminating waste with innovative strategies such as atom economy and reaction pathway minimization. By embracing catalytic reactions that enable higher yields and fewer byproducts, the chemical industry can not only decrease its environmental footprint but also boost economic efficiency. Applying these principles is able to transform conventional practices to more sustainable forms, thus incorporating broader ecological goals, to where waste minimization is the core in the modern chemical production activities (Achieng, 2025). Catalysis is then a primary function in the direction of reducing waste in chemical processes by optimizing reaction conditions as well as enhancing selectivity. Catalysts can have the ability to accelerate reaction rates by orders of magnitude and render chemical transformation more efficient, thereby reducing the quantity of raw material required and waste production. Furthermore, new catalytic systems rooted in green resources or nanotechnology could have the ability to reduce environmental impact even further. These advancements provide the means of manufacturing compounds with fewer operations and less energy input, which is vital in promoting sustainable industrial practice. As the industry continues to transition more and more towards greener alternatives, a focus on catalytic technologies will be instrumental in waste minimization while ensuring production of high-quality chemical products (Chen, 2025). In addition to the catalysis chemistry, part of an overall strategy for reducing waste in chemical processes is reconsidering the entire chemical life cycle, from feedstock acquisition through consumption. Encouraging the use of renewable resources as feedstocks not only aligns with the spirit of green chemistry but also lowers waste from non-renewable resources. The use of life cycle analyses (LCAs) would also make the environmental impacts of chemical processes clearer to companies so that they could identify most crucial areas of improvement. Through facilitation of industry-academic collaborations, the transition towards more sustainable approaches can be accelerated. By integrating these strategies, the chemical industry can minimize waste production greatly without contravening environmental regulations and meeting the demands of consumers for eco-friendly products, thereby paving the way to a more sustainable future of responsible production of chemicals (Achieng, 2025) (Figure 5.6).

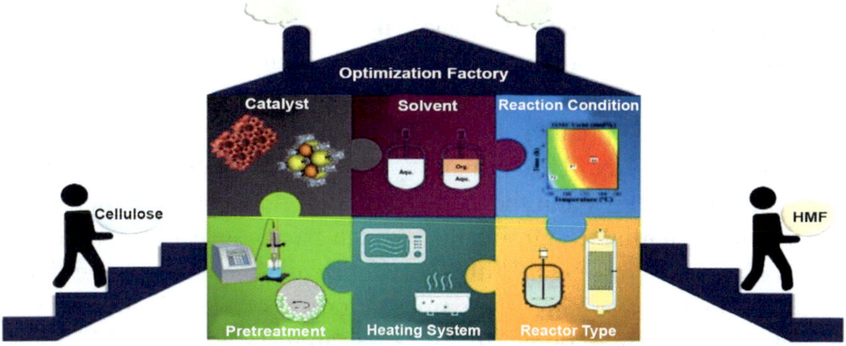

Figure 5.6. Catalysts can significantly accelerate reaction rates and increase the efficiency of chemical transformations, thereby reducing the quantity of raw materials required and limiting waste generation.

Source: https://chemistry-europe.onlinelibrary.wiley.com/cms/asset/5c0eeb3e-58e3-4832-a693-22aae273a2a1/cctc202300973-toc-0001-m.jpg.

5.2.2. Energy Efficiency in Catalytic Reactions

Catalytic reactions offer an important route to enhancing the energy efficiency of a vast array of chemical processes and are compatible with the philosophy of green chemistry. Catalysts facilitate reactions to be performed under less severe conditions by lowering the energy barriers for chemical transformations, and as a result, total energy expenditure is reduced. Such advancement is represented by optimizing catalytic systems, such as the integration of electrochemical methods. Surprisingly, recent research shows that the application of paired electrolysis in combination with nickel catalysis can be applied to achieve significant energy savings. The new approach involves simultaneous cathodic and anodic reactions, both of which maximize atom economy, and minimize wastage, serving to convey twin benefit in terms of enhanced efficiency and sustainability of catalytic reactions (Zhu et al., 2020). Moreover, the design of efficient energy catalysts is crucial to develop the hydrogen evolution reaction (HER) because it is a key path to sustainable hydrogen production. An understanding of HER catalysis principles has led to the development of novel noble and non-noble metal, as well as metal-free, catalyst designs. These designs focus on optimizing the correlation between catalytic activity and structure, morphology, and composition of the catalysts. Techniques for enhancing intrinsic activity and increasing active site density are now a major research priority over recent years, a reflection on the major trend towards the utilization of less energy-consuming catalysts for optimum performance (Zhu et al., 2019). This pioneering work not only facilitates new function in green chemistry but also holds out wider application across various industrial processes. The collaborative efforts to increase energy efficiency in catalytic reactions represent a broader commitment on the part of the chemical industry to sustainability. Through its application of such breakthroughs as electrochemical processes and the development of new catalytic materials, the industry has the potential to significantly reduce greenhouse gas emissions from traditional chemical processes. With evolving regulations and public demands for environmental stewardship, the emphasis on energy efficiency in catalysis will inevitably affect the course of future research. Long term, this emphasis is not only good for the environment but is economically advantageous as well, making energy-efficient catalytic processes more and more attractive to industries seeking to align with green chemistry principles. Development of these technologies will be paramount to their progress toward a sustainable chemical economy (Figure 5.7).

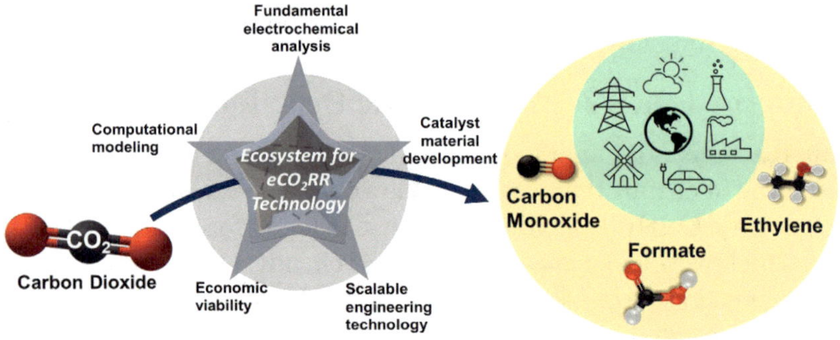

Figure 5.7. By leveraging innovations such as electrochemical methods and the development of advanced catalytic materials, the sector can significantly reduce greenhouse gas emissions associated with traditional chemical processes.

Source: https://chemistry-europe.onlinelibrary.wiley.com/cms/asset/ab25d657-14f2-409d-8336-c54054b3684f/celc202400512-toc-0001-m.jpg.

5.2.3. Renewable Resources in Catalysis

The application of renewable resources in catalytic processes is a milestone in green chemistry. As the world grapples with the implications of climate change and the depletion of resources, the use of sustainable approaches has become an urgent need than ever before. Catalysis, defined as the acceleration of chemical reactions with catalysts, has a great potential for optimization through the application of renewable feedstocks. Employing bio-based substances, such as biomass of plants or agricultural waste, not only decreases the dependency on fossil fuels but also minimizes the production of harmful waste. Through the incorporation of renewable resources in catalytic applications, industries are able to put more emphasis on achieving greater environmental sustainability in addition to alleviating economic issues related to raw material sourcing. It is an integrated approach that puts renewable resources at the center of modern catalysis, opening the doors to cleaner and more sustainable production processes (Bhaskar et al., 2025). Besides employing renewable feedstocks, the use of innovative catalytic system design and operation also enhances the sustainability of chemical processes. Green chemistry places special emphasis on catalysis as a way of minimizing harmful chemicals and optimizing reaction efficiency.

By adopting methods using biocatalysts—enzymes originating from living organisms—biologists can make reactions efficient and environmentally friendly. Biocatalysts have some advantages, including high specificity and lower energy requirements compared to traditional chemical catalysts. Furthermore, the potential for performing reactions at more moderate conditions provides greater possibilities for using renewable raw materials, which in many cases needs less stringent handling and storage. Such a shift not only addresses the environmental concerns most crucial in traditional chemistry but also answers the green chemistry principles that demand the development and employment of cleaner and better chemical processes (Babbar, 2024). The impact of employing renewable materials in catalysis is not only restricted to environmental effects but also to economic sustainability in chemistry. As business houses more and more adopt the principle of green chemistry, use of renewable materials and creative catalysis can lead to a drastic reduction in cost of production and liability in the future due to toxic waste. Companies that are sustainable in nature are also gaining immense popularity with consumers, thus demand for cleaner products is increasing. This union of market forces and regulatory attention reinforces the imperative for chemical firms to work with renewable sources in catalysis. By the redesign of traditional processes of chemical production, companies can stand out in an ever-growing global marketplace while facilitating a greener future. Therefore, the use of renewable feedstocks in catalysis is not only an environmental necessity but also an active economic advantage for the new chemistry sector (Bhaskar et al., 2025) (Figure 5.8).

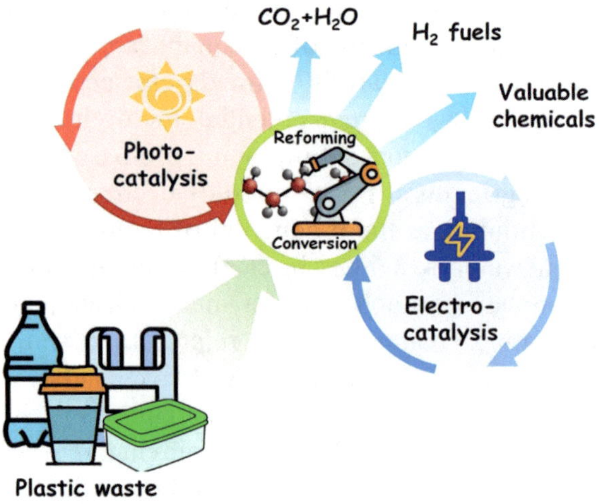

Figure 5.8. As businesses increasingly adopt green chemistry practices, the integration of renewable materials and innovative catalysis can lead to a significant reduction in production costs and potential liability associated with hazardous waste.

Source: https://chemistry-europe.onlinelibrary.wiley.com/cms/asset/7cf748f3-87f6-4d26-b193-4805e512200b/cssc202301352-toc-0001-m.jpg.

5.3. CATALYST DESIGN AND DEVELOPMENT

In the area of green chemistry, catalyst discovery and design have become crucial elements in making chemical processes greener as well as more efficient. Of key focus is the use of carbon dioxide (CO_2) as a renewable carbon source. The potential for employing CO_2 as a reagent in chemical synthesis, in addition to offering an eco-friendly alternative to traditional toxic reagents, also aligns with the principles of green chemistry in terms of waste minimization and maximizing resource utilization. However, the low reactivity of CO_2 presents daunting challenges which necessitate the development of new catalytic systems. Supercritical CO_2 as a reaction medium is a promising method, with favorable conditions for catalysis and potential for mitigation of VOCs emissions. In the investigation of combining catalytic processes with supercritical CO_2, catalyst engineering advances are pivotal for effective conversion of CO_2 to valuable products, such as dialkyl carbonates, making an imperative case for more research (Ballivet-Tkatchenko et al., 2005). The catalysis synthesis in green chemistry has progressed to encompass the ideas of modular synthesis and advanced material science. Specifically, layered double hydroxides (LDHs) have come into focus because of their unique structural and chemical properties, which make them good candidates for many catalytic processes. The potential to intercalate polyoxometalates (POMs) in LDHs allows for the construction of a new class of materials with functionalities of choice. Through precise control of the composition and shape of these catalysts, scientists can achieve desired catalytic behavior for specific reactions. This approach not only enhances catalyst efficiency but also enables closer control over reaction selectivity and kinetics. Hence, the application of modular design in catalyst construction is critical to the design of catalysts that operate effectively under environmentally benign conditions, thereby contributing to the objectives of green chemistry (Li et al., 2017). Furthermore, the link among catalyst design and environmental sustainability becomes increasingly evident as the science evolves. New catalyst synthesis technologies, where focus is placed on reducing toxic byproducts and increasing recyclability, are hot topics in this conversation. Employing catalysts that promote resource-efficient use, the need for harmful solvents and reagents is minimized, as the target is to minimize environmental impacts. By bringing together computer tools and experimental techniques to predict and simulate catalyst behavior, more targeted efforts in catalyst formulation can be made, enabling new catalysts to be discovered. Integration of design, efficiency, and environmental sensitivity is at the heart of green catalytic process development. With high-performance catalyst engineering combining with the green chemistry principles, the potential for making major contributions to industrial processes and environmental remediation is becoming increasingly possible, marking a revolution in the discipline (Ballivet-Tkatchenko et al., 2005; Li et al., 2017).

5.3.1. Strategies for Designing Efficient Catalysts

The structure of efficient catalysts is the foundation for building green chemistry, an area dedicated to chemical reaction sustainability. Among the strategies is the development of heterogeneous catalysts for increased reaction selectivity with minimized environmental impact. For instance, novel photocatalysts such as immobilizing 1,5-diaminoanthraquinone on silica nanoparticles prove the effectiveness of visible light use in organic synthesis. It not only sustains green reactions but also offers high recyclability, which allows for up to ten cycles of use without substantial degradation of performance. The development of such catalysts also illustrates a major intersection between green chemistry and innovative materials science to facilitate reactions that conserve energy and minimize waste, thereby aligning with the general objectives of green chemistry and improving overall efficiency in synthetic strategies (Martin et al., 2025). Furthermore, the use of ultrathin nanomaterials is also extremely significant in the operation of carbon-based catalysts, particularly in C1 catalysis, during which single-carbon compounds are transformed to chemicals and energy sources. The activation of inert C-H or C-O bonds turns out to be a problem needing rational design of the catalytic material with enhanced catalytic activity. Tandem catalytic strategies have the capacity to synergistically combine different reactions via multifunctional catalysts in order to maximize the conversion process. Additionally, taking advantage of confinement effects can regulate the distribution of product and improve activation efficiency. The new approach not only promotes better catalyst design but also enables significant progress towards carbon neutrality by enabling effective utilization of greenhouse gases (Wang et al., 2023). These strategies emphasize the pivotal role of strategic design in achieving carbon feedstock sustainable conversions. Lastly, catalytic system engineering plays a crucial role in achieving long-term stability and performance in catalytic operations. Optimizing metal-support interactions and modulating electronic character of active sites can lead to increased reactivity and selectivity in complex catalytic reactions. New synthesis technologies, like 3D printing and traditional wet-chemistry syntheses, provide opportunities for producing tailor-made catalytic materials for specific reactions. For instance, the formation of ammonia pools in zeolites and specifically designed capsule catalysts can effectively improve the efficiency of CO, CO_2, and CH_4 conversions and hence solve major issues in C1 catalysis. While investigations continue to explore such new materials and approaches, the future possibility of catalytic design has been observed to make the achievement of greener, more efficient chemical processes possible, reaffirming the significance of such advances in making the chemical industry sustainable (Wang et al., 2023) (Figure 5.9).

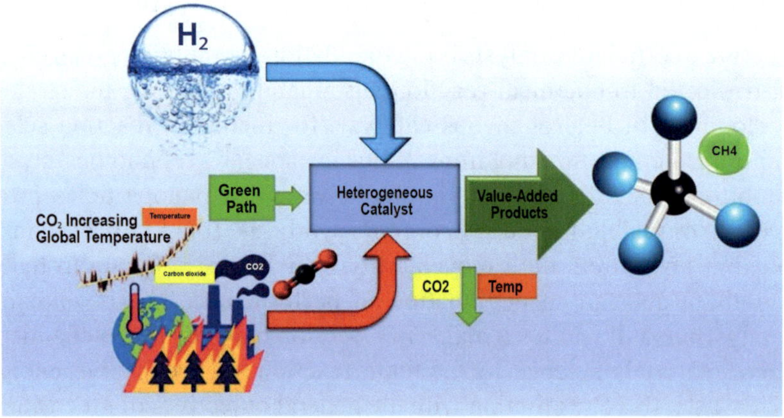

Figure 5.9. A pivotal strategy is the development of heterogeneous catalysts that enhance reaction selectivity while minimizing environmental impact.

Source: https://pub.mdpi-res.com/catalysts/catalysts-13-01514/article_deploy/html/images/catalysts-13-01514-ag.png?1702972830.

5.3.2. Role of Nanotechnology in Catalysis

Nanotechnology has initiated a revolutionary process in numerous disciplines, particularly catalysis in green chemistry. Nanomaterials have been introduced into research to dramatically enhance the efficiency, selectivity, and reusability of reactions. Nanoparticles, typically characterized by their high surface area-to-volume ratio, offer unique chemical properties that drastically improve catalytic activity. The development of the synthesis of carbon nanomaterials such as carbon quantum dots (CQDs) and graphene quantum dots (GQDs) points to their potential in green catalysis. They are not just catalysts but also exhibit favorable interaction with the reactants, providing higher yields and higher rates of reaction. According to the literature, green synthesis of CQDs and GQDs via laser-based methods highlights the combination between nanotechnology and green principles in catalysis, demonstrating the broader implications for forthcoming sustainable energy-efficient processes (Rey et al., 2025). Moreover, green synthesis routes of nanoparticles are key to developing green catalytic processes.

Traditional nanoparticle synthesis contains hazardous chemicals that are environmentally and health-related issues, whereas green synthesis uses natural extracts, rendering the process non-toxic and cost-effective. Using plant extracts to synthesize copper nanoparticles demonstrates how nanotechnology can capitalize on biological resources to develop efficient and safe manufacturing methods. The produced nanoparticles possess impressive catalytic properties, being able to catalytically facilitate degradation of organic pollutants under various environmental conditions efficiently. This demonstrates the double benefit of optimizing catalytic performance while maintaining adherence to green chemistry tenets—minimizing waste and reducing toxic byproducts. All these breakthroughs demonstrate the potential of nanotechnology in addressing critical environmental problems. The versatility of nanotechnology is also reflected in the fact that it can catalyze reactions for a broad range of chemical transformations. Nanoscale catalysts can be tailored to enhance some functionalities and imbue the material with desirable properties like photocatalytic activity to degrade pollutants or to catalyze reactions in mild conditions. For instance, metal nanoparticles and CNTs have been included in composite materials to create synergies capable of maximizing catalytic processes in as broad a range of applications as renewable energy and drug production. By the magnitude of the nanoscale, researchers can create more sophisticated catalysts that are environmentally sustainable and cost-effective. This plan not only foresees improved effectiveness in catalytic application but also represents a paradigm shift to sustainable practice at the heart of green chemistry. As such, the application of catalysis through nanotechnology represents a vital research axis, which stimulates innovation and spurs sustainable development (Rey et al., 2025).

5.3.3. Computational Methods in Catalyst Development

The development of computational methods in the design of catalysts is vital in propelling the concepts of green chemistry, and this is true with the optimization of catalytic processes. These methods apply computational software to model the efficacy and performance of catalysts under varying conditions of reaction, hence making it possible to discover new catalysts that are efficient and sustainable. Through the application of computational approaches, researchers can design reaction pathways, examine energy barriers, and make predictions about active site structures, ultimately resulting in the rational design of catalysts that minimize waste and energy consumption. This targeted strategy not only enhances the sustainability of chemical reactions but also ensures effective use of resources by eliminating the need for empirical trial-and-error approaches that are time- and resource-consuming. Of the most significant areas where computational methods have been instrumental is in the modeling of enzyme-like catalysts, which mimic biological reactions to facilitate essential chemical transformations. The power of molecular dynamics simulations and density functional theory has enabled scientists to study enzyme mechanisms at an atomic level, and discover structure-function relationships. This is especially crucial in order to design catalysts for renewable energy purposes, such as splitting water and reducing carbon dioxide, which require catalysts capable of lower activation energy reductions and improved reaction rates. By employing data from the Protein Data Bank, researchers can direct their designs by adjusting existing protein structures, as laid out in Pan et al. (2021). Such computer enhancement not only leads to the synthesis of more selective and efficient catalysts but also demonstrates how computational tools have become ever more indispensable to green chemistry. Moreover, the combination of machine learning techniques with computational ones has revolutionized catalyst design over the past few years. By examining vast quantities of data extracted through experiment and the literature, machine learning algorithms are capable of identifying patterns and predicting suitable catalyst candidates with extremely high accuracy. This predictive capability significantly accelerates new material screening, thereby conserving time and cost associated with catalyst discovery. Machine learning can also tune reaction conditions and performance parameters, further increasing the overall effectiveness of the catalytic systems under development for green chemistry uses. With growing demand for sustainable processes, demand for the application of computational approach to catalyst development continues to be vital since it captures the possibility towards innovative solutions that reflect paradigms towards reducing carbon footprints and increasing environmental sustainability in chemical manufacturing, as outlined in the discussion on (Song et al., 2020) (Figure 5.10).

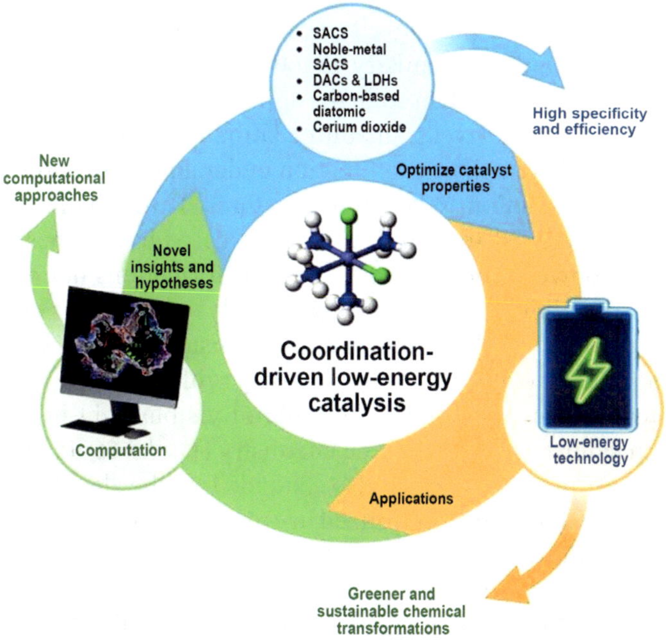

Figure 5.10. By utilizing computational tools, researchers can simulate reaction pathways, evaluate energy barriers, and assess active site characteristics, ultimately guiding the rational design of catalysts that can minimize waste and energy consumption.

Source: https://ars.els-cdn.com/content/image/1-s2.0-S0010854524002467-gr1.jpg.

5.4. APPLICATIONS OF CATALYSIS IN GREEN CHEMISTRY

Applications of catalysis in green chemistry are increasingly critical to the pursuit of sustainable chemical processes. One of the most significant innovations in recent years has been the marriage of electrochemical and catalytic methods, allowing for more effective and more green transformations. Electrochemical homogeneous catalysis, for instance, allows reactions to be conducted under less severe conditions, fewer toxic reagents and solvents being needed. This approach leverages the unique properties of transition-metal catalysts to facilitate a variety of organic transformations, such as C–H functionalization and cross-coupling reactions, in fewer steps with reduced waste. Furthermore, the union of electrochemistry with established catalytic approaches has opened doors for novel synthetic strategies, putting electrochemical processes at the forefront of the innovation of green chemistry practice (Cheng et al., 2021). Another important advance is achieved through the application of photocatalysis, further advancing chemical synthesis sustainability. Utilizing visible light as a source of energy, photocatalytic reactions can function under ambient conditions, thereby limiting energy consumption and carbon signatures. Through this process, diazo compounds are converted to useful intermediates with carbenes and radicals formed via solar-powered processes. These transformations not only provide a greener option to traditional transition metal-catalyzed reactions but also extend the range of reactions available to organic chemists. Photochemical reactions are able to increase reaction rates and selectivity without byproducts associated with traditional reactions, reiterating the role of visible light in redefining the future of green synthesis in sustainable chemistry (Durka et al., 2021). Further, focus on catalytic methodology is also extended to the development of new catalysts that are designed to be optimized in order to obtain maximum efficiency and selectivity at minimal environmental impact. For instance, the development of ligand-free catalysts and utilizing base metals and nontoxic metal have been prominent catalyst directions in green chemistry. Not only do these advances enhance the effectiveness of the reactions, but they also reduce resource use and waste generation. Additionally, the application of transition metals such as iron and cobalt in place of precious metals is a reflection of a divergence towards more sustainable practices. The shift towards utilizing sustainable material and optimizing catalytic systems is accompanied by the general goals of the green chemistry platform to promote green methods of synthesis of chemicals with applications in various industries and uses (Cheng et al., 2021).

5.4.1. Catalysis in Organic Synthesis

The catalytic function in organic synthesis plays a key role in promoting sustainable approaches within the scope of green chemistry. Catalysis greatly improves the efficiencies of reaction, allowing for conversion processes that are otherwise not practical or economically viable. This is especially significant as companies seek to avoid waste and maximize the use of resources. The integration of catalysis in organic synthesis accommodates the twelve green chemistry principles, which suggest waste avoidance, energy conservation, and use of renewable feedstocks. Micellar catalysis, for example, entails surfactants enabling reactions in aqueous systems such that reliance on poisonous organic solvents is circumvented. As Fabris et al. (2023) explains, this method is consistent with the tenets of green chemistry because it promotes greater reactant solubilization while minimizing negative effects on the environment and safety risks associated with the use of traditional solvents. In addition to micellar catalysis, the union of homogeneous catalysis with electrochemistry has emerged as a new strategy for organic synthesis. Electrochemical reactions enable efficient electron transfer that can trigger inert substrates under gentle conditions, thus providing a stable alternative to conventional chemical oxidants. Recent advances in this direction reveal the efficacy of noble transition metals as catalysts in catalyzing electrochemical reactions. As highlighted in Cheng et al. (2021), the integration of electrochemical approaches extends the array of organic synthesis means, with enhanced selectivity and reduced wastewater. Prominently, the strategies not only refine synthetic pathways but also pave the way towards realization of energy-efficient processes as the foundation for green chemistry objectives. The potential of catalyzing reactions at ambient conditions with less by-product is the reflection of the technological strength. The endeavors towards the assembly of catalytic systems based on renewable resources with high efficiency also underscore the importance of catalysis in green chemistry. Innovation in catalyst design aims at ensuring sustainability, particularly via the utilization of earth-abundant metals and biodegradable ligands. Changing directions from precious metals to more sustainable materials is a paradigm shift in organic synthesis protocols. Lastly, prioritizing the design of catalysts in accordance with green chemistry principles maximizes the possibility that reactions become as clean, efficient, and economically viable as possible. This integrated strategy in catalysis not only meets the present needs of organic synthesis but also paves the way to future advances that are centered on environmental and economic sustainability, complementing the greater vision of catalysis in green chemistry.

5.4.2. Catalysis in Biomass Conversion

Catalytic reactions, which are at the core of green chemistry thought, significantly aid the bioconversion of biomass to valuable chemicals and fuels. As renewable feedstocks, materials of biomass have the potential to offer biodegradable alternatives to fossil fuels, but effective conversion needs advanced catalytic processes. One of the promising thermochemical processes is hydrothermal carbonization (HTC), which can convert various forms of biomass to high-carbon-content products. This synthesis not only provides carbon precursors but also enables nitrogen-doped carbon material synthesis with satisfactory performance as support for catalytic applications. Utilization of these kinds of carbon materials is crucial since their physicochemical properties play an important role in determining the activity of supported catalysts, particularly in electrochemical reactions like the oxygen reduction reaction (ORR) to be used in fuel cells (Martin et al., 2024). The utilization of catalysts derived from biomass conversion has the potential to accelerate numerous chemical reactions and increase efficiency. A good example is the recent development of novel nitrogen-doped hydrothermal carbon materials (N-HTC) with low-cost organic precursors such as glucose, which are converted under mild conditions. The resultant materials can be thermally treated to modulate their surface properties, for example, nitrogen content and graphitization, significant parameters affecting catalytic activity.

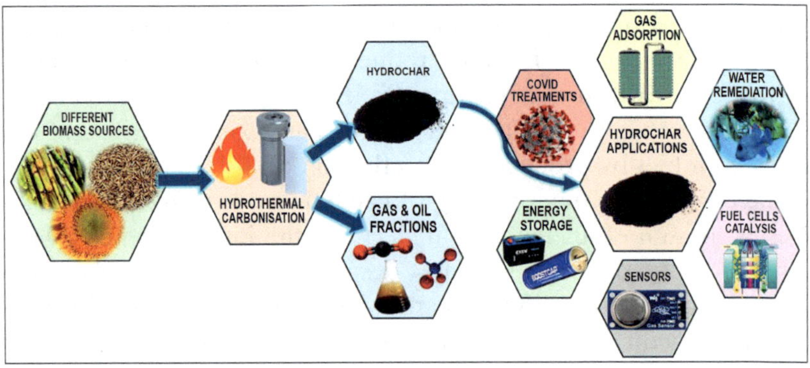

Figure 5.11. Hydrothermal carbonization (HTC), which allows for the transformation of various biomass types into carbon-rich materials.

Source: https://www.mdpi.com/polymers/polymers-16-02633/article_deploy/html/images/polymers-16-02633-g002.png.

The subsequent impregnation of these carbon supports with platinum nanoparticles presents an exciting route toward high-performance electrode construction. Apart from emphasizing the possibility of using biomass-derived materials in catalysis, this process also accentuates their possibility for use in sustainable energy, further elucidating concepts of green chemistry (Martin et al., 2024). The quest for novel catalytic approaches in the conversion of biomass has seen the application of innovative techniques such as microwave-assisted synthesis. The technique is unique in that it provides quick and even heating, which can drastically reduce reaction time and improve yields. As attention is drawn to optimizing the use of resources, microwave technology is priceless in its capacity to establish non-equilibrium conditions that induce fruitful reaction. Adopting this approach to catalytic processes enables researchers to achieve higher selectivity towards desired products and lower energy utilization. The significance of integrating new methodologies into paradigms for biomass conversion is highlighted by these advancements towards greener, more sustainable chemical processes. The synergy created by these catalytic developments puts biomass conversion squarely at the center of current green chemistry endeavors (Figure 5.11).

5.4.3. Catalysis in Waste Treatment

The intersection of waste treatment and catalysis is a milestone towards cleaner chemical processes. Catalysis enhances the efficiency of chemical reactions and, consequently, potentially reduces the amount of waste generated in the process of treating hazardous materials. Industries are able to reduce significantly the amount of energy used in waste treatment processes through the utilization of catalysts, more so nanocatalysts because of their greater surface area and distinct properties. Not only does this minimize operation costs but also through the primary principles of green chemistry in encouraging minimizing unwanted byproducts and energy consumption. The integration of catalysis into waste treatment processes is a typical example of the drive towards sustainable principles supporting environmental health and safety goals through the encouragement of a responsible industrial waste management and conversion process. The application of nanocatalysts within waste management systems has unveiled revolutionary potential for decontaminating wastewater through the elimination of pollutants. The nanocatalysts eliminate and degrade poisonous chemicals, such as heavy metals and organic pollutants, that are of utmost concern in modern environmental conservation. Research indicates that the application of green and bio-synthesized nanocatalysts can enhance the treatment of wastewater systems in order to curb environmental degradation in the most core region. As defined, the efficiency of the nanocatalysts has the potential to result in the effective degradation of complex pollutants without the highest level of secondary waste products. Through their application, thus, they not only follow the principle of green chemistry but also provide the scalable method of meeting wastewater treatment requirements, especially in countries with limited resources and infrastructure to conduct traditional waste disposal methods, towards providing cleaner water supplies. Furthermore, the advancement of catalysis for waste treatment is particularly critical in developing countries, where industrial pollution turns into a large-scale environmental and health issue. Emphasized in the context of global outreach for green chemistry, it is essential that these catalytic methods be incorporated into school curricula to stimulate innovation and awareness at an early level. Educational institutions can ascertain the learning of sustainable practices, and future generations can be prepared to manage current environmental issues. Furthermore, combined efforts from governments, schools, and industries are needed to establish good research systems that can aid in the application of catalytic approaches in waste treatment. By addressing these synergies first, developing countries are able to not only improve their waste management systems but also contribute enormously toward ushering the world toward sustainable chemical practice, thus in harmony with the tenets of green chemistry as championed by Guyo et al. (2024) and Eskandarinezhad et al. (2021) (Figure 5.12).

Figure 5.12. Educational initiatives can promote the understanding of sustainable practices, empowering future generations to solve existing environmental issues.

Source: https://usercontent.one/wp/www.winssolutions.org/wp-content/uploads/2024/09/students-working-on-sustainable-ideas-800x500.webp?media=1732096469.

5.5. CASE STUDIES IN GREEN CATALYSIS

The integration of green catalysis into chemical systems is a pillar to more sustainable practice in business. Case studies in green catalysis illustrate the successful integration of principles such as atom economy and reduced environmental impact, thereby bringing about the tangible benefit of green chemistry. The pharmaceutical industry, for instance, is usually criticized for environmental insensitivity through the creation of toxic byproducts and extensive generation of waste. Using green catalysis not only minimizes environmental damage but also promotes low-cost methods of drug production. Various projects have established that through catalysis, it is possible to enhance the efficiency of the reaction while keeping the amount of toxic byproducts generated during the process of pharmaceutical manufacture to a minimum.

This twofold benefit is in complete harmony with green chemistry philosophy, keeping in mind a reduction in waste and using safer processes to eventually yield a sustainable future for the industry, as evidenced through different case studies representative of such novel practices (Achieng, 2025). Moreover, the application of green catalysis is not exclusive to the pharmaceutical industry but can be applied to a wide range of chemical applications. For example, enzymatic catalysis has been at the forefront for its environmentally friendly profile to carry out reactions under mild conditions and with high selectivity. Examples of the enzymatic treatment of materials such as the production of bio-based butylene glycol illustrate the capacity of biocatalysts to reduce harmful environmental impacts and enhance process efficiency. The adoption of such green practices does not only minimize the usage of energy but also provides a vehicle whereby industries can transition to less polluting production techniques. The analysis of such instances shows that the adoption of environmentally friendly catalytic practices helps mitigate the negative impact of traditional processes, further entrenching the thematic focus of green chemistry, especially in pedagogical contexts that explore industrial redesign and minimization of waste (Vaz et al., 2025). Lastly, the move towards cleaner catalysis underscores the importance of education courses that prepare chemists of the future to manage the fineries of sustainable chemical operation. The incorporation of Problem-Based Learning (PBL) methods enables students to address case studies head-on, their grasp of green chemistry concepts under real-world conditions significantly enhanced. By going through various examples of green catalysis, like the preparation of acetanilide through varied processes, students are required to evaluate the greenness of each process. Such an approach enhances critical thinking and imposes the practical applicability of the sustainable approach in chemistry. However, as it appears from the research, certain concepts like atom economy and catalysis are problematic for students to understand, and thus special teaching material is required. The remediation of these obstacles plays a critical function in the evolutionary process of the integrated idea of green chemistry, thereby ultimately allowing students to effectively contribute towards sustainability in their respective fields (Achieng, 2025; Vaz et al., 2025).

5.5.1. Successful Industrial Applications

The application of green chemistry principles in manufacturing has brought a myriad of innovative and successful solutions in various industries. Through the focus on reducing the environment footprint through the application of catalysis, most industries have made greener efforts. For instance, the pharma manufacturing industry has employed greener ways in producing drugs, lessening waste and enhancing efficiency. This shift not only responds to the regulatory needs of lower emissions and waste but also to the growing customer-based need for sustainable products. The adoption of these practices is a turning point for how industries create profitability and environmental sustainability, demonstrating that environmental issues can equal economic benefit through improved processes and reduced operating costs (Ch et al., 2025). Among the notable accomplishments in catalysis use in industry are the results of advancements in nanocatalysis and biocatalysis. Such technologies utilize nanoscale catalysts or life forms to catalyze chemical reactions at an economical cost. In the process, industries have gained from changes in reaction rate, selectivity, and energy demands in addition to keeping in check main issues with respect to toxicity and waste formation. For example, biocatalysts facilitate greener production pathways in specialty chemical, pharmaceutical, and biofuel production. The coming together of such practices not only facilitates compliance with stringent environmental regulations but also presents prospects for new market opportunities driven by sustainability (Swamy, 2025). Furthermore, the petroleum refining industry has also gained significantly from the practices of green chemistry, particularly through catalytic process improvement. Catalysts have a vital role in the processing of crude oil to useful commodities, and the trend toward more environmental-friendly operations has spurred the development of catalysts that minimize harmful byproducts while maximizing yield. Progress in heterogeneous catalysis has introduced enhanced catalyst material separation and recovery, thus minimizing waste while maximizing production efficiency. As such, the industry not only experienced economic advantages but also substantial reduction in environmental degradation, going on to attest to the usefulness of catalysis in encouraging green practices in traditional industrial areas (Ch et al., 2025).

5.5.2. Academic Research Contributions

The integration of research work in academia is at the core of catalysis promotion in green chemistry. Researchers are not only setting new avenues but also closing the gap between theory and practical information through collaborations. Brazilian labs, for instance, have shown a strong interest in investigating catalysis and green chemistry through their strategic development of sustainable methods of organic synthesis. With rigorous citation analysis, the contributions have been made clear to be appreciated globally, thus indicating the importance of this research in the global scientific community. Full-time involvement by well-trained chemists guarantees a continuous flow of new ideas, which are critical for overcoming conventional chemical process limitations. This collaborative spirit is also set to underpin major advances in the scalability and efficacy of green catalytic reactions (Ronaldo et al., 2018).

As the research environment continues to evolve, the conversion of biomass into sustainable chemicals is an emerging area of investigation. The potential of lignocellulosic biomass as a renewably sourced feedstock is espoused through a number of innovative catalytic routes aiming to replace fossil fuels with sustainable chemicals. Academic publications in the field point to the trend in renewable resources today, emphasizing the significance of green chemistry towards achieving energy security and lighter environmental footprint. A number of methods, such as the catalytic processing of biomass to valuable chemicals, are still being researched. Product yields and efficiencies in processing remain issues, however. Active involvement of academically associated researchers has been instrumental in shaping groundbreaking methodologies that avoid these constraints, culminating in the successful application of biomass in various chemical processes (Paresh et al., 2017). Academic institutions are also significant players in the technological advancements necessary for the optimization of catalytic systems. Research work not only aims at optimizing the present processes but also finding new catalytic processes and materials. The collaborative environment among the research scholars enables the sharing of ideas and techniques, and thus innovation that might not have been attainable under restricted conditions. With the emphasis on sustainability, the researchers are forced to revisit traditional paradigms and employ an interdisciplinary strategy to catalysis. This development is crucial to the realization of the potential benefit of green chemistry for application in industry, where the combination of renewable resources becomes increasingly determinate. Thus, research contributions from academia not only push knowledge frontiers but also become driving forces for sustainable chemical industry growth in the broader industry, as well as further strengthening the position of academia in bringing about change (Ronaldo et al., 2018).

5.5.3. Innovations in Catalytic Processes

Innovative catalytic reaction strategies are key to the advancement of green chemistry, with aims at the development of sustainable routes for chemical conversion. Perhaps the most groundbreaking advance has been the application of nanostructured catalysts. These nanocatalysts, with their increased surface area and tunable reactivity, facilitate chemical conversion under less rigorous conditions with significantly enhanced efficiency. A close-up analysis highlights that by modifying such materials at atomic and molecular levels, their performance can be improved in a wide variety of applications varying from biomass conversion and pollutant degradation to energy applications and catalytic applications. Lastly, the use of nanocatalysts in industrial processes not only accelerates reaction rates but also decreases the environmental footprint, thus achieving green chemistry objectives. However, these must surmount challenges such as nanoparticle agglomeration and scalability to utilize best their potential to sustainable applications (Chahal, 2025). In addition to nanostructured materials, organometallic catalysis is another field of innovative catalysis, offering efficient, selective solutions to vital reactions. Palladium, nickel, and copper transition metals are increasingly dominant in various transformations, including hydrogenation and utilization of CO_2. A critical examination of the literature illuminates how modification to ligand systems enhances reactivity and selectivity of such catalysts, a huge step towards optimizing chemical processes. Furthermore, examination of earth-abundant metals like iron and cobalt as replacement catalysts is in line with principles of sustainability by reducing precious metal usage. These advances indicate a complete understanding of catalytic processes, which is essential for the generalizability of organometallic catalysts in green chemistry (Singh, 2025). Searching for innovations in catalytic processes also involves considering environmental efficiency and sustainability through the catalytic material lifecycle. Even though huge strides have been achieved in the design of efficient catalysts, the shift to industrial scalability is fraught with concerns about cost and use-viability of newer catalytic systems. Life cycle assessment and modeling techniques play a significant role in estimating long-term performance and environmental outcomes of new catalytic processes. For instance, giving precedence to valorization of CO_2 facilitates the transformation of greenhouse gases to valuable commodities, something that is advantageous to a circular economy. A fine balance between technological progress and environmental sustainability shall be critical in ensuring these catalytic innovations find successful integration into conventional uses, and interdisciplinary research and cross-collaboration in green chemistry become even more pertinent (Singh, 2025).

5.6. CHALLENGES IN GREEN CATALYSIS

The development of green catalysis is an important focus in the branch of green chemistry, through which environmental footprints can be minimized while chemical processes can be made more efficient. But there are certain challenges in its large-scale implementation. One of the main issues is the scalability of green catalytic operations for practical industrial applications. While most methods are promising on a lab scale, their transfer into mass production generally reveals defects in their cost, efficiency, and operation stability. Further, integrating such processes within existing industrial frameworks may face resistance due to the need to drastically alter existing infrastructure or invest capital in new technologies. Therefore, the optimization of the balance between efficiency and environmental issues remains a significant challenge towards sustainable catalysis solutions that demands ongoing interdisciplinary research and collaboration in crossing laboratory breakthroughs and industrial applications. An additional key problem in green catalysis is the selection and optimization of effective catalysts meeting sustainability needs. Traditional catalysts, some of which consist of precious metals or have large environmental footprints, cannot easily be replaced by greener catalysts without extensive research. The search for long-lasting and low-cost catalysts, especially those derived from plentiful feedstocks, is full of promise and daunting challenges. For instance, while metal-organic frameworks (MOFs) and covalent organic frameworks (COFs) have been recognized to be highly functional and tunable, questions about these materials' production scalability as well as stability under real-world operating conditions linger in their implementation into commercial applications (Freund et al., 2021). Ensuring that certain catalysts can be effective under a range of environmental conditions and feedstocks, but remain economic, is imperative in the challenge of overcoming the current limitation of applying them on larger scales in the chemical industry. Moreover, the growth of biocatalysis has been an interesting area for green catalysis, but not without its own set of challenges. As biotechnology comes to enable the design of enzymes for specific reactions, economic viability for such a process is still uncertain. While biocatalysts bring the advantages of milder reaction conditions and reduced byproduct formation, their utility in the commercial context can be negatively impacted by constraints such as enzyme stability and substrate specificity. Enzyme immobilization, as a performance enhancement and reuse strategy, increases the complexity of the design phase (Roger et al., 2017). With advancing science, further research needs to be conducted to scale up biocatalytic processes, expand the substrate range that can be used, and ensure compatibility with traditional methods of manufacture, in order to boost the role of biocatalysis toward sustainable chemical conversion.

5.6.1. Economic Viability of Green Catalysts

Green catalysts are also an economic factor in pushing industries toward greener operations. Green catalysts not only enhance reaction efficiencies but also reduce the environment's impact, providing a dual advantage of cost savings and environmental responsibility. By refraining from using harmful substances and waste production, green catalysts complement the concepts of green chemistry to ensure cleaner production. The application of novel materials such as nanomaterials and biocatalysts provides enhanced catalytic activity, thus leading to lowered operating costs in the long run. Moreover, the conservation of energy expenditures through greener catalytic processes is equivalent to one big saving on utility bills for businesses heretofore reliant on more energy-costly processes. This economic foundation emphasizes the potential of investment in the research and development of green catalysts for promoting adoption across industrial sectors, further proving that sustainability can also be economically viable. As industries are subjected to tighter regulation and increasing pressure from the public for sustainable practices, the economic interest of green catalysts is further justified. The potential for *in situ* synthesis of vital reagents, as noted in recent advancements, offers a innovative means of achieving optimal catalytic activity at minimum expenditure. For example, *in situ* hydrogen peroxide (H_2O_2) synthesis rather than by typical energy-intensive procedures illustrates how green catalysts reduce chemical processes. This approach not only yields greater atom efficiency but also curtails greenhouse gas emissions markedly, favoring international sustainability.

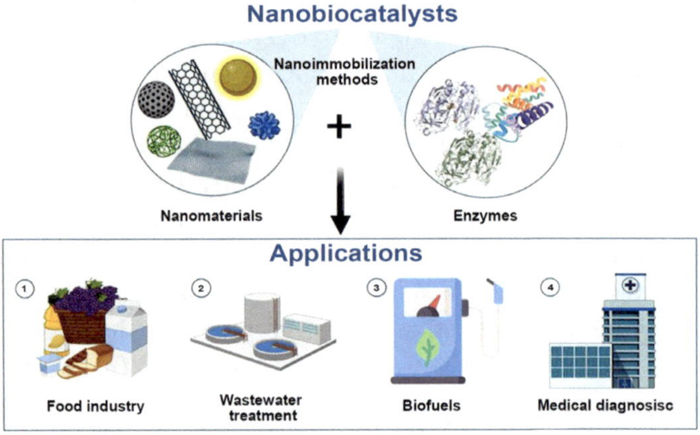

Figure 5.13. The integration of innovative materials, such as biocatalysts and nanomaterials, enhances catalytic efficiency, thus leading to lower operational costs in the long run.

Source: https://pubs.acs.org/cms/10.1021/acsomega.2c03155/asset/images/large/ao2c03155_0006.jpeg.

A switch to such newer technologies may have initial capital outlays; however, long-term operating cost savings and environmental standards compliance dramatically enhance economic feasibility. Therefore, green catalyst research and deployment are likely to be primary drivers for sustainable industrial growth concurrently with economic needs. Furthermore, use of decentralized systems for green catalyst usage presents additional economic opportunities in the industry. By allowing production of chemicals and materials locally, these systems lower shipping costs and associated emissions and therefore the sustainability of chemical production overall. Such developments are vital to enhancing circular economy principles, which foster efficiency in use of resources and waste reduction. Being able to produce fine and bulk chemicals via innovative, ecofriendly approaches—like chemo-catalytic/enzymatic one-pot synthesis highlighted in the latest science—has the promise of transforming traditional manufacturing paradigms. Notably, their economic effects run deep: not merely lowering production costs but generating new markets and employment for green technology and green jobs. As such, continued investment in green catalytic process research is certain to yield fantastic economic dividends as well as preserve the environment, further supporting their place in green chemistry's future (Figure 5.13).

5.6.2. Scalability of Catalytic Processes

Scale-up of catalytic processes is a big challenge in the field of green chemistry, as it has direct impact on the feasibility of adopting new catalytic reactions on a commercial scale. Catalysts play a central role in terms of maximizing reaction efficiency and minimizing waste and hence ensuring sustainable operations. Nevertheless, the switch from laboratory-scale tests to large-scale manufacture tends to introduce several challenges, some of which include the complexity of the reaction environments and the necessity for efficient catalyst recovery and recycle. These points highlight the need for developing catalysts that not only show high activity and selectivity but also show stability under industrial operation conditions. This flexibility is crucial, as it achieves the eco-friendly objectives of green chemistry while also solving problems in large-scale production, ultimately minimizing the environmental impact of chemical processes. Further, innovation through new techniques like the employment of metal-organic frameworks (MOFs) shows the scalability of catalytic processes in green chemistry. MOFs possess unique structural features, i.e., high surface areas and adjustable surface areas, that can elevate catalytic performances to a significant degree. However, their effective application is constrained by scale-up synthesis and forming techniques, which could discourage their overall commercial exploitation. Researchers have been attempting to optimize synthetic pathways—e.g., electrochemistry and environmentally friendly reflux solvents—to make the materials more viable for process applications on the large scale. According to recent literature, addressing such scale-up challenges is essential in a bid to tap the full potential of MOFs in catalysis, and forms part of a broader movement towards the use of advanced materials within sustainable chemistry approaches (Li et al., 2022). Lastly, scalability in catalytic process must be achieved through a multi-pronged approach involving not only innovative catalyst designs but also integration with efficient and eco-friendly methodologies. The application of green metrics to quantify the efficiency of various catalytic processes can guide researchers and industries on how to optimize their approaches in terms of efficiency. A case in point is focusing on process intensification and developing reactive distillation approaches, which can significantly contribute to the overall sustainability of catalytic processes through energy saving and minimizing the production of waste. The ability to analyze and enhance the scalability of catalytic processes in an orderly fashion is important for enhancing the field of green chemistry, so new catalytic technologies are not only effective in a laboratory environment but also practical and economical in actual applications (Andraos et al., 2022).

5.6.3. Regulatory and Safety Concerns

The use of catalytic processes in green chemistry is not without challenges, the regulatory and safety aspects being of great concern. While green chemistry aims to develop sustainable processes and minimize environmental effects, regulation must also see to it that such new modes of doing things do not undermine public health and safety. Regulatory bodies fall behind the rapid advancement in catalysis technology development, leaving loopholes that can pose potential threats. For example, although catalysis can significantly reduce energy expenses and waste output, the new materials and processes introduce as yet unforeseen dangers not yet fully accounted for by existing regulation. Regulatory bodies themselves need to keep pace and adapt along with these advances in order to facilitate innovation without compromising against adverse effects on human health and the environment. Commercialization of catalytic processes is also made more challenging by safety concerns. Effective safety assessments require close scrutiny of both the byproducts and catalyst materials to establish that they are safe for consumers and workers. Particularly, the shift to renewable feedstocks and biodegradable feedstocks can lead to regulatory attentions towards their safety profiles. A large number of new chemicals can appear on the market as catalysts or products from catalytic processes in green chemistry, which calls for comprehensive evaluations and guidelines. For instance, storage of hydrogen in porous media, as potential as it is for energy efficiency, poses questions of safe handling and transport of hydrogen, especially at large scale. Such solutions to these challenges might protect against accidents and instill public trust in new green technologies (Heinemann et al., 2021). Moreover, legislative interventions like the proposed carbon neutrality targets demonstrate the imperative of establishing robust regulatory systems. Global cooperation is necessary, but the absence of coordination between nations in the creation of adaptation and mitigation policies hinders progress towards collective environmental objectives (Chen et al., 2022). As catalysis plays a vital role in the achievement of these goals, policymaking organizations have a mandate to provide congruence in safety standards and environmental policies across borders. It involves creating harmonized policies that maintain safety while promoting the safe development and usage of catalytic technologies. By so doing, stakeholders are able to support a smoother shift toward environmentally friendly practices that are not only scientifically credible and economically feasible but end up promoting the world's response to climate change and environmental degradation.

5.7. FUTURE TRENDS IN CATALYSIS

As the demand for sustainable chemical processes increases, the catalytic science is poised to undergo radical transformations. Most hopeful is the development of bioelectrocatalysis, particularly ammonia synthesis. The novel approach employs enzyme-based processes in electrochemical systems as an alternative to the conventional Haber-Bosch process. Improvements in nitrogenase-mimetic enzymes and genetically modified microorganisms have exhibited higher efficiency in nitrogen reduction under mild conditions, not only decreasing energy intake but also favoring the objectives of green chemistry. Bioelectrocatalyst scalability offers a pathway to industrial-scale usage, making them focal points of future research and development activities in sustainable chemistry. As companies increasingly seek greener strategies, such advancements are able to redefine traditional catalytic processes to be more environmentally friendly and sustainable (Jayakumar et al., 2025). Another amazing catalysis trend lies in using photocatalytic technology that effectively utilizes solar radiation in producing chemical changes. Such a process holds much promise for green chemistry, just because it only runs at mild reaction conditions and low energy demands compared to normal catalytic reactions. Photocatalysis has increasingly found applications in organic synthesis with innovative pathways that minimize environmental impact at maximum efficiency. Current research highlights its effectiveness in building up complicated organic molecules with minimized risk of secondary pollution. By further refining the mechanisms and applications of photocatalytic synthesis, scientists are addressing critical issues in the area and paving the way to a greener future. This technology will play a vital role in fostering innovation and achieving sustainable development globally, making it part of the discussion for green chemistry (Li et al., 2025). The future of catalysis is not only about technological advancement; it is also a paradigm change in how chemical processes are envisioned and applied. Sustainability will involve the integration of multidisciplinary methods, bridging ideas from biology, materials science, and environmental science. Nanomaterials and purposeful catalysts advance themselves at the forefront of catalysis, capable of maximizing the efficiency and selectivity of reactions while minimizing environmental effects. Additionally, the establishment of research collaborations among universities and industry will facilitate the rapid translation of these new methodologies to commercial significance. As tomorrow's catalysts emerge, it will become necessary to monitor their performance and institute guidelines that encourage sustainable practice across the chemical production sector. Ultimately, a worldwide commitment to more sustainable catalytic processes will drive the future of catalysis in green chemistry (Figure 5.14).

Figure 5.14. As industries increasingly seek greener practices, such advancements may redefine traditional catalytic methodologies, rendering them more sustainable and environmentally friendly.

Source: https://www.deloitte.com/content/dam/insights/articles/2024/us187651_e-i_2025_outlook-chemicals/primary-images/us187651_banner.png.

5.7.1. Advances in Catalytic Materials

Development of catalytic materials has witnessed tremendous progress, particularly in the area of green chemistry. One of the notable developments is the employment of graphene oxide as a metal-free catalyst, and it is a great representative of the trend toward sustainable alternatives in catalytic reactions. Graphene oxide (GO), benefiting from its unique two-dimensional structure and high density of functional groups, offers a new platform for various organic conversions. Its utility has been discovered in oxidative coupling, oxydehydrogenation reactions, and group functional transformations and thereby broadening the scope of precious metal-free catalytic reactions. Not only does this innovation address the environmental problem that is posed by conventional catalytic materials, but it is also in line with the principles of green chemistry as it provides efficiency while keeping toxic byproducts at a minimum. The constructed synthetic methods of GO are simple and eco-friendly, and it is a good future prospect as a catalyst for various catalytic applications in sustainable processes (Alka et al., 2025). Besides this, the introduction of silver nanoparticles (AgNP) with azo reagents is another remarkable advancement in synthesizing catalytic material. These nanostructures have gained significant attention due to their high catalytic activity and potential for application in analytical chemistry. The green synthesis of AgNP has exhibited enhanced selectivity and sensitivity, particularly in UV spectroscopy cases. The complexation between AgNP and azo compounds reflects the versatility of these materials; they can be engineered for specific reactions, thus optimizing yield while minimizing waste. By resonating with the unique properties of silver nanoparticles, researchers can engineer greener pathways for catalysis that conform to green chemistry principles. This new marriage of materials is a reflection of the ongoing push towards more efficient and sustainable catalytic systems. Ultimately, advances in catalytic materials portend a landmark change towards sustainability in chemical processing. Materials such as silver nanoparticles and graphene oxide are a reflection of what can be achieved in terms of making catalysts both cost-effective and sustainable. Such technologies contribute to increasing knowledge of catalytic processes that are capable of reducing energy consumption and limiting the production of harmful waste. With the progress in research in this direction, integration of new materials with traditional approaches is continuously paving the way for better catalytic applications in green chemistry. This holistic approach relies on joining the newest technology with established science in a bid to foster sustainable processes within the chemical industry and offer a greener tomorrow for scientists and society at large (Alka et al., 2025).

5.7.2. Integration of Artificial Intelligence in Catalysis

The integration of artificial intelligence (AI) into catalysis is a significant step towards green chemistry, particularly more sustainable catalytic processes. Traditional catalytic methods are a time-consuming trial-and-error method with high levels of manual experimentation and resource utilization. In contrast, AI-based methods such as machine learning algorithms can identify extensive amounts of catalytic performance data to identify patterns and make predictions more efficiently. This paradigm shift lowers not just the time taken to find effective catalysts but also enable scientists to focus on earth-abundant materials in line with the green chemistry philosophy. By employing AI, scientists can design rationally catalysts of certain specifications reducing the environmental impact while enhancing performance, thus developing a greener and cheaper way of catalysis for other industrial operations (Guo et al., 2024). In addition, AI goes beyond catalyst discovery; it is also central to optimization of previously existing catalytic processes. For instance, optimization of reaction conditions can be achieved using machine learning algorithms, with radical improvements in yields and enantioselectivity in asymmetric synthesis, an area that is of significant interest to the pharmaceutical industry. Given that chemical reactions typically involve complex interaction between catalysts and substrates, AI can systematically explore the reaction space with a focus on conditions to create higher enantiomeric excess and overall higher efficiency. New enzymatic catalysis breakthroughs, also outlined as more sustainable solutions, also find beneficial use through integration by means of enzyme engineering. Techniques such as microfluidics-based ultrahigh-throughput screening combined with AI-guided directed evolution allow one to design and create super-efficient enzymes to conduct sustainable cascade reactions (Yi et al., 2024). The future of catalysis in green chemistry is definitely interconnected with the ongoing convergence of AI technologies to further transform the field. With the advances in the sophistication of algorithms for AI and their capacity to analyze complex chemical information rapidly, researchers are now able to tackle the pressing challenges that control catalyst design and optimization. The emergence of creative strategies, such as the utilization of biomimicry, demonstrates the promise of AI to create innovative biomimetic catalysts from nature's own mechanisms. Through the move towards a more data-driven approach, the scientific community can facilitate the development of high-performing catalysts from abundant and sustainable materials. Ultimately, the confluence of AI and catalysis not only offers solutions to extant challenges but also holds the potential to reveal new paths to cleaner chemical procedures that align with worldwide sustainability goals (Guo et al., 2024).

5.7.3. Global Initiatives for Sustainable Catalysis

The impending imperative for sustainability has necessitated global action towards new catalysis technologies in green chemistry. Global action increasingly centers on the imperative to stimulate sustainable processes with lower environmental footprints and higher efficiency. These initiatives strive to lead industries away from traditional petrochemical dependency through encouraging biobased alternatives. Notably, lignocellulosic biomass has emerged as a principal resource, with different sustainable applications such as biofuels and bioplastics. The latest developments in integrated biorefineries are illustrative examples of how farm waste could be successfully converted into high-value bioproducts and hence help spur the circular economy.

Facilitating the use of lignocellulosic fractions, such schemes underscore the central role of green catalysis to mitigate overall climate emergencies and advance environmental responsibility (Mujtaba et al., 2023). Furthermore, the complex contribution of catalysis to sustainability is now recognized across various industries and necessitates an interdisciplinary approach towards green chemistry. Global programs aim at not just optimizing catalytic reaction efficiency but also exploring renewable reactants and waste minimization. For instance, low-temperature plasma (LTP) technology gains increasing popularity as it provides new perspectives on chemical transformations and pollution protection.

Figure 5.15. Global initiatives increasingly emphasize the importance of developing sustainable processes that minimize environmental impacts while enhancing efficiency.

Source: https://kaizen.com/wp-content/uploads/2023/11/corporate-sustainability-2-scaled.webp.

The 2022 LTP Roadmap establishes outstanding challenges and opportunities and calls out interdisciplinary collaboration and greater investment in research and development. Utilizing soft materials and data-driven approaches, these developments are examples of how scientific ingenuity is facilitating dovetailing with sustainable development goals to make industrial processes more environmentally friendly (Adamovich et al., 2022). Combined, these global initiatives towards sustainable catalysis provide a base for cross-disciplinary cooperation between academia, business, and policy-making. Through collaborative learning and sharing of knowledge, partners can accelerate the enforcement of green chemistry regulations, which will have transformative impacts on global sustainability. Moreover, the integrated processing of lignocellulosic biomass not only enhances waste management but also promotes the production of second-generation materials in concordance with climate targets. With the world facing critical environmental issues, the embracement of sustainable catalysis in traditional practices will be an important factor in shaping a future that is less reliant on fossil fuels and where there is higher utilization of resources. Innovation and collaboration in such endeavors will be most vital in an effort to achieve the ambitious targets set by global climate policies and sustainability agendas (Mujtaba et al., 2023) (Figure 5.15).

5.8. CONCLUSION

In short, green chemistry catalysis is an innovative approach that reduces environmental problems while maximizing the efficiency of chemical reactions. Focusing on waste reduction and energy consumption, green chemistry seeks to create sustainable processes with low environmental impact. Through catalytic processes implemented by industries and scientists, innovation can introduce a paradigm shift in pharmacy, agriculture, and materials engineering. These techniques are not only hypothetical practices; they play sophisticated roles in regulatory frameworks and technological innovation, as conferred by the requirement of environmental regulations that promote green operations among varied industries (Maïder et al., 2025). Ultimately, the integration of catalysis into green chemistry presents a way to respond to the increasing worldwide demand for sustainable products without compromising efficiency or safety. Further, looking at the meaning of green chemistry catalysis, the proof indicates its capacity to offer new procedures that effectively minimize ecological dangers. The use of catalysts during chemical reactions not only catalyzes the reaction but also facilitates milder conditions, which are energy-conserving. This compatibility with sustainability goals emphasizes the necessity of continuous investment in research and development efforts to find good catalysts which are both financially and environmentally feasible. Importantly, regulatory frameworks that promote innovation in green chemistry can encourage the automobile and chemical sectors to adopt cleaner technology, thereby aligning manufacturing activities in accordance with social expectations of environmental stewardship (Maïder et al., 2025). Thus, the significance of catalysis is not merely a matter of its function; it represents an ideology of uniting science with environmental responsibility. The future of catalysis in green chemistry appears bright, given the steady dedication of academia and industry to set up more environmentally responsible chemical processes. It will be necessary for scientists, policymakers, and business executives to come together in overcoming existing hurdles in catalytic processes gaining widespread usage. This collaboration can facilitate the sharing of knowledge and resources such that discoveries in catalysis can be common and be applied to industries. Green chemistry principles education and awareness at the public level will also generate public support for legislation favorable to a transition toward a sustainable manner of conducting affairs. As highlighted by recent discussions within the chemical community, the role of catalysis in the development of a sustainable world is not only inevitable but also achievable through co-operation and research innovation. Therefore, adopting catalysis as a key catalyst of green chemistry not only solves the existing environmental problems but also provides a base for future generations to thrive in a clean environment.

5.8.1. Summary of Key Points

The field of green chemistry has brought large momentum due to its concentration on sustainable operation in chemical reactions, particularly catalysis. One key point is that catalysis is an intrinsic mechanism to enhance the efficiency of chemical reactions with reduced waste and energy consumption. Such effectiveness is characterized by the ability of catalysts to accelerate chemical reactions without consumption, allowing for decreased energy requirements and increased rates of reaction. The integration of catalysis in green chemistry principles effectively converts traditional approaches to sustainable alternatives, thus responding to the urgent need for sustainability in the chemical industry. As mandated by current studies, the use of catalytic processes has been linked to beneficial outcomes economically as well as environmentally, with potential to revolutionize paradigms in industrial production and chemical synthesis. The role of innovation in advancing towards new catalytic processes is also central to advancing towards a more environmentally friendly approach. The research and development of heterogeneous catalysts, particularly those from renewable resources, offer opportunities towards more environmentally friendly processes in chemical production. These are significant to industries such as the automotive sector, where the application of green chemistry has transformed methods of operation and regulatory frameworks. Such catalysis technologies are not merely in harmony with environmental norms but also promote a technology regime that upholds sustainability, in consonance with the transition to a more responsible industrial paradigm. The collective sum of the scientist society's response to the environmental question serves as a watershed in the approach to chemical synthesis and process efficiency, extending the contribution of catalysis to achieving these results (Maïder et al., 2025). Moreover, the role of education and research in catalysis in green chemistry cannot be overstated. Academic studies and student and expert training programs ensure deeper understanding of sustainable processes as well as environmental impacts. It is a critical area of focus in equipping future scientists and engineers with the ability to innovate in catalysis. Collaboration among research centers, research organizations, and industry giants will assist in enhancing the efficiency of catalytic processes, thereby facilitating greater application of green chemistry principles. Last but not least, integration of theoretical knowledge and practical knowledge will lead to new paradigms that represent the spirit of sustainability and effectiveness. As the field advances, increased recognition of the interconnectedness of catalysis, research, and education in green chemistry will be critical to the creation of solutions that support the environmental priorities of our times.

5.8.2. The Future of Catalysis in Green Chemistry

The future of catalysis in green chemistry promises not only to revolutionize the manner in which chemical reactions are induced but also how industry interacts with sustainability. As pressure to become greener escalates, catalysts that minimize energy consumption and waste generation are quickly emerging as a central focal point for research and development. Carbon nanotubes (CNTs), for instance, are emerging as multipurpose materials with immense potential to transform catalytic processes beyond recognition. Their outstanding qualities, including high surface area and better conductivity, ensure greater charge capacity and energy conversion and storage efficiency. By integrating CNTs into catalytic systems, researchers aim to maximize processes such as hydrogen production and biomass conversion, thus facilitating grand visions of green chemistry, which aims to be resource conservative and environmentally benign (Zafar et al., 2025). Aside from that, both nanocatalysis and biocatalysis bring into focus the multiverse strategies implemented to green catalysis. Nanocatalysts have their distinctive size-to-surface ratio that enhances the reaction rate while reducing the content of noble and rare metals involved, which are generally costly and environmentally destructive to extract. Biocatalysts, on the other hand, such as synthetic enzymes, bring with them a greener solution by conducting catalysis at mild conditions and producing fewer by-products. These emerging strategies, besides lowering the carbon footprint of chemical processes, also align with regulatory trends towards sustainability. Hence, the deployment of these next-generation catalytic systems is vital to industries aiming to satisfy increasingly stringent environmental regulations without compromising competitiveness based on productivity (Fatemah, 2025). In the future, scalability and affordability issues of green catalysts need to be addressed to facilitate wide-scale use. Existing research reveals the need for efficient synthesis methods that can produce high-performance catalysts at a lower cost, hence made accessible to various industries. Interconnection between industry and academia will have a pivotal role in overcoming such issues, formulating new solutions that have the potential to be applied in the real world. In addition, creating supportive regulatory environments and investment incentives in green technology has the capability to encourage companies to go greener with greater ease. As catalysis keeps evolving in the context of green chemistry, it also holds out the potential to significantly minimize the environmental impact of chemical manufacturing, ultimately ushering in a cleaner and greener tomorrow (Zafar et al., 2025; Fatemah, 2025).

5.8.3. Call to Action for Researchers and Industry

In the changing green chemistry landscape, scientists and industrial players have to come together in a common endeavor to adopt innovative catalytic techniques that critically minimize environmental burdens. Continuous-flow photochemistry is an example of a vital shift towards sustainable practice, highlighting the greater efficiency and safety provided by such technologies. It offers more control over reaction conditions and allows processes to be scaled up that were previously constrained by conventional batch methods. As highlighted in recent literature, application of continuous-flow technology not only makes discovery and optimization of reactions easier but also facilitates the development of multiphasic reactions that are essential to the pharmaceutical industry and other uses (Sambiagio et al., 2019). A focus on cooperation between fundamental research and industrial application will allow for a more integrated approach to catalysis in green chemistry. At the same time, merging electrochemical conversions provides a fascinating option towards sustainability that requires urgent action from researchers and industry. While the increased interest in electrochemical synthesis is focused primarily on its ability to reduce the production of waste, there is also potential for it to simplify synthetic pathways and apply milder reaction conditions. With the use of electricity as a reagent, electroorganic conversions can avoid the use of stoichiometric reagents that cause less efficient processes that are also safer and less toxic material-dependent (Wiebe et al., 2018). A such innovative strategy can set a new standard in green chemistry that renders it imperative for scientists to exhaustively explore the potential of electrochemical systems and develop scalable processes that business can enact promptly. An interdisciplinary call to action is imperative for this transition to sustainable operations. Scientists must collaborate with industry players to develop practical solutions that address urgent environmental issues while advancing economic growth. By utilizing continuous-flow systems and electrochemical processes, the platform for sustainable chemistry is enhanced. There must be a revamp of education, so that future scientists possess the tools and information required to innovate in this field, able to comprehend theoretically and implement practically. Eventually, the development of a culture of collaboration among academia, industry, and regulatory bodies will accelerate the implementation of green chemistry concepts into real practice, thereby expanding their overall value contribution to the environment and society.

5.8.4. Final Thoughts on Sustainability and Innovation

In closing remarks about sustainability and innovation, the potential of catalysis in green chemistry should be acknowledged as a key building block in addressing contemporary environmental issues. Catalytic process innovation is central to developing sustainable production technologies of chemicals and materials previously reliant on fossil feedstocks. Possibly the most prominent example of such a shift is towards biobased replacements, i.e., using lignocellulosic biomass of agricultural waste. Not only does this reduce dependence on petrochemicals but also supports environmental conservation by lowering waste and emissions. An integrated biorefinery approach can allow for the conversion of such biomass into a wide range of desired commodities, thus illustrating the potential for fusion of innovation and sustainability to solve increasing demands in a world of dwindling resources, highlighting research and continuous enhancement of catalytic processes (Mujtaba et al., 2023). Furthermore, the development of chiral organocatalysts brings a revolutionary role of green chemistry into play, which focuses on efficiency and eco-friendliness. Enantioselective radical reactions have seen the world of chemistry witness a revolution over the last two decades, driven by novel approaches and catalysis. These new-age methods not only reduce toxic byproducts characteristic of conventional chemical reactions but also lead to increased functionality and complexity in the resultant compounds. The use of photocatalysis in organocatalytic systems is an excellent example of the innovative techniques that are being used to diversify radical precursors and achieve maximum reaction yield. This interplay between innovation and sustainability in green chemistry is a reflection of our capability not only to redesign conventional methods but to adopt greener processes towards industry application, which is crucial to attain sustainability targets (Mondal et al., 2022). As the dialogue on sustainability advances, it can be seen that interdisciplinarity will be key to fueling greater innovation in green chemistry. By encouraging interdisciplinary research, scientists and researchers can harness new catalytic pathways by employing the knowledge from different disciplines, such as engineering, biology, and materials science. The potential for biotechnological use, such as the use of enzyme-catalyzed reactions in radical chemistry, illustrates the level of creativity that can be reached through the use of interdisciplinary combinations of knowledge. Such synergies lead to an era where environmentally positive practices can be simply added into industrial processes to help ease severe environmental problems while stimulating economic growth. Lastly, the future is in the hands not only of scientific ingenuity but of a shared determination to lead sustainability through new solutions, with catalysis at the center of both environmental responsibility and economic benefit.

REFERENCES

1. Achieng, N. M. (2025). Green chemistry in pharmaceuticals: Reducing environmental impact. *Newport International Journal of Public Health and Pharmacy*. https://www.semanticscholar.org/paper/fcf7c9c0751460e055a99f3481f52db214e8d25d (accessed on 24th October 2025).
2. Adamovich, I., Agarwal, S., Ahedo, E., Alves, L. L., Baalrud, S., Babaeva, N. Y., Bogaerts, A., et al. (2022). The 2022 plasma roadmap: Low temperature plasma science and technology. *Journal of Physics D: Applied Physics, 55*, 373001–373001. https://doi.org/10.1088/1361-6463/ac5e1c (accessed on 24th October 2025).
3. Alka, A., Patel, U. K., & Agarwal, A. (2025). Carbonatitic activity of graphene oxide in organic transformations: A comprehensive review. *Chemistry Select*. https://www.semanticscholar.org/paper/ed199629d520198eca69d53074c0aa74887f76eb (accessed on 24th October 2025).
4. Andraos, J., & Matlack, A. S. (2022). *Introduction to Green Chemistry*. https://doi.org/10.1201/9781003033615 (accessed on 24th October 2025).
5. Babbar, R. (2024). Green chemistry and its applications. *Journal of Pharmaceutical Technology, Research and Management*. https://www.semanticscholar.org/paper/22a2ff31f002a034fa93718f2f5d1593eb55e08f (accessed on 24th October 2025).
6. Ballivet-Tkatchenko, D., Camy, S., & Condoret, J. S. (2005). Carbon dioxide, a solvent and synthon for green chemistry. https://core.ac.uk/download/12039955.pdf (accessed on 24th October 2025).
7. Bhaskar, S., & Kishore, V. (2025). Advancing green and sustainable chemistry: Importance and applications in human life. *Shodh Sari–An International Multidisciplinary Journal*. https://www.semanticscholar.org/paper/f34231f2c4756fb632c58e18c200fbdbbf366db7 (accessed on 24th October 2025).
8. Chahal, P. (2025). Nanostructured catalysts in green chemistry: Enhancing efficiency and sustainability through nanotechnology. *Resmilitaris*. https://www.semanticscholar.org/paper/3b9f5b6ffd5b3bb2899e93e965ff7d8c8121701f (accessed on 24th October 2025).
9. Chen, E. (2025). Exploring green chemistry: Sustainable approaches to chemical synthesis. *Revita Latinoamericana de Química*. https://www.semanticscholar.org/paper/37ef2209f3997399352c45ee568c965c16604a1a (accessed on 24th October 2025).

10. Chen, L., Msigwa, G., Yang, M., Osman, A. I., Fawzy, S., Rooney, D. W., & Yap, P. S. (2022). Strategies to achieve a carbon neutral society: A review. *Environmental Chemistry Letters, 20*, 2277–2310. https://doi.org/10.1007/s10311-022-01435-8 (accessed on 24th October 2025).

11. Cheng, X., Lei, A., Mei, T. S., Xu, H. C., Xu, K., & Zeng, C. C. (2021). Recent applications of homogeneous catalysis in electrochemical organic synthesis. *CCS Chemistry, 4*, 1120–1152. https://doi.org/10.31635/ccschem.021.202101451 (accessed on 24th October 2025).

12. Cinca-Fernando, P., Vázquez-Rodríguez, A., Mangas-Sánchez, J., & Ferreira, P. (2025). Aryl-alcohol oxidases: Catalysis, diversity, structure–function and emerging biotechnological applications. *Applied Microbiology and Biotechnology, 109*. https://www.semanticscholar.org/paper/a270c6611a21e32855bfee95599d3db968d7f601 (accessed on 24th October 2025).

13. Cortes, F. R. U., Falomir, E., Doñate-Buendía, C., & Mínguez-Vega, G. (2025). A review on pulsed laser-based synthesis of carbon and graphene quantum dots in liquids: From fundamentals, chemistry to bio applications and beyond. *The Journal of Physical Chemistry C: Nanomaterials and Interfaces, 129*, 10378–10414. https://www.semanticscholar.org/paper/96e9dcc1c5f32145e3a5600a2784c6c725029cac (accessed on 24th October 2025).

14. Dhepe, P. L., Tomishige, K., & Wu, K. C. W. (2017). Catalytic conversion of biomass. *ChemCatChem, 9*, 2613–2614. https://doi.org/10.1002/cctc.201700884 (accessed on 24th October 2025).

15. Durka, J., Turkowska, J., & Gryko, D. (2021). Lightening diazo compounds? *ACS Sustainable Chemistry & Engineering, 9*, 8895–8918. https://doi.org/10.1021/acssuschemeng.1c01976 (accessed on 24th October 2025).

16. Eskandarinezhad, S., Khosravi, R., Amarzadeh, M., Mondal, P., & Filho, F. (2021). Application of different nano catalysts in industrial effluent treatment: A review. *Journal of Composites and Compounds*. https://www.semanticscholar.org/paper/a5c3f3468aed0bbccb5d253128f783476d75c649 (accessed on 24th October 2025).

17. Fabris, F., Illner, M., Repke, J. U., Scarso, A., & Schwarze, M. (2023). Is micellar catalysis green chemistry? *Molecules, 28*, 4809–4809. https://doi.org/10.3390/molecules28124809 (accessed on 24th October 2025).

18. Freund, R., Zaremba, O., Arnauts, G., Ameloot, R., Skorupskii, G., Dincă, M., Bavykina, A., et al. (2021). The current status of MOF and

COF applications. *Angewandte Chemie International Edition, 60*, 23975–24001. https://doi.org/10.1002/anie.202106259 (accessed on 24th October 2025).

19. Guo, J., Haghshenas, Y., Jiao, Y., Kumar, P. V., Yakobson, B., Roy, A., Jiao, Y., et al. (2024). Rational design of earth-abundant catalysts toward sustainability. *Advanced Materials, 36*. https://www.semanticscholar.org/paper/b2b454a5afa7811605b13eded14d3ccee57cb84b (accessed on 24th October 2025).
20. Guyo, U., Chigondo, F., Nyoni, S., Muchanyereyi, N., Zinyama, N. P., Kusena, W., & Nyoni, D. (2024). Green chemistry: Current status and challenges in Zimbabwe. *Physical Sciences Reviews, 9*, 3305–3322. https://www.semanticscholar.org/paper/d95e1ffdc006006814deffbcd4323d9b488dbdb9 (accessed on 24th October 2025).
21. Heinemann, N., Alcalde, J., Miocic, J., Hangx, S., Kallmeyer, J., Ostertag-Henning, C., Hassanpouryouzband, A., et al. (2021). Enabling large-scale hydrogen storage in porous media–The scientific challenges. *Energy & Environmental Science, 14*, 853–864. https://doi.org/10.1039/d0ee03536j (accessed on 24th October 2025).
22. Hisatomi, T., Kubota, J., & Domen, K. (2014). Recent advances in semiconductors for photocatalytic and photoelectrochemical water splitting. *Chemical Society Reviews, 43*, 7520–7535. https://doi.org/10.1039/c3cs60378d (accessed on 24th October 2025).
23. Jayakumar, K., Sakthilatha, D., & Venkatesan, R. (2025). Advancements in bio electrocatalysis for electrochemical ammonia production: Integrating enzyme-based catalysis. *Chemistry Select*. https://www.semanticscholar.org/paper/e9032f1608dcc7a1ad3d9a1255e898927887cebc (accessed on 24th October 2025).
24. Lewis, R., & Hutchings, G. J. (2023). Selective oxidation using *in situ*-generated hydrogen peroxide. *Accounts of Chemical Research, 57*, 106–119. https://www.semanticscholar.org/paper/e98ebd1ab434494ba59ae753a135c420b7685150 (accessed on 24th October 2025).
25. Li, J., & Li, S. (2025). Photocatalytic synthesis in green chemistry: Methods andapplications. *MATEC WebofConferences*. https://www.semanticscholar.org/paper/ee1717f5e7dadd12a119ed737d7d844e41f7f429 (accessed on 24th October 2025).
26. Li, T., Miras, H. N., & Song, Y. F. (2017). Polyoxometalate (POM)-layered double hydroxides (LDH) composite materials: Design and catalytic

applications. https://core.ac.uk/download/96883278.pdf (accessed on 24th October 2025).

27. Li, Y., Wen, G., Li, J., Li, Q., Zhang, H. X., Tao, B., & Zhang, J. (2022). Synthesis and shaping of metal–organic frameworks: A review. *Chemical Communications, 58*, 11488–11506. https://doi.org/10.1039/d2cc04190a (accessed on 24th October 2025).

28. Martin, J., Mitrovic, J., & Fischer, A. (2024). Scalable and tunable synthesis of sustainable nitrogen-doped hydrothermal carbon materials (N-HTC) as support material for platinum/N-HTC ORR catalysts. *ECS Meeting Abstracts*. https://www.semanticscholar.org/paper/80686088156 85f42331fe4b2335ebe7d66ca52e4 (accessed on 24th October 2025).

29. Martin, M. G., Clavero, L. R., Gomez, C. S., Aiassa, V., Lázaro-Martínez, J., Martín, S. E., Uberman, P., et al. (2025). Design of aminoanthraquinone-based heterogeneous photocatalysts for visible-light-driven reactions and antibacterial applications. *RSC Advances, 15*, 24393–24405. https://www.semanticscholar.org/paper/7d3c8826e503941dfec6c65776b7b953b1f eb451 (accessed on 24th October 2025).

30. Mattiello, S., Lucarelli, G., Calascibetta, A., Polastri, L., Ghiglietti, E., Podapangi, S., Brown, T., et al. (2022). Sustainable, efficient, and scalable preparation of pure and performing Spiro-OMeTAD for perovskite solar cells. *ACS Sustainable Chemistry & Engineering*. https://www.semanticscholar.org/paper/522775d808e7d53f7475f599c8abd0359a6104 ee (accessed on 24th October 2025).

31. Mendioroz, P., Casoni, A. I., Volpe, M. A., & Gerbino, D. C. (2024). Xanthone synthesis through catalysis: Exploring the green limits of homogeneous and heterogeneous methods. *European Journal of Organic Chemistry*. https://www.semanticscholar.org/paper/bbe977bee5cd64fcbf16b04ccbadfe2f08d844b8 (accessed on 24th October 2025).

32. Mondal, R., Ghanta, R., & Chattopadhyay, T. (2025). Applications of surface modified Fe_3O_4 based nanoparticles for the oxidation of various alcohols: A review. *Journal of Coordination Chemistry, 78*, 841–936. https://www.semanticscholar.org/paper/7a88e6ca686054a71ed59891158 88a3e2a0ccdbc (accessed on 24th October 2025).

33. Mondal, S., Dumur, F., Gigmès, D., Sibi, M. P., Bertrand, M. P., & Nechab, M. (2022). Enantioselective radical reactions using chiral catalysts. *Chemical Reviews, 122*, 5842–5976. https://doi.org/10.1021/acs.chemrev.1c00582 (accessed on 24th October 2025).

34. Mujtaba, M., Fraceto, L. F., Fazeli, M., Mukherjee, S., Savassa, S. M., de

Medeiros, G. A., Pereira, A. E. S., et al. (2023). Lignocellulosic biomass from agricultural waste to the circular economy: A review with focus on biofuels, bio composites and bioplastics. *Journal of Cleaner Production, 402*, 136815–136815. https://doi.org/10.1016/j.jclepro.2023.136815 (accessed on 24th October 2025).

35. Mukherjee, R. (2024). A revealing study on metals and metal salts impregnated silica: An efficient and versatile catalyst in green chemistry. *RASAYAN Journal of Chemistry*. https://www.semanticscholar.org/paper/ae0e440f7c12aa40527d7ab67696ac6b42508b0f (accessed on 24th October 2025).

36. Pan, X., & Kortemme, T. (2021). Recent advances in de novo protein design: Principles, methods, and applications. *Journal of Biological Chemistry, 296*, 100558. https://doi.org/10.1016/j.jbc.2021.100558 (accessed on 24th October 2025).

37. Parthasarathy, K., Jayakumar, J., & Sanjana, R. (2025). Metal-catalyzed C–H activation for the synthesis and applications of polysubstituted quaternary ammonium salts. *Asian Journal of Organic Chemistry*. https://www.semanticscholar.org/paper/f09efd6bca6fa5ebdcfa1a4ba717136ae3887ca5 (accessed on 24th October 2025).

38. Pawar, S. S., Chatterjee, R., & Dandela, R. (2024). Comparative study of catalysis strategies: Friedel-Crafts alkylation vs. Michael's addition of indoles to nitroalkenes. *Current Organic Chemistry*. https://www.semanticscholar.org/paper/698027b9c7e2e4e47d022bf9a5852497c5d2fa53 (accessed on 24th October 2025).

39. Pilli, R. A., & de Assis, F. F. (2018). Organic synthesis: New vistas in the Brazilian landscape. *Anais da Academia Brasileira de Ciências, 90*, 895–941. https://doi.org/10.1590/0001-3765201820170564 (accessed on 24th October 2025).

40. Sai Laya, C. H., Saroja, K. S. N., Aswini, V. L., Sameera, M., Jayasree, M., Rani, E. U., Kamala, G., et al. (2025). A review of industrial applications of green chemistry. *Journal of Pharma Insights and Research*. https://www.semanticscholar.org/paper/26f8b58af8118b4f6accb051efc5beaaddeeb245 (accessed on 24th October 2025).

41. Saint Jean, M., & Oltra, V. (2025). Incrementalism of environmental innovations versus paradigmatic change: A comparative study of the automotive and chemical industries. https://core.ac.uk/download/pdf/6297038.pdf (accessed on 24th October 2025).

42. Sakhahari, K. V. (2025). Green chemistry and catalysis: Innovations toward a cleaner and sustainable world. *International Journal of Pharmaceutical*

Research and Applications. https://www.semanticscholar.org/paper/4a3292276b1c806584dc1e557279ec5e81c0f3e8 (accessed on 24th October 2025).

43. Sambiagio, C., & Noël, T. (2019). Flow photochemistry: Shine some light on those tubes! *Trends in Chemistry, 2*, 92–106. https://doi.org/10.1016/j.trechm.2019.09.003 (accessed on 24th October 2025).

44. Sheldon, R. A., & Woodley, J. M. (2017). Role of biocatalysis in sustainable chemistry. *Chemical Reviews, 118*, 801–838. https://doi.org/10.1021/acs.chemrev.7b00203 (accessed on 24th October 2025).

45. Singh, R. (2025). Organometallic complexes in catalysis for sustainable chemistry: A review. *International Journal for Multidisciplinary Research*. https://www.semanticscholar.org/paper/0281edc46bddf1d70b2fd402909ee2c5c2556f18 (accessed on 24th October 2025).

46. Song, J., Wei, C., Huang, Z. F., Liu, C., Zeng, L., Wang, X., & Xu, Z. J. (2020). A review on fundamentals for designing oxygen evolution electrocatalysts. *Chemical Society Reviews, 49*, 2196–2214. https://doi.org/10.1039/c9cs00607a (accessed on 24th October 2025).

47. Swamy, S. (2025). Advances and applications of catalysis in industry: From nanocatalysis to environmental sustainability. *International Journal of Research Publication and Reviews*. https://www.semanticscholar.org/paper/f02ae92d6acfa888bc15b2aa11050573216cc33e (accessed on 24th October 2025).

48. Taqi, F. A. (2025). Impact of green catalysis on reducing industrial pollution. *International Journal of Chemistry*. https://www.semanticscholar.org/paper/f0e19ae71b7f6400e386fc2b82cb2eb630d85881 (accessed on 24th October 2025).

49. Vaz, C. R. S., Morais, C., Pastre, J., & Girotto Júnior, G. (2025). Teaching green chemistry in higher education: Contributions of a problem-based learning proposal for understanding the principles of green chemistry. *Sustainability*. https://www.semanticscholar.org/paper/dbed2d8ecd249a504a76ba03c24b0092e8574aa8 (accessed on 24th October 2025).

50. Wada, Y. (2018). On the occasion of publication of the special issue on "Now and future of technology of microwave chemical processes." *Journal of the Japan Petroleum Institute, 61*, 87. https://www.semanticscholar.org/paper/57775071842d482213b9af0b8592c75789141081 (accessed on 24th October 2025).

51. Wang, Y., Sun, J., & Tsubaki, N. (2023). Clever nanomaterials fabrication techniques encounter sustainable C1 catalysis. *Accounts of Chemical*

Research. https://www.semanticscholar.org/paper/030ebffca54b12a7e0cc c23838cfe25e0156cb2f (accessed on 24th October 2025).

52. Wattanakit, C., Fan, X., Mukti, R. R., & Yip, A. C. K. (2024). Green chemistry, catalysis, and waste valorization for a circular economy. *ChemPlusChem*. https://www.semanticscholar.org/paper/9a6322d7f1799 a3a25198b63042e91bce798311b (accessed on 24th October 2025).
53. Wiebe, A., Gieshoff, T., Möhle, S., Rodrigo, E., Zirbes, M., & Waldvogel, S. R. (2018). Electrifying organic synthesis. *Angewandte Chemie International Edition, 57*, 5594–5619. https://doi.org/10.1002/anie.201711060 (accessed on 24th October 2025).
54. Yan, H., Song, J., Zhu, S., & Xu, H. C. (2021). Synthesis of acridinium photocatalysts via site-selective C–H alkylation. *CCS Chemistry, 3*, 317–325. https://doi.org/10.31635/ccschem.021.202000743 (accessed on 24th October 2025).
55. Yi, J., Goh, N. J. J., & Li, Z. (2024). Green and enantioselective synthesis via cascade biotransformations: From simple racemic substrates to high-value chiral chemicals. *Chemistry–An Asian Journal*. https://www.semanticscholar.org/paper/74056f5bfb3af405dc755dd4cdba82921ee103 1d (accessed on 24th October 2025).
56. Zafar, F., Sultan, M. A., Nasir, A., Kiani, A. F., Ghouri, M. S., Ahmad, S., Yousaf, M., et al. (2025). Carbon nanotubes as multifunctional tools advancing batteries and catalysis for sustainable solutions. *Scholars Academic Journal of Biosciences*. https://www.semanticscholar.org/paper/fe4b7e2dd5707e59feb6301e689c794b26f083a8 (accessed on 24th October 2025).
57. Zhu, C., Yue, H., Nikolaienko, P., & Rueping, M. (2020). Merging electrolysis and nickel catalysis in redox neutral cross-coupling reactions: Experiment and computation for electrochemically induced C–P and C–Se bonds formation. *CCS Chemistry, 2*, 179–190. https://doi.org/10.31635/ccschem.020.201900112 (accessed on 24th October 2025).
58. Zhu, J., Hu, L., Zhao, P., Lee, L. Y. S., & Wong, K. Y. (2019). Recent advances in electrocatalytic hydrogen evolution using nanoparticles. *Chemical Reviews, 120*, 851–918. https://doi.org/10.1021/acs.chemrev.9b00248 (accessed on 24th October 2025).

CHAPTER 6

RENEWABLE FEEDSTOCKS AND MATERIALS

Contents

INTRODUCTION

Renewable materials and feedstocks are gaining prominence as the globe tackles the twin challenge of resource depletion and environmental degradation. Green chemistry practice is a model for formulating sustainable approaches that tackle resource efficiency and environmental stewardship. The utilization of non-renewable fossil fuels not only exacerbates energy crises but also contributes substantially to the production of toxic waste and air pollution.

Here, the use of renewable feedstocks—biologically derived materials—becomes imperative. By transitioning from conventional chemical processes to ones based on renewable feedstocks, industries can save the environment while simultaneously satisfying growing demand for green chemicals. It is not just desirable but imperative for a greener age of chemical production, a threshold in raw materials sourcing and utilization. The integration of renewable feedstocks with chemical processes is inextricably interlinked with catalysis and reaction methodology progress. Progress in biocatalysis, along with the development of improved, safer reaction conditions, are crucial in furthering green chemistry. For instance, scientists are employing various technologies, including supercritical fluids and ionic liquids, to minimize toxic waste and energy use without compromising product quality. With the employment of more benign solvents and other synthesis routes, the chemical industry can integrate more sustainable processes with the global sustainability objectives. Single-site heterogeneous catalysts are also being explored for their ability to facilitate efficient reactions with renewable substrates, thus practicing green chemistry principles on a day-to-day basis. Such innovations introduce a paradigm shift towards more ethically based chemical practice integral to future advancement. Besides, the OECDs 12 principles form the foundation of the green chemistry framework, guiding chemists to approach the development of environmentally friendly chemical processes. The principles demand reduced unwanted substances, improved atom economy, and maximum energy efficiency, among others. Emphasizing the use of renewable feedstocks, these guidelines facilitate ongoing innovation in chemical synthesis and foster an industry-wide commitment to sustainability. Besides, the research of microporous and mesoporous systems in catalyst synthesis has led to sophisticated catalytic performance in chemical reactions using renewable feedstocks, such as chemicals production from renewable feedstock-derived chemicals. With the trend towards a sustainable economy, these innovations in materials and methods will be central to furthering the goals of green chemistry and establishing a more environmentally friendly chemical industry (Babbar, 2024; Thomas, 2012).

Definition of Renewable Feedstocks

Renewable feedstocks is the term given to raw materials of biological origin that can be naturally replenished within a time frame, distinguishing them from limited but eco-damaging fossil fuels. Vital examples of renewable feedstocks are plant biomass, agricultural waste, and specific algae, all of which can be converted into chemicals, fuels, and materials without the danger of depletion. The transition to the utilization of renewable feedstocks is significant in the area of green chemistry because it makes production systems capable of supporting sustainable development and mitigating carbon emissions. By being harmonious with the philosophy of the application of renewable resources, the aforementioned feedstocks play a significant role in facilitating innovations that are based on biological substrates, thereby enabling industries to reduce their consumption of fossil-based inputs in preference for cleaner inputs towards environmental protection. By being harmonious with the goals of green chemistry, renewable feedstocks not only minimize ecological harm but also boost the circular economy. The use of biomass in production promotes design for resource and waste reduction, helping to displace traditional linear economic systems with take-make-use-dispose attitudes.

Figure 6.1. Renewable feedstocks refers to raw materials derived from biological sources that can be replenished naturally over time, distinguishing them from fossil fuels that are finite and environmentally detrimental.

Source: https://cdn.britannica.com/90/222690-050-41E9A975/installing-solar-panels.jpg.

Further, renewable feedstocks allow for several uses like pharmaceuticals, biofuels, and biodegradable products and thus promote innovation across industries. However, their sustainability extends beyond mere availability of biomass; they demand holistic evaluation by means of metrics such as process mass intensity and life cycle assessment. Their construction and optimization guarantee that what exists by way of environmental burden during the course of the products life cycle is quantified, offering a basis for sound comparability to fossil comparators. Therefore, renewable feedstocks are not just a fix for the resource shortage, but also a force for more widespread alteration in manufacturing systems. In order to gain an understanding of the implications of renewable feedstocks for green chemistry, it is important to consider the difficulties with their adoption. While promising, the transition to renewable feedstocks can entail surmounting technical and economic hurdles. Industries would have to make investments in research and development to optimize extraction and conversion processes such that efficiency matches traditional fossil fuels. Market acceptability and regulatory frameworks also have the ability to discourage full assimilation of such sustainable resources. The interdependency between policy support and technological innovation emphasizes the dynamic nature of renewable feedstock use. As green chemistry continues, it is evident that new strategies, such as those focused on hydrogen as a future clean energy carrier, are necessary in the pursuit of advancing the pursuit of sustainability in chemical production in order to demonstrate that renewable feedstocks are the solution to a decarbonized future (Staffell et al., 2018; Roger, 2017) (Figure 6.1).

6.1. THE ROLE OF RENEWABLE FEEDSTOCKS

The application of renewable feedstocks in green chemistry has significantly improved the sustainability of chemical processes. Renewable feedstocks, which are sourced from biological materials such as plants, agricultural waste products, and waste biomass, reduce the dependency on fossil fuels. This step is of greater importance in combatting global warming and decreasing environmental pollution. Use of renewable sources not only mitigates greenhouse gas emissions but also promotes applications of circular economy by leveraging materials that are regrowth or recyclable. Furthermore, advances in biotechnology and waste treatment have provided avenues for improved conversion of the biological materials into valuable chemicals and fuels. Thus, renewable feedstocks must have a pivotal role in establishing a more sustainable and resilient chemical industry that is compatible with planetary environmental goals. One striking example of the potential of renewable feedstocks is the development of bio-refining processes, which convert biomass into a range of bio-based products. These processes can produce not only biofuels but also a range of chemicals, polymers, and materials traditionally obtained from petroleum. The Roadmap Bioraffinage introduces possible routes of the bio-based economy, e.g., the establishment of commercial biorefineries and related pilot plants. In addition to contributing to regional economies, these operations create jobs and encourage research into sustainable technologies. Through renewable feedstocks and upgrading technologies, the chemical industry can become more efficient with regard to resources, hence addressing issues of waste management and resource availability. Therefore, the transition to a bio-based economy not only matters for environmental sustainability but also for economic viability, as indicated in initiatives like the Roadmap Bioraffinage. In addition, the employment of renewable feedstocks has the potential to drive innovation in green chemistry, and its result would be enhancing the properties of new materials. Researchers are now more looking for the use of lignocellulosic biomass, including wood and crop waste, as potential feedstocks for the manufacture of bio-based polymer and composite materials. Studying these materials aligns with a greater tendency to place emphasis on sustainability in the chemical industry. The diversity and importance of renewable feedstocks as raw materials for innumerable products affirm their relevance to contemporary chemical research. By being dedicated to renewable feedstocks, the industry not just meets current demand for sustainable materials, but plays an important role in a richer understanding of the lifecycle and possibilities of bio-derived products, ultimately to more sustainable approaches to working within and beyond the chemical industry.

6.1.1. Types of Renewable Feedstocks

It is required to be well-versed in renewable feedstocks to design green chemistry because it enables one to make sustainable processes with minimal environmental impacts. Renewable feedstocks like biomass, agricultural residues, and waste streams emerge as the players of prime significance in the transition. Biomass derived from organic materials such as plants and animals is a highly significant source of renewable feedstocks. It can be converted into energy or chemicals through gasification, which transforms solid biomass into syngas that can be utilized in different uses. The ongoing advancement of gasification technology underscores the necessity for optimization of feedstock types and conditions so that efficiency and sustainability of the process is enhanced, as noted by Anthony et al. (2016). In addition to biomass, another significant category of renewable feedstocks is agricultural residues, such as straw, husk, and sugarcane bagasse by-products. These residues are typically underutilized although they possess immense potential to be used in the form of biofuels, biochemicals, and other bioproducts. Application of agricultural residues in biorefineries not only aligns with circular economy principles but also reduces agricultural waste. This approach has been utilized in numerous strategic roadmaps that are meant to promote a bio-based economy, as in Annevelink et al. (2009). Using these materials effectively, it is possible to achieve enhanced resource efficiency alongside eliminating issues related to wastage of food and environmental sustainability. Municipal solid waste and industrial by-products are waste materials that form another significant class of renewable feedstocks. These feedstocks are converted into valuable chemicals or energy with novel recycling and conversion technologies. Waste valorization in green chemistry achieves environmental and economic aims by reducing waste going to landfills and providing room for circular use of resources. Further, effective waste management can allow the building of biorefineries based on diverse feedstock streams so that chemical conversions are integrated with sustainability targets. By embracing the full set of renewable feedstocks, industries and chemists can make a great leap in the principles of green chemistry and make the future greener (Figure 6.2).

Figure 6.2. Biomass.

Source: https://www.energy.gov/sites/default/files/styles/full_article_width/public/Biomass%20Resources.png?itok=7msmb1_6.

6.1.2. Comparison with Non-Renewable Feedstocks

Among the most important considerations in evaluating the relative advantages of renewable feedstocks over non-renewable feedstocks is sustainability. Non-renewable feedstocks, most notably fossil fuels, play a large role in causing environmental damage in the form of greenhouse gas emissions and other toxins. The production and consumption of these materials are frequently pursued at the cost of harsh ecological effects, such as habitat loss and water pollution. Contrarily, renewable biomass feedstocks offer a cleaner alternative by recycling carbon through biological systems rather than letting it go into the environment. By utilizing renewable resources, green chemistry aligns with reducing environmental degradation and encouraging resource conservation. Such a transition is required, as indicated by the need to limit environmental catastrophes due to reliance on fossil fuel, as highlighted by green chemistry practices that favor less toxic chemical reactions (Babbar, 2024). Secondly, the economic effect of renewable over non-renewable feedstocks cannot be overstated. Non-renewable feedstocks are prone to price volatility in large measure as a result of market and geopolitical uncertainty. This volatility poses threats to those sectors that rely on stable supply bases, having a tendency to cause increased production costs which can ultimately be transferred to consumers. Alternatively, renewable feedstocks are usually more price-stable as they rely on agricultural cycles and can be locally produced, thus ending the dependence on foreign sources. For instance, sustainable platform chemicals such as levoglucosenone (LGO) possess multifaceted applications in creating a variety of desired products, expanding manufacturers' space for innovation. This not only enhances the economic viability but also forms the resilience of the supply chain, driving the shift towards a more sustainable economy (Caio et al., 2024). Lastly, the technology improvements of renewable feedstocks have a major advantage over their non-renewable counterparts. Improvements in green chemistry, including biocatalysis and the use of alternative solvents, allow for the effective conversion of biomass into high-value products with less energy used and little waste generated. Contrary to non-renewable processes that usually result in hazardous waste and are subject to high energy inputs, renewable feedstock technologies focus on reducing environmental impact while ensuring maximum efficiency. The ability of chemicals such as Cyrene™ to be sourced from LGO is testament to the possibility of renewable feedstocks as non-toxic and biodegradable substitutes in compliance with the principles of green chemistry. This transition not only supports product sustainability but also maintains the objective of maintaining ecological balance, underlining the advantages of renewable feedstocks in contemporary chemical practices (Caio et al., 2024; Babbar, 2024).

6.1.3. Benefits of Using Renewable Feedstocks

The transition to renewable feedstocks for the manufacture of chemicals has significant environmental benefits, above all the lowering of greenhouse gases. Compared to conventional feedstocks that are fossil-based, renewable feedstocks such as biomass and crop residues may be replenished naturally and tend to have lower emissions throughout their complete lifecycle. With these cleaner materials, industries can balance their carbon footprint and assist in moderating climate change. This is consistent with the general goals of green chemistry, which value the importance of designing processes not just efficiently but also sustainably. The utilization of renewable feedstocks is thus crucial in minimizing the impact of chemical manufacturing on the environment, further emphasizing the shift of industrial practice towards sustainability (Ch et al., 2025). Another important advantage of renewable feedstocks is their possible economic resilience and innovation in the green chemistry strategy. By diversifying raw materials sources, industries can reduce dependency on unpredictable fossil energy markets, thereby increasing economic security. Cultivation and consumption of renewable feedstocks can also create job expansion in sectors such as agriculture, forestry, and waste treatment. In addition, product formulation and use innovation is promoted by innovation in the development and commercialization of new biobased materials.

Figure 6.3. The cultivation and use of renewable feedstocks can also stimulate job creation in sectors such as agriculture, forestry, and waste management.

Source: https://martech.com.vn/pictures/catalog/products/biomass/Biomass-la-gi.jpg.

For instance, consumer goods and packaging markets are being revolutionized by the use of bioplastics produced from renewable raw materials, and these replacement products are environmentally less degrading. This not only enhances brand image for companies engaged in sustainable practices but also attracts environmentally friendly consumers, gaining a market competitive edge (Ch et al., 2025). Finally, the utilization of renewable feedstocks in chemical processes completes green chemistry concepts through optimizing overall efficiency and product quality. For instance, reactions that utilize renewable resources are found to have improved atom economy and higher yield, reducing waste and resource consumption. This is particularly critical in fields like the pharmaceutical sector, where efficient synthesis can lead to lower cost and less toxic waste production. Moreover, renewable feedstocks can lead to the development of safer chemical alternatives, thereby enhancing product safety and health protection for consumers. With the concept of sustainability gaining prominence all across the globe, utilization of renewable feedstocks is not an environmental imperative but a strategic industrial advantage that aligns with contemporary social values (Deepika et al., 2024) (Figure 6.3).

6.2. SOURCES OF RENEWABLE FEEDSTOCKS

The creation of renewable feedstocks is paramount to the future of green chemistry because it promotes the utilization of sustainable production methods while reducing the reliance on fossil fuels. Of the numerous possibilities for renewable feedstocks, carbohydrate biomass is one that encompasses diverse varieties of agricultural residues and energy crops. These can be converted to valuable intermediates such as hydroxymethylfurfural (HMF), a chemical that has a vital role to play in the manufacture of biofuels and platform chemicals. The catalytic conversion system allows efficient conversion of hexose substrates to HMF, thus paving the way to generate biofuels that could replace conventional diesel and kerosene. According to recent studies, new catalytic strategies are required to achieve maximum conversions for such a direction to ensure high-yield renewable fuel production (Amarasekara et al., 2010). Another significant source of renewable feedstocks is biomass gasification, which is a thermochemical conversion process that uses organic materials to produce syngas, or a combination of hydrogen and carbon monoxide. This can be converted into various chemicals, including fuels and industrial chemicals. Technological advancements in gasification emphasize the necessity of optimizing efficiency and innovative solutions for decreasing costs in order to make biomass gasification viable. It has been proven in research that production yields and quality of produced gases can be enhanced by optimizing feedstock types and by modifying operation parameters. The process enhances the financial worth of raw biomass, besides enhancing waste minimization and environmental sustainability, through the use of by-products that would otherwise be wasted (Anthony et al., 2016). The integration of different renewable feedstocks in the creation of green materials plays a pivotal role in achieving sustainability in chemical processes. Processes that promote the utilization of forestry and agricultural residues, along with energy crops, help towards the development of a circular economy. The creation of bioplastics, biofuels, and other green materials from renewable feedstocks not only minimizes the emission of greenhouse gases but also reduces dependence on petroleum-based products. Additionally, the integration of conversion technologies, including enzymatic hydrolysis and thermochemical conversions, depicts the variety of renewable feedstocks in generating a variety of products. Research and development of such technologies continue to play an imperative role in achieving the full potential of renewable feedstocks, towards a brighter future in green chemistry.

6.2.1. Agricultural Products

The significance of agricultural commodities as renewable feedstocks in green chemistry cannot be overstressed. Agricultural by-products, in general, are underutilized and represent a vast resource that can be employed to enhance environmental sustainability and economic feasibility. For instance, raw materials such as crop residues, fruit peels, and biomass are not only abundant but inexpensive. These materials can be transformed into valuable commodities, reducing the utilization of non-renewable material while, at the same time, solving the problem of waste disposal. By integrating farm products into chemical processes, industries are able to downsize their impact on the environment, facts above principles that guide green chemistry. This is done by shifting paradigms in material sourcing, whereby waste becomes a valuable input in the production of new and greener solutions to environmental problems to enhance agriculture's symbiotic existence with sustainable industrial processes. The application of agricultural products extends beyond waste valorization for its utility purpose; it also encompasses the development of efficient adsorbents for environmental remediation. From the majority of studies, it has been observed that such agricultural organic waste materials can be used as efficient adsorbents for pollutants in aqueous solutions, such as Congo Red dye. This is due to their high surface area and the presence of many functional groups, which increase their efficiency for adsorption. With the utilization of these renewable materials in adsorption, industries will be able to reduce the use of synthetic adsorbents that may cause negative effects on the environment. The dual benefit of utilizing agricultural products as adsorbents can be noted in their water purification quality improvement and contribution towards the circular economy, thus projecting the significance of green chemistry. As green chemistry emerges, interest in renewable feedstocks, particularly from agriculture, increases vigor across a variety of industries, certainly the pharmaceutical industry. The shift away from fossil fuel-derived materials is part of a broader trend to sustainability in society. Research into sustainable synthesis pathways is proving that agricultural chemicals can be employed as drug and therapeutic building blocks with low environmental impact typically associated with pharmaceutical production. This emphasis on renewable resources not only works towards enhancing the safety and efficacy of drugs but also serves the greater environmental agenda by accelerating innovation in green chemistry.

Through facilitating the move from petrochemical to bio-based feedstocks, the pharmaceutical industry can lower greenhouse gas emissions significantly, thus contributing to global efforts towards satisfying climate change and assuring sustainability (Purushottam et al., 2024) (Figure 6.4).

Figure 6.4. Agricultural waste, often underutilized, represents a vast resource capable of fostering environmental sustainability and economic viability.

Source: https://shapiroe.com/wp-content/uploads/2023/08/uses-of-agricultural-waste-1.png.

6.2.2. Waste Biomass

The utilization of waste biomass is an important milestone in the quest for sustainable energy resources in the green chemistry framework. Waste biomass includes organic material such as agricultural residues, food wastes, and wood products, which are typically disposed of in landfills and lead to environmental deterioration. The recycling of these materials into useful applications is capable of reducing greenhouse gas emissions and mitigating waste disposal problems. Additionally, this method is congruent with the principles of circular economies, where waste is not considered refuse but a resource. Adopting integral biorefine strategies may increase the recovery of compounds of value, such that waste biomass is converted to renewable biofuels, chemicals, and materials feedstocks (Annevelink et al., 2009). This shift in paradigm not only inspires innovation but enhances local economies through employment generation and sustainability development (Annevelink et al., 2009). It is feasible to operate the process of transforming waste biomass to valuable products utilizing gasification, a thermochemical process that has been lately increasing in popularity. Under the condition of oxygen deficiency, biomass is subjected to high temperatures and syngas is produced, which may be transformed into fuel or chemicals. This is a process with several benefits, including high efficiency and feasibility for application using diverse feedstocks, and presents an interesting prospect for biomass waste conversion (Anthony et al., 2016). Moreover, new gasification technologies have experienced significant advancements, and this places a need for ongoing research and development in the field. With proper control, gasification not only minimizes waste biomass disposal issues but also ensures energy security by providing a clean alternative to fossil fuels. By integrating such technologies in integrated waste management systems, sustainability advantage increases.

Figure 6.5. Waste biomass includes organic materials such as agricultural residues, food waste, and forestry by-products, which are often disposed of in landfills, contributing to environmental degradation.

Source: https://www.homebiogas.com/wp-content/uploads/2023/09/shutterstock_2057386766.jpg.

Besides energy production, conversion of waste biomass into biodegradable materials presents new possibilities for sustainable material science. These substances can replace conventional plastics, which are major environmental risks owing to their long-lasting occurrence in ecological systems. Waste biomass bioplastics offer an opportunity to minimize reliance on petrochemical feeds while meeting consumer demands for environmentally friendly products. Expanded economic gain from these bio-based products can be obtained by investing in research focused on optimizing the extraction and conversion processes. Process innovations in biorefineries can also enhance further the efficiency of waste biomass utilization and serve as the basis for future green economies (Annevelink et al., 2009). The transition, thus, to renewable feedstocks from waste biomass is not only a technology transition but a critical step toward a greener and more responsible management of materials in green chemistry (Figure 6.5).

6.2.3. Algal Biomass

Algal biomass is a very promising contender in the search for green chemistry's renewable materials and feedstocks due primarily to its ability to be employed to generate biofuels as well as other useful bioproducts. Unlike traditional energy sources, algal biomass growth does not require arable land and therefore does not compete with food crops. This makes it a sustainable option for alleviating pressure on food supply chains and increasing energy diversity. Research has indicated that algal biodiesel produced from the lipids of various microalgae is a promising green fuel because of its better emissions profile and energy density compared to fossil fuels. Involving investments in research and development of algal biodiesel technology can result in transition toward a more sustainable energy economy as suggested by Kowalski et al. (2010), based on the argument that it's innovation that makes it commercially viable. The isolation and breeding of lipid-rich algal strains are central to maximizing the yield efficiency of algal biomass for biofuel purposes. Such strains must be tolerant to stress and adapted to grow under extreme conditions, particularly in regions with climatic conditions that may be demanding for growing. New advances in techniques like fluorescence activated cell sorting (FACS) have enabled researchers to sort autochthonous microalgae well established and competitive in the natural habitat.

For example, studies have discovered various strains from extreme habitats, e.g., those isolated in the Red Sea, and determined that these microorganisms possess lipid contents between 20% and 25% of their biomass dry weight, thereby establishing their potential for use in biodiesel production. These targeted screening protocols are critical in the guarantee of the chosen strains furnishing the necessary fatty acid compositions for productive biofuel production, as noted in Mouffouk et al. (2013). Commercialization of algal biomass as a renewable product is not free of challenges, but it is also a significant economic development and environmental security opportunity. Construction of robust intellectual property frameworks, through strategic patenting of algal biodiesel technology in particular, can enhance innovation and investment in this nascent industry. With continued advancement of research on algal biomass beyond the conceptual stages to practical technologies, public education and support for algal biofuels' potential must be enhanced. Subsidy and policy by the government would play a critical role in initiating further research on algal biomass applications, thereby accelerating its viability in the world energy market. Ultimately, the establishment of this industry has the capability to facilitate a circular economy where waste material is used as a valuable resource, which is basically the primary agenda of green chemistry (Figure 6.6).

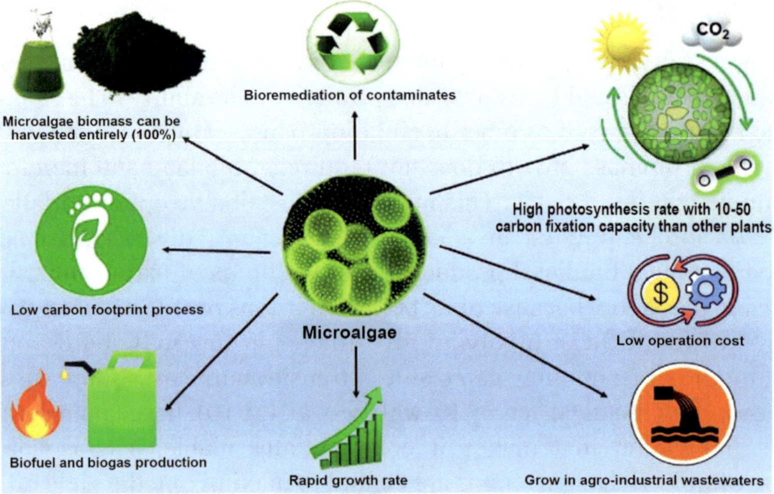

Figure 6.6. Algal biomass emerges as a compelling candidate in the pursuit of renewable feedstocks and materials within green chemistry, primarily due to its potential to produce biofuels and other valuable bioproducts.

Source: https://media.springernature.com/lw1200/springer-static/image/art%3A10.1007%2Fs10311-022-01458-1/MediaObjects/10311_2022_1458_Fig7_HTML.png.

6.3. CHEMICAL PROCESSES IN GREEN CHEMISTRY

Application of the principles of green chemistry is crucial in transforming traditional processes into more sustainable ones by emphasizing renewable feedstocks. The transformation to using biomass resources can significantly reduce our reliance on fossil fuels and make chemical production more sustainable. With the implementation of approaches that bring together the circular economy, green chemistry drives resource utilization efficiency and waste minimization through the design strategies that render the product recyclable or reusable. The 2030 Agenda for Sustainable Development captures the significance of such concepts by asserting that chemistry can facilitate a transition to more sustainable systems by the ingenuity of product design for minimizing environmental footprints with efficient resource conservation. This interdependence underscores the imperative need for industrial process and policy reformations that promote the growth of green chemistry and the utilization of renewable feedstocks as the norm (Ncube et al., 2023). In reviewing the chemical reactions in green chemistry, the new modes of synthesis of the important monomers from biomass need to be addressed. For instance, production of traditionally petrochemical-derived chemicals, such as acrylic acid and maleic anhydride, can be obtained from biomass-derived furfural. Utilizing green chemistry methods, such as green photooxygenation and aerobic oxidation, ensures the synthesis process to be atom-efficient and low in waste. This approach not only addresses the pressing need for sustainable materials but also illustrates the promise of biobased synthesis to provide competitive substitutes for conventional petrochemical routes. Above all, these processes highlight the strategic role of green chemistry in providing the means to create sustainable industrial pathways that are harmonious with ecological and economic goals, which once again identifies renewable feedstocks as the key to avoiding pollution (Johannes et al., 2021). Moreover, the increased footprint of including green chemistry processes in industrial applications cannot be overstated. The synergistic effect of academia, industry, and government on one another can lead to significant improvements in chemical processes with the upgrading of sustainable practices on a broad scale. By addressing critical issues, such as regulatory frameworks, investment needs, and environmental education, stakeholders can guarantee the dimensions of a circular economy while spurring the application of renewable materials. Interplay among these forces can drive innovations that are not only sustainable economically but also environmentally friendly. It is thus the mandate of green chemistry, founded on chemical operations that guarantee sustainability, to lead the way in designing a greener and more resilient future while supporting global sustainability strategies and opening doors for new materials and technologies (Ncube et al., 2023).

6.3.1. Biocatalysis

The integration of biocatalysis with renewable materials and feedstock technology is a giant step towards green chemistry. Biocatalysis uses nature's catalysts, primarily enzymes, to catalyze chemical reactions under generally mild conditions and minimize the energy used. Traditional synthesis methods in most industries—fragrance and pharmaceutical among them—are usually toughest by chemicals and secondary environmental degradation. By contrast, biocatalytic processes work with renewable feedstocks, leading to enhanced atom economy and lower by-product formation. This change not only enhances the sustainability of industrial processes but also aligns with the global goal of waste and pollution reduction, thus demonstrating an eco-stewardship drive by the manufacturing sectors. With the implementation of biocatalysis, industries can make greener products without compromising the principles of green chemistry, which are centered on safety, efficiency, and reduced ecological footprint. Recent years have witnessed major strides in biocatalysis in the synthesis of fragrance products and medicine drugs, thus establishing its flexibility and efficacy.

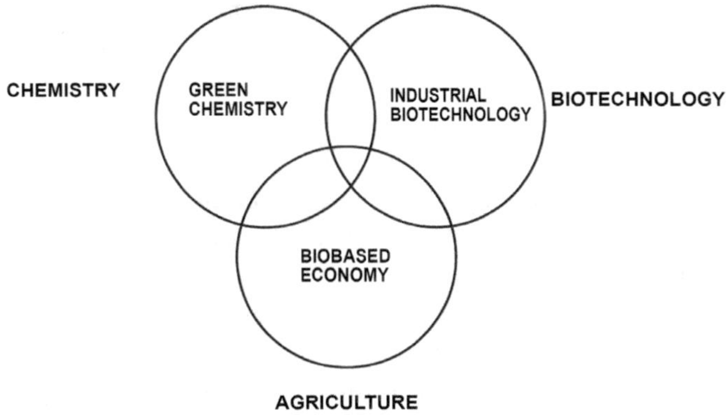

Figure 6.7. Biocatalysis utilizes natural catalysts, primarily enzymes, to facilitate chemical reactions under mild conditions, thereby minimizing the energy input required.

Source: https://www.mdpi.com/catalysts/catalysts-07-00009/article_deploy/html/images/catalysts-07-00009-g001.png.

Through the action of enzymes, the reactions are more specific and selective, which significantly reduces the formation of unwanted by-products. This holds particular relevance in the field of perfumes, where natural and environmentally friendly products are increasingly sought by consumers. As stated, traditional processes have a tendency to lead to huge environmental pollution since harsh conditions are applied during the process of synthesis. By employing biocatalysis, companies can switch to cleaner processes that utilize renewable raw materials and yield high-quality products. Moreover, biocatalysis also holds potential for novel applications in pharmaceutical synthesis, thus validating the role of enzymes towards developing cost-effective and sustainable processes for industries with a bearing on public health (Ojukwu et al., 2025). With growing momentum in the transition towards biocatalysis, policy and regulatory frameworks are obviously critical to further this groundbreaking process. The pharmaceutical industry, for example, stands to reap a great deal by embracing green chemistry principles, as epitomized by the emergent importance of sustainable manufacturing practices. Regulators can enhance this transition by promoting standards that ensure safety while offering an environment conducive to innovation. Encouragement of partnerships between academic institutions, industrial participants, and government agencies will be essential in maximizing biocatalytic processes and integrating them with sustainable feedstocks effectively. Overall, biocatalysis's utilization does not merely promise radical reduction of the environmental impact but also economic viability and innovation in the pharma and fragrance sectors. Here, the advancements in biocatalysis contribute to a more environmentally friendly future, eventually coupling business objectives with a sensitivity for the environment in an effort to benefit society as a whole (Yang, 2025) (Figure 6.7).

6.3.2. Solvent-Free Reactions

In the context of green chemistry, reactions carried out without solvents are a milestone towards minimizing environmental impact and resource consumption. Solvent-free reactions eliminate traditional solvents, which typically contribute the lion's share of hazardous waste as well as non-renewable resource consumption. By excluding solvents, atom economy in chemical reactions is dramatically improved with an optimal yield of target products while waste formation is minimized to the barest minimum. Green chemistry principles stress such methods, calling for reactions that weigh in favor of ecological sustainability as well as synthetic efficiency. This trend toward solvent-free methods falls into line with an increasing dedication by chemists to the creation of greener options that are both successful and environmentally friendly. As it has been noted in recent book reviews, solvent-free reaction developments have the potential to radically influence the overall environmental footprint of chemical production, further emphasizing the need for continuing research in the area (Arshiya et al., 2025).

The application of solvent-free reactions underscores the role of renewable feedstocks and substances to revolutionize green chemistry. As increasingly more industries are being compelled to adopt sustainable processes, these methods incorporate biomass and other renewable reagents as primary reagents, reducing the need for petroleum chemicals. Employing natural material promotes the kind of circular economy in which waste is avoided and materials are recycled systematically. For example, lignin and cellulose, which are renewable building blocks, can be used as the starting materials for large-scale chemical conversions to produce solvent-free processes that align with the greater sustainability agenda. This collaboration not only renders chemical processes economically feasible but also plays a crucial role in countering global environmental challenges. Sufficient strategies for their application in solvent-free reactions highlight the new avenues required for sustainable chemistry of the future (Tiwari et al., 2021). Further, other catalytic innovations, which belong to this paradigm, enable the development of solvent-free reactions and make the reactions more efficient while being consistent with the principles of green chemistry. Methods like mechanochemistry and microwave synthesis are examples of how outside energies can be used to facilitate chemical reactions efficiently without solvent use. These technologies not only simplify processes but also minimize energy use by significant degrees, which is a major issue in assessing the sustainability of chemical processes. In addition, the identification of alternative reaction conditions also illustrates the versatility of solvent-free methodology, allowing chemists to manipulate reaction pathways formerly reliant on solvents for efficiency. Lastly, the development and fine-tuning of solvent-free reactions represent a paradigm shift in chemical practice, wherein sustainability is now built into methodology, paving the way towards a greener, cleaner future for chemical research and industry (Arshiya et al., 2025).

6.3.3. Energy-Efficient Processes

Implementation of energy-saving processes is essential towards promoting sustainability in the case of green chemistry, particularly when handling renewable feedstocks and materials. Such processes revolve around effectiveness in energy utilization while optimizing output efficiency to minimize the overall carbon footprint of chemical synthesis. Conventional processes require high temperature and pressure, resulting in huge energy demands. On the contrary, green chemistry encourages new approaches, such as microwave-driven synthesis and continuous flow processing, that can considerably reduce energy consumption at the same or even greater rate of reaction. This change not only aligns chemical manufacture with sustainable development goals but also supports the broader industry trend toward environmentally sustainable production approaches. As realized, the pursuit of chemical process energy efficiency is of particular significance in developing a sustainable future that encompasses economic profitability as well as environmental preservation. Moreover, the deployment of renewable feedstocks is another integral component of energy-efficient processes in green chemistry. The utilization of biomass and other responsibly sourced materials decreases reliance on fossil fuels and therefore addresses some of the sector's most critical environmental challenges today. Utilizing these renewable inputs, the processes are less energy intensive and produce fewer greenhouse gas emissions. No less important is the role of catalysis in energy efficiency; using biocatalysts or solid catalysts not only makes reactions easier but also lowers energy input by a huge amount as well as waste generation. Through these catalytic reactions, like using renewable feedstocks, green chemistry finds the door open to more sustainable pathways in organic synthesis. The complementarity between renewable resources and energy-saving processes amplifies the green chemistry potential for revolutionizing the chemical industry.

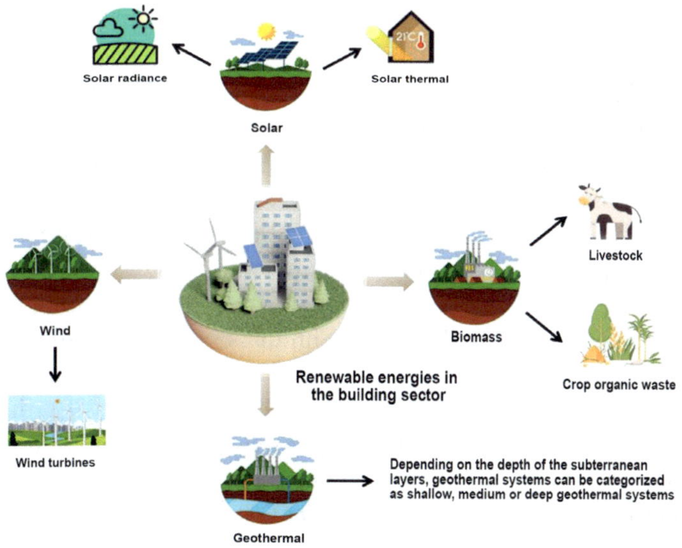

Figure 6.8. The utilization of biomass and other sustainably sourced materials minimizes reliance on fossil fuels, thereby addressing some of the most pressing environmental issues faced by the chemical industry today.

Source: https://media.springernature.com/lw1200/springer-static/image/art%3A10.1007%2Fs10311-023-01675-2/MediaObjects/10311_2023_1675_Fig1_HTML.png.

The issues of transitioning to energy-saving processes are multifaceted and require synchronized input from academia, industry, and the regulatory environment. While the benefits are clear—reduced use of resources, reduced environmental impact, and improved economic competitiveness—the ground realization remains hampered by technology as well as costs. For example, the first-time costs of implementing new technologies might be too high and keep most companies from implementing them, especially those in energy-demanding industries accustomed to conventional processes. Furthermore, magnifying new energy-efficient processes at large scale has logistic issues to address in order to achieve large-scale implementation. Nevertheless, as the world economy becomes increasingly centered on sustainable development, it will be necessary to design and streamline energy-efficient, environmentally friendly chemical processes. With the implementation of these technologies, the industry not only becomes in line with environmental regulations but also leads the way toward a greener future for chemical manufacturing, as attested to in recent studies with renewable feedstocks and materials in green chemistry (Kumar et al., 2024; Dixit et al., 2024) (Figure 6.8).

6.4. APPLICATIONS OF RENEWABLE MATERIALS

The combination of renewable materials in green chemistry processes is critical in the development of sustainable technologies that carry low environmental consequences. One of the widespread applications of renewable resources can be seen in the synthesis of ligands in coordination chemistry. With the utilization of biomass compounds as precursor chemicals for the synthesis of ligands, researchers have been able to synthesize new chemical compounds while adhering to the sustainability dogma. Techniques such as organic synthesis and coordination chemistry have been employed to build novel ligand precursors that not only exhibit the desired chemical diversity but also reduce reliance on fossil resources. The technique emphasizes the potential for renewable materials to maximize the efficiency of chemical conversions while promoting environmentally friendly approaches (Chauhan et al., 2024). As these new ligands have favorable properties in various coordination reactions, they are strong demonstrations of the groundbreaking potential of renewable materials in green chemistry. In addition to the stimulation of ligand synthesis, renewable materials are significant for the formation of platform chemicals that can be the building blocks of a wide variety of applications. Levoglucosenone (LGO), derived from pyrolysis of cellulose and hemicellulose, is a representative platform chemical with unparalleled structural diversity. Chemical conversion of LGO allows researchers to produce valuable derivatives such as levulinic acid, furan derivatives, and intermediates essential for agrochemicals and pharmaceuticals. This aspect not only reflects the economic viability of using renewable feedstocks but also enhances the sustainability of processes for chemical production (Caio et al., 2024). As the industries move to become greener, their quest for renewable material like LGO opens up a path towards less reliance on fossil fuels as they meet the growing needs of eco-friendly products in various industries. Beyond that, the incorporation of renewable materials makes its way into the development of biodegradable solvents in line with the aims of green chemistry. Cyrene™, which results from hydrogenation of LGO, is a revolutionary dipolar aprotic solvent that happens to be biodegradable and non-toxic, with less potential to cause harm to the environment than traditional solvents. Its use in chemical processing not only minimizes toxic waste but also results in a cleaner solvent system. As the LGO cascade transformation to Cyrene™ shows on a continuous-flow scale, effective synthetic pathways can be constructed that embrace both efficiency and sustainability (Caio et al., 2024). This new trend of using renewable materials in solvent manufacture reflects the growing value placed on their role in energy saving, greenhouse gas emission reduction, and chemical hazards associated with the use of solvents in certain industry (Figure 6.9).

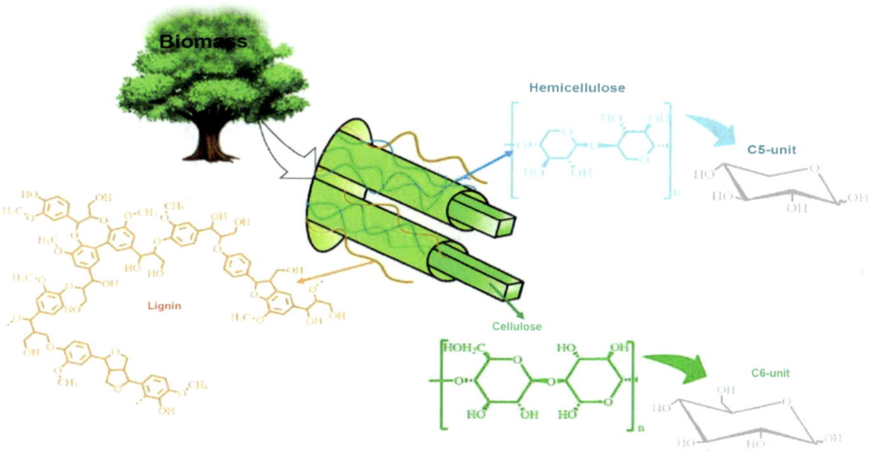

Figure 6.9. Levoglucosenone (LGO), derived from the pyrolysis of cellulose and hemicellulose, exemplifies a platform chemical with remarkable structural versatility.

Source: https://www.mdpi.com/catalysts/catalysts-12-01524/article_deploy/html/images/catalysts-12-01524-g002.png.

6.4.1. Bioplastics

The new field of bioplastics is the best illustration of an outstanding advance in the search for sustainable materials on the green chemistry stage. Being produced from renewable resources such as plant-based materials, bioplastics provide a replacement for conventional petroleum-based plastics, which have been reported to be environmentally persistent and polluting. This transition to bioplastics is not only technological; it is a broader social awareness of environmental degradation and the necessity of sustainable material science practices. While industries and governments increasingly recognize plastic waste's damage, bioplastics are a compelling solution capable of lessening ecological load without sacrificing functional requirements of various applications, from packaging to auto components. The ambition to share bioplastics as a central element of the circular economy is also supported by enabling regulatory systems and funding initiatives geared towards stimulating innovative biotechnological strategies (Cukalovic et al., 2009). The versatility of bioplastics allows them to be classified into various types, including biodegradable and non-biodegradable polymers, each tailored to distinct applications and final life management. Biodegradable materials such as polylactic acid (PLA) and polyhydroxyalkanoates (PHA) are also gaining fame because of the higher rate at which they can biodegrade than traditional plastics, adhering to the sustainable principles and reducing waste. On the other hand, non-biodegradable bioplastics, although still sourced from renewable feedstocks, are made to last longer periods, thus ideal for applications where durability and strength are required. This dual categorization underscores the flexibility of bioplastics in fulfilling a wide variety of industrial and consumer needs, a changing reality where bioplastics are likely to replace a significant portion of commonly used plastics. As studies continue, focus remains on creating attributes of bioplastics, simplifying production processes, and reducing the environmental impact across the entire life cycle of the product (Cukalovic et al., 2009). Emerging from joint ventures and research programs, such as those funded by the European Community, the bioplastics sector is rapidly evolving to cover new end uses and materials. These programs are inclined to concentrate on making bio-based polymers, composites, and additives that transfer a mix of eco-friendliness and functionality. For instance, projects like Biostruct, Forbioplast, and Woody encourage cooperation between university institutions and industry leaders in searching for alternative ways of employing wood-based materials towards bioplastic production. Knowledge sharing through the transfer of outcomes from conferences, like the International Conference on Bio-based Polymers and Composites (BiPoCo), helps foster research in the field further. By exchange of ideas among experts, the industry is able to tap into sustainable processes and potential applications across industries from packaging to biomaterials in biomedical uses. This continuous feedback of research, development, and exchange plays a critical role in establishing bioplastics as a cornerstone in green chemistry and in developing a greener tomorrow (Imre et al., 2013).

6.4.2. Biofuels

The quest for biofuels as a substitute renewable energy source has gained much attention on the basis of reducing the dependence on fossil fuels and reducing greenhouse gases. Biofuels, which are derived from biological feedstocks, make use of renewable feedstocks such as agricultural residues, energy crops, and waste streams. These renewable resources not only minimize the environmental impact of traditional fuels but also align with circular economy principles by reusing waste as useful energy. Biodiesel production from *Jatropha curcas* oil using waste eggshell as a catalyst is just one of the innovative approaches in green chemistry, with a viable means of converting non-food feedstocks into high-yielding biodiesel that is eco-friendly through microwave heating approaches (Amesho et al., 2022). Through this process, it showcases the value of integrating renewable feedstocks in the production of biofuels in delivering significant benefits to energy security and environmental well-being. The growth of technology in the synthesis of biofuel is fueled by the discovery of new catalysts that enhance the efficiency of biofuel synthesis. The advances made have focused on one-dimensional titanate nanotube materials, such as protonated titanate nanotubes and sodium titanate nanotubes, as new low-cost catalysts.

Figure 6.10. Biofuels, derived from biological materials, harness renewable feedstocks such as agricultural residues, dedicated energy crops, and waste materials.

Source: https://bioenergytime.com/wp-content/uploads/2024/12/Photo-1.jpg.

The materials exhibit better catalytic activity compared to zeolites employed widely and other solid catalysts, thereby showcasing their applicability in green chemistry studies (Adeleye et al., 2021). The application of such new catalysts within biofuel production processes can potentially lead to greener practices through lowering the demand for hazardous chemicals and more efficient biomass conversion in general. Through the use of such green catalysts, the biofuel industry can further support the United Nations Sustainable Development Goals, especially for accessible and sustainable energy solutions. The inclusion of biofuels in the energy mix is a multi-faceted strategy towards meeting climate change, energy security, and environmental degradation. The future of biofuels is dependent on continuous breakthroughs in feedstock-to-fuel conversion technologies, including the innovation of new catalysts and processes to optimize yields. For example, using waste materials not only reduces raw material expense but also promotes sound waste management methods—transforming unwanted biomass to valuable energy resources. Identification of biofuel innovations as central to the larger context of green chemistry reinforces the importance of interdisciplinary solutions to designing sustainable energy alternatives. Lastly, the shift to biofuels is a reflection of a willingness to make low-carbon futures feasible and highlights the important role that research has in advancing innovations in renewable feedstocks and material compatible with environmental sustainability goals (Figure 6.10).

6.4.3. Green Surfactants

Under the field of green chemistry, the development of green surfactants is also one step towards green processes in industries. Unlike traditional surfactants that tend to be made from petrochemical sources and pose serious environmental hazards, green surfactants are typically produced from renewable raw materials. This shift, besides minimizing environmental damage that accompanies artificial surfactants, leads to more overarching goals of sustainability. Because green surfactants are synthesizable from renewable sources such as plant oils and fats, they encompass improved biodegradability, non-toxicity, and reduced volatility factors, which cumulatively reduce harmful environmental impacts. The utilization of renewable resources in the manufacturing of such surfactants thus promotes the development of a circular economy in which waste is minimized and resources used with high efficiency so that ecological balance is improved in industrial application. In addition, their functional attributes have seen them applied in varied industries including agriculture, personal care, and pharmaceuticals. Their capacity to lower surface and interfacial tension makes it possible for them to be utilized in procedures like enhanced oil recovery, emulsification, and formulation stabilization. Biobased surfactants, for instance, those derived from renewable sources, exhibit comparable performance with synthetic surfactants yet they are less toxic and environmentally harmful. Such research effort keeps improving the properties of such surfactants so they can be tailored with characteristics suited for specific industrial applications. Such natural versatility fits perfectly with the principles of green chemistry, enabling them to become more accepted in industries that advocate for sustainability and environmental stewardship. Green surfactant development is also pushed further by advancements in biotechnological processes that include enzyme engineering and biocatalysis. These technologies enable the synthesis of surfactants through enzymatic pathways and gain control over the conditions of the reaction as well as the properties of the product. Biocatalytic routes streamline surfactant synthesis, without unnecessary energy usage and generation of waste material, thereby guaranteeing optimum process efficiency. Moreover, immobilization processes in biocatalysis enhance the stability and recyclability of enzymes, thereby significantly lowering costs of production while allowing sustainable operations. Growing convergence between green chemistry and biotechnology is a reflection of the potential of new technologies that satisfy market demands while upholding environmental demands. As industries become more attentive to green practices than ever, the role of green surfactants grows more critical, solidifying their role as a cornerstone in seeking greener materials and processes in the chemical industry (Trache et al., 2020; Roger et al., 2017) (Figure 6.11).

Figure 6.11. The development of green surfactants represents a critical advancement towards sustainable industrial practices.

Source: https://signicent.com/wp-content/uploads/2025/05/digital-screen-with-environment-day-1024x819.jpg.

6.5. CHALLENGES IN IMPLEMENTING RENEWABLE FEEDSTOCKS

Transition to renewable feedstocks poses numerous challenges in the realm of green chemistry, particularly when industries try to replace traditional fossil resources. One of the significant challenges is related to the availability and scalability of renewable raw material. While renewable feedstocks such as biomass and agricultural commodities bring with them the environmental advantages of sustainability, their supply chain is typically tainted by seasonality, competition for land use against food, and intensive inputs required for agriculture. As a result, the uncertainty of supply can interrupt and encourage manufacturers to revert to more reliable, though environmentally not so sound, fossil-based feedstocks. Sustaining a steady and sustainable supply chain is thus vital to driving the application of renewable feedstocks in process industries, for driving environmental benefits as well as for addressing increasing consumer demand for ecofriendly products. Technical feasibility in applying renewable feedstocks is a problem of unprecedented scale, in addition to supply stability. Variability in feedstock composition can influence process efficiency in chemicals, leading to yields that vary and are more expensive to produce. Ongoing development of new catalytic processes that can effectively convert heterogeneous biomass feedstocks to products of value remains a topic of very active research interest. The majority of existing industrial processes are currently optimized for petrochemical feedstocks and are not necessarily directly transferable to renewable feedstocks. This involves not only new research in catalysis and process technology but also significant investment in new technologies that render the economic utilization of renewable feedstocks more viable. Hence, aligning technical capability with green product demand is the means to further employing renewable materials in green chemistry initiatives. Lastly, economic incentive and policy regulation also play an important role in the successful implementation of renewable feedstocks. In the majority of regions across the globe, present policy can inadvertently promote conventional fossil fuels through subsidies and tax breaks, creating a cost barrier for industries wishing to transition into renewable-based alternatives. Also, the lack of controlled procedures for assessing the sustainability of renewable feedstocks creates a challenge to comply with and confuses manufacturers. Strategies like embedding green chemistry principles in higher education programs, as in the case of most nations including Zimbabwe, are essential towards embedding knowledge of sustainable methods and reducing such regulatory hindrances (Guyo et al., 2024). In short, conquering the hurdles of renewable feedstocks' challenges will require concerted efforts on science research, policy reformulation, and civic action to see a green chemistry sustainable future.

6.5.1. Economic Viability

The economic viability of green chemistry feedstocks is a significant consideration that involves cost of production, competitiveness in the market, and sustainability over time. As conventional fossil fuels continue to come in and out of favor, investment in renewable bioproducts has gained increasing interest from stakeholders. The economic appeal of utilizing renewable feedstocks, such as agricultural residues and non-food biomass, is additionally accompanied by the concomitant decrease in greenhouse emissions and footprint. Moreover, with technological advances in production processes, particularly with the evolution of bio-refinery technologies, the profitability of these green alternatives will enhance. This provides a setting where investments in renewable feedstocks not only reduce environmental degradation but also contribute affirmatively to economics, re-tuning the industrial base towards a sustainable future that fosters conservation and growth. The production of biofuels, one of the major industries in the renewable feedstock sector, outlines the economic issues that are associated with moving towards sustainable material. The first-generation biofuels produced from food crops are already struggling with resource competition and environmental sustainability. The second-generation biofuels produced from lignocellulosic biomass is an economically viable alternative according to the use of non-food biomass and agricultural residues. As Walker et al. (2011) describe, advancement in technologies for effective treatment of lignocellulosic feedstocks is key to lowering the cost and enhancing the availability of such biofuels. By utilizing waste materials, the bioprocessing industry not only avoids waste but also creates a circular economy that optimizes overall economic resilience through employment and efficiency of resources, further making the economic appeal of renewable feedstocks clear. Moreover, policy interventions by the state and market incentives play an important role in shaping the economic conditions around renewable materials. Subsidies, grants, and biofuel requirements for utilization create a favorable regulatory environment that encourages investment in green chemistry initiatives. The rising importance of policymaking in decarbonization and sustainability, Cukalovic et al. (2009) notes, puts renewable feedstocks at the heart of efforts to fulfill climate goals. Thus, business in renewable materials today has the advantage of more consumer demand for green products, encouraging innovation and ensuring economic security. All this market demand synergy and policy systems together finally make economic business in green chemistry more attractive, providing a wide business case for transitioning to renewable feedstocks and developing a sustainable industrial foundation.

6.5.2. Technological Barriers

The transition to green chemistry's renewable feedstocks and materials is followed by many technological challenges that restrict widespread adoption. These are usually the nature of poor infrastructure, lack of technical know-how, and the innate complexity of bio-refinery systems. For instance, incorporating bioraffinage processes within traditional chemical production facilities is a challenge since enterprises must change to new forms of production that essentially differ from traditional fossil fuel-based systems. The roadmap presented in Annevelink et al. (2009) emphasizes that there must be short- and long-term strategies to address these challenges, and that successful adoption of renewable feedstocks will require collective effort in the development of technology. Unless these obstacles are cleared, the value of green practice transformation is lost, limiting broader societal and environmental impacts of green chemistry initiatives. Beyond infrastructural concerns, there are numerous measurement and assessment challenges to complicate analyses of bio-energy pathways. The complex nature of comparing alternative bio-energy systems has the potential to create ambiguity that deters investment and innovation in green chemistry. As identified by Bauen et al. (2010), such measures of evaluating such pathways vary very widely between the studies, making data comparison and interpretation hard and eventually leading to industry confusion. The multiplicity of factors such as capital cost, emissions profiles, and use of resources complicates it for designing a conclusive framework for technological development. Such uncertainty not only restrains development but also distorts the vision of renewable material potential. Therefore, establishing common evaluation criteria is crucial in order to overcome these technology barriers successfully. Also, opposition to chemical industry technological changes usually stems from a strongly rooted habituation with traditional processes. Many existing firms may be reluctant to invest in new, unproven technology despite the growing amount of evidence favoring the utilization of renewable feedstocks. Such reluctance is, to a certain degree, instigated by the fear of economic loss or operational disruption on changes. Furthermore, the existing regulatory climate may punish newer technologies that do not have an evident, well-defined route to take for regulatory acceptance. A robust roadmap, as discussed by Annevelink et al. (2009), is thus required to lead stakeholders through these changes while addressing the natural resistance in adopting new methods. It is only through creating an innovation culture and encouraging cross-sector cooperation that the industry can look to break down the technological barriers currently hindering the development of renewable feedstocks and materials in green chemistry.

6.5.3. Regulatory Issues

The regulatory landscape for the application of renewable feedstocks and materials within green chemistry is dynamic and intricate. With countries seeking to be sustainable, governments are more and more creating frameworks to facilitate the development and utilization of bio-based materials. These regulatory frameworks address emissions reductions, waste management, and resource efficiency, which are central in encouraging innovation while conserving the environment. Notably, the transformation to a bio-based economy in regions such as the Netherlands is well articulated in strategic policy reports outlining short and long-term opportunities for biorefinery development, with a clear pointer to a robust regulatory framework for facilitating such initiatives (Annevelink et al., 2009). Consequently, navigation in this complex regulatory environment is essential for businesses that want to become engaged with sustainable activities, and it requires them to be continuously competent at compliance while consolidating their experimental and operational efficiencies. In addition to providing supportive frameworks for renewable feeds, regulatory issues also include stringent safety and health standards.

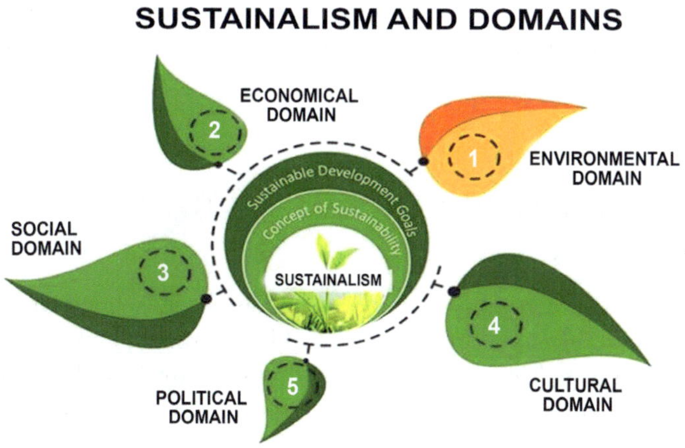

Figure 6.12. As countries strive to meet sustainability targets, governments are increasingly establishing frameworks to encourage the development and application of bio-based materials.

Source: https://www.mdpi.com/sustainability/sustainability-15-10682/article_deploy/html/images/sustainability-15-10682-g009-550.jpg.

The chemical production industry, particularly in Europe, is increasingly under pressure to rethink its operating practices in the face of pressures for environmental sustainability. This involves the requirement of transparent reporting against sustainability measurements, also known as the Triple Bottom Line approach, where success is measured not just in terms of profit but also in ecological and social terms (Allen et al., 2010). This type of regulation necessitates that companies not only be creative in the production of products but also embrace processes that align with safety, health, and the protection of the environment. Thus, green chemistry companies are compelled to make these regulatory matters part of their research and development plan, solidifying the worth of an integrated set of regulations. Development of regulations on renewable feedstocks provides challenges as well as opportunities for green chemistry. As the industry accepts the consequences of this change in regulation, there is the possibility of a transformation in chemical engineering practice towards sustainable methods. Continued refinement in regulatory demands calls for a reorientation of curricula within educational institutions to prepare future engineers to handle this complexity and support sustainable innovation. In the process, future professionals should not only have a strong foundation in scientific fundamentals but also knowledge of regulatory frameworks that shape the chemical manufacturing sector. The nexus among regulatory compliance, education, and industrial practice shapes how sustainable solutions are developed in chemistry and making it imperative to incorporate these into an integrated strategy that outlines the destiny of green chemistry programs (Figure 6.12).

6.6. CASE STUDIES IN RENEWABLE FEEDSTOCKS

In recent years, discovery of renewable feedstocks has been at the forefront of driving sustainable methods in the field of green chemistry. Most case studies prove the transition from traditional petroleum-based materials to bio-based materials, significantly reducing the environmental impact of chemical processes. The pharmaceutical industry, for instance, is beginning to embrace green chemistry principles, incorporating renewable feedstocks into their manufacturing processes. These innovations lead to a tangible decrease in the quantity of poisonous wastes generated and energy consumption, making it a sustainable line of action. Through the application of the concept of atom economy and catalysis, companies are capable of decreasing not just their ecological footprint but also the process of synthesizing drugs, a good example to their peers in the industry to follow, as stipulated in Achieng (2025). Furthermore, the work of biodegradable plastics produced with renewable feedstocks illustrates the immense potential of green chemistry to address challenges of conventional plastics. Emerging case studies have also brought out successful transformations of the materials science industry, with biopolymers being employed to create sustainable packaging solutions in line with consumer demand for ecologically sustainable products. These advances not only minimize pollution concerning plastic waste but also attest to the innovative uses of renewable feedstocks, which may be obtained from agricultural wastes and organic wastes. Through comprehensive research and collaboration among scientists, the manufacture of such bio-based products considerably contributes towards the reduction of greenhouse gas emissions, which is in line with the general objectives of green chemistry as exemplified in Geetha (2025). In addition, the use of renewable feedstocks is not limited to pharma and materials, and is found to permeate other industrial sectors like the production of energy. As an example, biodiesel from plant oils and waste lipids has been found to be a very effective substitute for fossil fuels, and several case studies have shown its potential. These case studies indicate the successful use of renewable feedstocks in energy, resulting in decreased carbon emissions and less dependency on non-renewable sources. The integration of renewable feedstocks into energy infrastructure not only moves the agenda of green chemistry forward but also injects strength and security into the energy system. With the world confronting the grim realities of climate change, these technologies reinforce the need to adopt sustainable practices on the basis of renewable feedstocks towards a better, more sustainable, and environmentally-responsible future.

6.6.1. Success Stories in Industry

One of the greatest success stories of the renewable feedstocks business is the transition from traditional chemical processes to processes based on bio-based materials. Companies worldwide are quickly realizing the value of utilizing renewable materials so as to minimize waste generation and environmental degradation. This paradigm shift is driven by the growing acknowledgment of the environmental consequences of traditional chemical manufacturing, as encapsulated by the realization that waste has a key contribution to make to environmental problems like climate change and plastic pollution (Roger, 2023). For instance, several of the large chemical firms have invested in technologies that transform agricultural waste and biomass into valuable chemicals. These technologies not only reduce fossil fuel dependency but also foster a circular economy through recycling of materials that would otherwise contribute to landfill waste. Another brilliant example of industrial achievement is found in the new suite of battery technologies where renewable materials are central to the production and lifecycle management of energy storage devices. With growing demand for electric vehicles and renewable energy, industries have realized the necessity to produce sustainable battery systems. This is an indicator of increased recycling intensity and reuse opportunities for batteries because of the looming waste management problems with end-of-life products (Zhao et al., 2021). The companies have begun to seek out alternative approaches that incorporate the utilization of renewable feedstocks in the production of batteries, thus aligning their manufacturing processes with more sustainable principles.

Besides, these advances not only address existing environmental concerns but also promote a sustainable pattern of consumption through the integration of the sustainability and efficiency of resources. Lastly, the integration of renewable feedstocks into consumer goods is a very attractive narrative in the area of green chemistry. Industries have also been able to create alternatives to conventional plastics from bioplastics based on renewable feedstocks such as corn starch or sugarcane. These technologies have gained significantly due to increased consumer demand for environmentally friendly products and the imperative of reducing plastic waste. Furthermore, initiatives towards life cycle analysis of the products indicate that these are capable of radically minimizing environmental impact relative to fossil fuel-based counterparts (Roger, 2023). By promoting the usage of renewable resources in the production of everyday commodities, companies not only achieve commercial success but also contribute to the bigger agenda of sustainable development, generating a future where sustainable products are the standard and not the exception.

6.6.2. Research Innovations

The frontier field of research advancements in green chemistry has been at the forefront of renewable feedstock and material development. The change is driven by growing needs for reducing fossil fuel reliance and environmental pollution. Researchers are committed to the design of substitute feedstocks from biomass that not only perform for industrial applications but are also sustainable. These technologies have an emphasis on material synthesis with lowered carbon content, along with those which are progressively biodegradable. In doing so, it is seen that interdisciplinary coordination between chemists, environmental scientists, and engineers plays a key role, promoting the creation of systems with high capacity to convert renewable resources into valuable chemicals and materials. This transdisciplinary research approach provokes technological innovations and regulatory models that promote environmental sustainability, which is paramount in the transition towards cleaner production processes (Maïder et al., 2025). Green chemistry technological innovations have also influenced the design of innovative processes aimed at optimizing the use of renewable feedstocks. For instance, the integration of biocatalysis and enzymatic systems facilitates more energy-efficient conversion of biomass into useful chemicals with decreased waste generation. Researchers are working to enhance selectivity and efficiency of such processes to yield high-value products with fewer environmental costs associated with traditional technologies. Also, advances in analytical techniques enable scientists to witness the conversion processes more intensively, and further understanding of the chemical pathways taken. Such developments are facilitated through policy and regulatory frameworks that encourage clean technology development and use across sectors, such as the automotive sector, which has a significant impact on greenhouse gas emissions. These regulatory measures are significant in eliciting research innovations that resonate with sustainability and environmental responsibility (Maïder et al., 2025). Moreover, the effects of research innovations extend beyond the exclusive scientific arena; they include socioeconomic implications that can revamp industries and communities. By embracing renewable feedstocks, businesses can diversify their product lines and create new jobs in rural areas where biomass is harvested. This transition underscores the economic sense of embracing greener production that can help stimulate regional economies and propel sustainable development. Additionally, education is the key to making this transformation happen, as it prepares successive generations of engineers and scientists to seek solutions to environmental problems. As society increasingly recognizes the importance of sustainability, the integration of the principles of green chemistry into breakthroughs in research will begin to influence policy-making and investment priorities, finally setting industries on the path of a more sustainable and planet-friendly trajectory. These technologies' development and use signify a significant paradigm change in understanding the interdependence of economy, science, and sustainability (Maïder et al., 2025).

6.6.3. Lessons Learned

In the pursuit of green chemistry feedstocks and materials using renewable resources, so much can be learned of the importance of waste utilization in sustainable process. The shift towards biodiesel production from waste products highlights the innovative method of producing power and the pragmatic approach to the environmental concerns of fossil fuels. By utilizing waste cooking oils and other biomass, researchers have demonstrated biodiesel can be produced with phenomenal efficiency, garnering up to 100% from feedstock materials such as food waste lipids. This underscores the significant lesson that the use of available waste locally not only reduces pressure on natural resources but also reduces pollution and health risk from conventional energy sources, as evident by the rising contribution of waste feedstocks to the biodiesel market (Ahmed et al., 2024). Equally important are advances in computational chemistry and machine learning as tools to enhance biodiesel processes. Lessons learned through use of these new technologies promise improvement in catalyst design and reaction efficiency, critical for scaling up biodiesel production from renewable feedstocks. Such computational methods permit improved understanding of reaction mechanisms as well as transesterification process optimization needed in the conversion of the feedstocks to biodiesel. In addition, the use of advanced computational approaches to green chemistry processes is a perfect illustration of how interdisciplinarity can stimulate innovation and optimization. This intersection of biology, chemistry, and technology is essential to solve the immediate energy crisis but also in keeping with sustainable development goals (Ahmed et al., 2024). Aside from biofuels, investigation on the application of hydrogen as a sustainable energy carrier offers valuable lessons on the future of green chemistry. The transition to a long-term hydrogen economy highlights its possible role in the decarbonization of transport and industry. By analyzing various strategies for hydrogen production, scientists have not only identified the problems, but also the opportunities, of incorporating it into renewable energy infrastructure. This school of thought highlights the necessity of collaboration between academia, business, and policymakers in facilitating the development of robust hydrogen technologies. The key lesson here is that collaboration across disciplines is vital in overcoming challenges and realizing the full potential of hydrogen to deliver a sustainable future. As the debate on hydrogen unfolds, it is clear that critical partnerships will be central to building the technology and applications.

6.7. CONCLUSION

In bringing together the discourse on renewable materials and feedstocks to green chemistry, it is evident that the move away from fossil resources to sustainable resources is not only preferable but also imperative. With the population of the world growing and environmental challenges rising, reliance on petroleum products becomes increasingly unsustainable. The vast exploration of renewable feedstocks suggests the promise of plant-based materials, namely lignocellulosic biomass, which is expansive and untapped. These resources offer a promising path toward mitigating the harmful impacts of fossil fuel consumption and are essential to the future of more sustainable chemical operations. By investing in and working towards the development of these renewable resources, society can envision a stronger and more sustainable chemical sector, in line with broader sustainability goals, while promoting economic growth. Besides, the multi-product versatility of chemicals and materials derived from lignocellulosic biomass further solidifies its status as the premier resource in modern green chemistry. Conversion of organic matter into valuable chemicals and materials presents a two-pronged benefit: it addresses the near-term challenge of sustainable commodities while simultaneously reducing carbon footprint and waste. With over 200 value-added compounds already obtained from this biomass, according to recent studies, potential for innovation and commercialization is high. The relevance of lignocellulosic biomass extends beyond production levels; it is also the knowledge about the circular economy where waste products are transformed into high-value materials. This reflects a paradigm shift in addressing natural resource limitations with a sense of drama, projecting the flexibility and robustness of green chemistry as a driver towards the development of sustainable solutions with a future outlook. Significantly, the shift in renewable energy resources, such as hydrogen, also presents additional opportunities towards decarbonization and the production of green materials. The potential of hydrogen in facilitating clean energy solutions can hardly be overstated since it offers cogent alternatives to the decarbonization of world heat, power, and transportation systems. Coupled with maximization of lignocellulosic biomass, the complementarity of renewable energy systems suggests a three-dimensional strategy towards sustainable objectives. The future must be about maximally exploiting such complementary resources so that the transition towards renewable feedstocks is not only feasible but also scalable. This comprehensive plan will play a pivotal role in the attainment of a sustainable future, with emphasis on the central role of innovation in green chemistry as companies shift from fossil fuel dependency to a more sustainable future.

6.7.1. Summary of Key Points

The movement to renewable materials and feedstocks in green chemistry is typified by a dramatic transformation of industrial processes with the aim of reducing the impact on the environment. The trend highlights the use of biomass and other renewable resources to produce chemicals, materials, and fuels instead of relying on finite fossil fuels. Renewable feedstocks reduce the greenhouse gas emissions and promote the circular economy, in which waste is redirected for transformation into new products. The Intergovernmental Panel on Climate Change (IPCC) emphasizes policy strategies that promote utilization of such sustainable processes to combat climate change effectively (Calvin et al., 2023). This highlights the multi-faceted benefits of transitioning to renewable resources, including improved sustainability markers and future innovation in materials science. Responsible recycling and sourcing practices create a platform for cleaner, greener industrial settings. Overcoming the issues of managing waste, green chemistry uses renewable feedstocks also entails evolving better recycling technologies. The current situation regarding plastics is beset by great challenges, particularly regarding their environmental impact through random disposal and perpetual pollution. Comparative plastic life cycles show the potential of industrial chemical recycling to solve these issues and develop a green material use approach (Li et al., 2022). Effective recycling methods can be employed to transform wastes into feedstocks for the manufacture of new products, thereby minimizing virgin plastics usage. Additional research directed towards discovering new approaches that enhance chemical recycling technologies will guide the use of blending renewable materials in driving the world to a more sustainable consumption pattern. The blend of recycling and renewable feedstocks displays the variety of initiatives required to address the plastic problem in the current world. Furthermore, renewable feedstock potential does not only extend to industrial processes but also to other industries such as agriculture and the pharmaceuticals industry. Green chemistry procedures allow for the manufacture of biodegradable products and bio-based chemicals that are just as good as the traditional counterparts but with reduced ecological footprints. The conventional chemical industry has long been criticized for its large carbon footprint and exploitation of resources, and an overhaul to embrace renewable sources is necessary on an urgent basis. It aligns with global sustainable goals and lends weight to the necessity of extensive education campaigns and policy measures to spur acceptance and awareness among the stakeholders. Lastly, supporting a broader recognition of renewable feedstocks is central to inspiring change across industries such that both economic and environmental goals are met with the same intent.

6.7.2. Implications for Future Research

The investigation of renewable feedstocks and materials in green chemistry is of huge future significance, particularly in light of the massive push for a transition away from fossil fuel dependency. There is an urgent need to re-examine energy usage patterns as well as agricultural practices to achieve carbon neutrality. As mentioned, few countries have achieved carbon neutrality, with others having targets to become so by 2050–2070. Thus, research must focus on designing pro-active strategies toward the use of renewable resources like novel farming practices and substitute feedstocks that limit reliance on petroleum products. In aiding these changes, tools such as input-output models and spatial mapping techniques must be used to develop knowledge on carbon emissions and potential cutbacks. Achieving a carbon-neutral economy implies the need for a multidisciplinary approach that conciliates economic, environmental, and technologic knowledge (Chen et al., 2022). Further, research on hydrogen-producing technology offers fertile ground for in-depth exploration in the field of renewable resources. Given the findings of current research, rising emphasis is being put on the role of hydrogen as a clean energy carrier, especially when produced from renewable resources. Future research must focus on enhancing the technology of biomass gasification and steam reforming, which have demonstrated high yields and economic viability. Diving deep into these technologies can optimize hydrogen production while reducing operation costs. In addition, the combination of renewable feedstocks with clean energy can make it possible to produce green hydrogen, which can transform energy systems worldwide. This type of work can be of great assistance in achieving broader climate goals by reducing carbon dioxide emissions and encouraging a hydrogen economy (Pedro et al., 2021). In addition, the implications of shifting to renewable feedstocks extend beyond technology and electricity generation and encompass socio-economic and policy factors as well. The transition requires not only technological development but also public engagement and governmental support to create sustainable conduct. Subsequent research should investigate effective policy structures that facilitate the uptake of green chemistry principles in a manner that caters to diverse stakeholders. This includes determining the socio-economic impacts of transition to renewable materials on different communities, with a focus on job creation and economic development in new sectors. Lastly, collective initiatives involving governments, business executives, and researchers can ensure that policies fostering sustainability in the chemicals industry are more efficient, such that everyone reaps the benefits and they are equitable (Pedro et al., 2021).

6.7.3. Call to Action for Stakeholders

The transition to green chemistry using renewable materials and feedstocks entails a call for action from stakeholders across numerous sectors, including industry, academia, and government. Engagement of these various stakeholders is pivotal to setting up a collaborative environment in which innovative ideas can develop. Stakeholders must bring about the sustainable approaches as a primary integration in their operations, most importantly in chemical production, where the change from fossil-based feedstocks to renewable feedstocks is necessary. Achieving cross-sectoral collaborations can simplify knowledge exchange and the utilization of resources, thus accelerating the innovation of green technologies. Notably, workshops, such as that organized by the European Commission's Joint Research Centre, have emphasized the necessity of stakeholder dialogues in meeting the multi-faceted issues of scaling up renewable feedstock technologies effectively (Andrei et al., 2013). Furthermore, universities and schools are also accountable in this system by teaching the future generation of scientists and engineers to innovate within the field of green chemistry. Through expanded curricula to emphasize renewable resources and sustainable chemical processes, universities are able to create a workforce trained in these areas. Collaboration in the industry can facilitate research studies that bring together theoretical understanding and application realities, thereby improving the quality of learning and opportunities for successful technology transfer. Significantly, this can lead to products that not only serve market needs but also are environmentally sustainable. Regional strategies, such as those examined through case studies in various places, are indicative of the importance of aligning educational programs with regional bioeconomy goals to demonstrate how local engagement can lead to real transformation of the industry (Bianchini et al., 2016). Finally, regulatory frameworks must adjust to facilitate the new market for renewable materials and feedstocks. Stakeholders, especially policymakers, should create a setting that pays back research and development of sustainable practices and provides clear guidelines for implementation. This could include subsidies to companies that have invested in clean technologies, along with more stringent carbon emissions regulation that demands a shift towards cleaner feedstock sources. Stakeholder involvement in the regulatory process helps assure policy is based on a knowledge of industry capability and science. By creating a space in which all can be included in policy development, the shift to renewable feedstocks can gain traction, leading to a green future in green chemistry and beyond.

6.7.4. Final Thoughts on Sustainability

The principle of sustainability has come to the forefront in addressing the global problems of environmental degradation and climate change. It has to take into account a three-dimensional model based on economic sustainability, ecological sustainability, and social justice. As industrialization continues to exert pressure on our planet, adopting renewable feedstocks in green chemistry is an immediate solution. When we transition from traditional fossil fuels to biobased products, we not only reduce carbon footprints but also promote a circular economy. This approach focuses on resource efficiency and waste management while keeping energy and material loops unbroken, thus aiming for carbon neutrality in 2050. This is a transformative process that involves looking at industrial practices and aligning them with sustainability values that appreciate the finite nature of our resources (Yang et al., 2022). Also, pursuing sustainable solvents for application in industries points towards the need to harmonize renewable values into the practice of everyday chemistry. From the start of green chemistry, efforts have been made to minimize the negative consequences of conventional solvents, which are the source of major emissions and environmental problems. The advent of neoteric solvents is a step in the right direction but whose partial use hints at the complexities of developing their true performance in practice. Successful utilization should take into account not only the intrinsic properties of solvents but also their origin, particularly non-fossil organic carbon sources. This perspective adds to the sustainability debate beyond merely the green character of a material to a wider thought of its lifecycle and environmental effects in order to be compatible with the overall aims of green chemistry today (Winterton, 2021). Lastly, transitioning to a sustainable future relies upon collaboration from numerous varied sectors utilizing expertise from business, academia, and policymaking. The transition to renewable feedstocks and sustainable materials cannot be left to one body; instead, cooperation is necessary in order to have facilitatory frameworks, research support, and public consideration. Life cycle assessments will come into play in achieving maximum resource efficiency and minimizing adverse impacts throughout the supply chain. With increasing climate pressures, the development and implementation of sustainable practices in chemistry must focus on collaboration and innovation to bring forth a stronger and more sustainable industrial climate. Education on the difficulties of implementing a circular economy across industries will be critical to foster real change, and optimizing such systems is vital to ensure sustainable industrial, agricultural, and business horizons (Yang et al., 2022).

REFERENCES

1. Achieng, N. M. (2025). Green chemistry in pharmaceuticals: Reducing environmental impact. *Newport International Journal of Public Health and Pharmacy*. https://www.semanticscholar.org/paper/fcf7c9c0751460e055a99f3481f52db214e8d25d (accessed on 24th October 2025).
2. Adeleye, A. T., John, K. I., Adeleye, P. G., Akande, A., & Banjoko, O. (2021). One-dimensional titanate nanotube materials: Heterogeneous solid catalysts for sustainable synthesis of biofuel precursors/value-added chemicals—A review. *Journal of Materials Science, 56*, 18391–18416. https://www.semanticscholar.org/paper/7e9fb715117ea6f1548ea97032d28eea3b362669 (accessed on 24th October 2025).
3. Allen, A., Allenby, B., American Chemistry Society, Aragon-Correa, J. A., Armstrong, C., Batterham, R., Benoit, J. P., et al. (2010). Sustainability of the chemical manufacturing industry–Towards a new paradigm? https://core.ac.uk/download/339962.pdf (accessed on 24th October 2025).
4. Amarasekara, S., Antal, M. J., Balaban, A. T., Bao, X., Bicker, M., Bilgen, S., Binder, J. B., et al. (2010). Towards the conversion of carbohydrate biomass feedstocks to biofuels via hydroxylmethylfurfural. https://core.ac.uk/download/12356021.pdf (accessed on 24th October 2025).
5. Amesho, K., Lin, Y. C., Chen, C. E., Cheng, P. C., & Ponnusamy, V. (2022). Optimization and kinetics studies of biodiesel synthesis from *Jatropha curcas* oil under the application of eco-friendly microwave heating technique: An environmentally benign and sustainable bio-waste management approach. *Sustainable Environment Research, 32*, 1–13. https://www.semanticscholar.org/paper/d7aa6c40c2b8f6be64a88bc658fcfa68063f57e2 (accessed on 24th October 2025).
6. Andrei, B. D., Maria Del Mar, P. F., Thea, S., & Tzimas, E. (2013). Carbon capture and utilization workshop: Background and proceedings. https://core.ac.uk/download/38626968.pdf (accessed on 24th October 2025).
7. Annevelink, E., Broeze, J., van Ree, R., et al. (2009). Opportunities for Dutch biorefineries. https://core.ac.uk/download/pdf/29248632.pdf (accessed on 24th October 2025).
8. Anthony, E. J., Clough, P., Fennell, P. S., Memon, M., et al. (2016). An overview of advances in biomass gasification. https://core.ac.uk/download/42144485.pdf (accessed on 24th October 2025).
9. Arshiya, A. S., Challa, M. S., Majji, M. K., Umme, K., Suneetha, A., & Patibandla, J. (2025). ECO-logic reactions: Shaping the future of sustainable molecular science. *World Journal of Current Medical and*

Pharmaceutical Research. https://www.semanticscholar.org/paper/47de7be2a84c33aa2ba7ebf2a55bcaffe902d9b5 (accessed on 24th October 2025).

10. Babbar, R. (2024). Green chemistry and its applications. *Journal of Pharmaceutical Technology, Research and Management*. https://www.semanticscholar.org/paper/22a2ff31f002a034fa93718f2f5d1593eb55e08f (accessed on 24th October 2025).
11. Bauen, A., Gross, R., & Slade, R. (2010). Prioritizing the best use of biomass resources: Conceptualizing trade-offs. https://core.ac.uk/download/144577367.pdf (accessed on 24th October 2025).
12. Bianchini, C., Charles, D., Clement, K., Davies, J., et al. (2016). Case studies of regional bioeconomy strategies across Europe. https://core.ac.uk/download/76998041.pdf (accessed on 24th October 2025).
13. Calvin, K., Dasgupta, D., Krinner, G., Mukherji, A., Thorne, P., Trisos, C. H., Romero, J., et al. (2023). IPCC, 2023: Climate change 2023: Synthesis report. Contribution of Working Groups I, II and III to the Sixth Assessment Report of the Intergovernmental Panel on Climate Change [Core Writing Team, H. Lee and J. Romero (Eds.)]. IPCC, Geneva, Switzerland. https://doi.org/10.59327/ipcc/ar6-9789291691647 (accessed on 24th October 2025).
14. Chauhan, P., Mathur, P., & Sharma, A. K. (2024). Study on sustainable resource utilization as ligand precursors in coordination chemistry. *International Journal for Research in Applied Science and Engineering Technology*. https://www.semanticscholar.org/paper/0c474778bdc49131398f7e2beff4f0c7d21807ea (accessed on 24th October 2025).
15. Cheekatamarla, P. (2024). Hydrogen and the global energy transition—Path to sustainability and adoption across all economic sectors. *Energies, 17*, 807. https://doi.org/10.3390/en17040807 (accessed on 24th October 2025).
16. Chen, L., Msigwa, G., Yang, M., Osman, A. I., Fawzy, S., Rooney, D. W., & Yap, P. S. (2022). Strategies to achieve a carbon neutral society: A review. *Environmental Chemistry Letters, 20*, 2277–2310. https://doi.org/10.1007/s10311-022-01435-8 (accessed on 24th October 2025).
17. Clark, J. H., Farmer, T. J., Hunt, A. J., Sherwood, J., et al. (2015). Opportunities for bio-based solvents created as petrochemical and fuel products transition towards renewable resources. https://core.ac.uk/download/83940849.pdf (accessed on 24th October 2025).
18. Cukalovic, A. (2009). Fifth International Conference on Renewable Resources and Biorefineries, 10–12 June 2009, Ghent, Belgium. https://

core.ac.uk/download/55849830.pdf (accessed on 24th October 2025).

19. Deepika, G., Unnisa, Z., Bhavani, P. G., Ashwitha, Y., & Manaswini, V. (2024). Future trends in green chemistry. *IJPAR Journal.* https://www.semanticscholar.org/paper/1856854a43c3c0fbcdf7c56aa19f5cac0d8eeeaf (accessed on 24th October 2025).

20. Dixit, C., Dixit, K. L., Dixit, C. K., Pandey, P. K., & Siddiqui, S. A. (2024). Advances in green chemistry: Sustainable synthesis of novel organic compounds. *International Journal of Modern Achievement in Science, Engineering and Technology.* https://www.semanticscholar.org/paper/43ff090ce9950f7fde97b3493ea211b03c86ffd6 (accessed on 24th October 2025).

21. Geetha, M. P. (2025). Green chemistry for a greener planet: Combating climate change through innovation. *International Journal on Science and Technology.* https://www.semanticscholar.org/paper/dabbb33f37a8f6570e26417f69c88c9f24aa9417 (accessed on 24th October 2025).

22. Guyo, U., Chigondo, F., Nyoni, S., Muchanyereyi, N., Zinyama, N. P., Kusena, W., & Nyoni, D. (2024). Green chemistry: Current status and challenges in Zimbabwe. *Physical Sciences Reviews, 9,* 3305–3322. https://www.semanticscholar.org/paper/d95e1ffdc006006814deffbcd4323d9b488dbdb9 (accessed on 24th October 2025).

23. Hermens, J. G. H., Jensma, A., & Feringa, B. L. (2021). Highly efficient biobased synthesis of acrylic acid. *Angewandte Chemie International Edition, 61.* https://doi.org/10.1002/anie.202112618 (accessed on 24th October 2025).

24. Imre, B., & Pukánszky, B. (2013). Recent advances in bio-based polymers and composites: Preface to the BiPoCo 2012 special section. https://core.ac.uk/download/18405899.pdf (accessed on 24th October 2025).

25. Isikgor, F. H., & Becer, C. R. (2015). Lignocellulosic biomass: A sustainable platform for the production of bio-based chemicals and polymers. *Polymer Chemistry, 6,* 4497–4559. https://doi.org/10.1039/c5py00263j (accessed on 24th October 2025).

26. Jeerh, G., Zhang, M., & Tao, S. (2020). Recent progress in ammonia fuel cells and their potential applications. *Journal of Materials Chemistry A, 9,* 727–752. https://doi.org/10.1039/d0ta08810b (accessed on 24th October 2025).

27. Khan, M. I., Shin, J. H., & Kim, J. D. (2018). The promising future of microalgae: Current status, challenges, and optimization of a sustainable

and renewable industry for biofuels, feed, and other products. *Microbial Cell Factories, 17*. https://doi.org/10.1186/s12934-018-0879-x (accessed on 24th October 2025).

28. Kim, D. S., Kumar, A., Kuppili, S. K., Omidvarborna, H., et al. (2016). Experimental modeling of NOx and PM generation from combustion of various biodiesel blends for urban transport buses. https://core.ac.uk/download/70427234.pdf (accessed on 24th October 2025).
29. Kowalski, S. P., & Priess, M. R. (2010). Algae and biodiesel: Patenting energized as green goes commercial. https://core.ac.uk/download/72052027.pdf (accessed on 24th October 2025).
30. Kumar, R., & Maurya, A. (2024). Green chemistry techniques for sustainable pharmaceutical synthesis. *Journal of Drug Discovery and Health Sciences*. https://www.semanticscholar.org/paper/5a8589fb8f69f6010c18833e91cf6d3d83a8a2bc (accessed on 24th October 2025).
31. Laddha, P. R., Sitaphale, G. R., Charhate, K. B., & Tathe, P. R. (2024). Green chemistry in pharmaceutical synthesis: Sustainable strategies for drug production. *International Journal of Advanced Research in Science, Communication and Technology*. https://www.semanticscholar.org/paper/0b10f3c32499241c0564c879bfc236d0d9c8ab4a (accessed on 24th October 2025).
32. Li, H., Aguirre-Villegas, H. A., Allen, R. D., Bai, X., Benson, C. H., Beckham, G. T., Bradshaw, S. L., et al. (2022). Expanding plastics recycling technologies: Chemical aspects, technology status and challenges. *Green Chemistry, 24*, 8899–9002. https://doi.org/10.1039/d2gc02588d (accessed on 24th October 2025).
33. Maity, S. K. (2015). Opportunities, recent trends and challenges of integrated biorefinery: Part I. https://core.ac.uk/download/38678810.pdf (accessed on 24th October 2025).
34. Megía, P. J., Vizcaíno, A. J., Calles, J. A., & Carrero, A. (2021). Hydrogen production technologies: From fossil fuels toward renewable sources. A mini review. *Energy & Fuels, 35*, 16403–16415. https://doi.org/10.1021/acs.energyfuels.1c02501 (accessed on 24th October 2025).
35. Mouffouk, F., Pereira, H., Varela, J., Abu-Salah, K., Lepage, M., Barreira, L., Custódio, L., et al. (2013). Isolation and fatty acid profile of selected microalgae strains from the Red Sea for biofuel production. https://core.ac.uk/download/225162963.pdf (accessed on 24th October 2025).
36. Ncube, A., Mtetwa, S., Bukhari, M., Fiorentino, G., & Passaro, R. (2023). Circular economy and green chemistry: The need for radical innovative

approaches in the design for new products. *Energies, 16*, 1752. https://doi.org/10.3390/en16041752 (accessed on 24th October 2025).

37. Nnabuife, S. G., Ugbeh-Johnson, J., Okeke, N. E., & Ogbonnaya, C. (2022). Present and projected developments in hydrogen production: A technological review. *Carbon Capture Science & Technology, 3*, 100042. https://doi.org/10.1016/j.ccst.2022.100042 (accessed on 24th October 2025).
38. Ogodo, U. P., & Abosede, O. (2025). The role of chemistry in achieving sustainable development goals: Green chemistry perspective. *International Research Journal of Pure and Applied Chemistry*. https://www.semanticscholar.org/paper/07af873696e6d8aad8a3273d34bfe3dd60458101 (accessed on 24th October 2025).
39. Ojukwu, U., Arthur, C., Jonathan, W., Ehigiegba, O., Yussuf, M., & Olawuyi, M. (2025). Green chemistry in pharmaceutical synthesis: Implications for sustainable public health. *European Journal of Scientific Research and Reviews*. https://www.semanticscholar.org/paper/3ea7b2d35d9abe52e2b253c0329f127725ff5baf (accessed on 24th October 2025).
40. Osman, A. I., Nasr, M., Farghali, M., Rashwan, A. K., Abdelkader, A., Al-Muhtaseb, A. H., Ihara, I., et al. (2024). Optimizing biodiesel production from waste with computational chemistry, machine learning and policy insights: A review. *Environmental Chemistry Letters, 22*, 1005–1071. https://doi.org/10.1007/s10311-024-01700-y (accessed on 24th October 2025).
41. Pacheco, C. M., Lima, W., Lima, F. A., Gomez, M., da Silva, I. G., Miranda, L. S. M., Esteves, P. M., et al. (2024). Levoglucosenone as a starting material for cascade continuous-flow synthesis of (R)-γ-carboxy-γ-butyrolactone. *RSC Advances, 14*, 34611–34619. https://www.semanticscholar.org/paper/634405c810a1f9c13251023179f3b28e3d412b60 (accessed on 24th October 2025).
42. Sai Laya, C. H., Saroja, K. S. N., Aswini, V. L., Sameera, M., Jayasree, M., Rani, E. U., Kamala, G., et al. (2025). A review of industrial applications of green chemistry. *Journal of Pharma Insights and Research*. https://www.semanticscholar.org/paper/26f8b58af8118b4f6accb051efc5beaaddeeb245 (accessed on 24th October 2025).
43. Saint Jean, M., & Oltra, V. (2025). Incrementalism of environmental innovations versus paradigmatic change: A comparative study of the automotive and chemical industries. https://core.ac.uk/download/pdf/6297038.pdf (accessed on 24th October 2025).
44. Sheldon, R. A. (2017). Metrics of green chemistry and sustainability: Past,

present, and future. *ACS Sustainable Chemistry & Engineering, 6*, 32–48. https://doi.org/10.1021/acssuschemeng.7b03505 (accessed on 24th October 2025).

45. Sheldon, R. A. (2023). The E factor at 30: A passion for pollution prevention. *Green Chemistry, 25*, 1704–1728. https://doi.org/10.1039/d2gc04747k (accessed on 24th October 2025).

46. Sheldon, R. A., & Woodley, J. M. (2017). Role of biocatalysis in sustainable chemistry. *Chemical Reviews, 118*, 801–838. https://doi.org/10.1021/acs.chemrev.7b00203 (accessed on 24th October 2025).

47. Staffell, I., Scamman, D., Abad, A. V., Balcombe, P., Dodds, P. E., Ekins, P., Shah, N., et al. (2018). The role of hydrogen and fuel cells in the global energy system. *Energy & Environmental Science, 12*, 463–491. https://doi.org/10.1039/c8ee01157e (accessed on 24th October 2025).

48. Thomas, J. M. (2012). Design and applications of single-site heterogeneous catalysts: Contributions to green chemistry, clean technology and sustainability. https://www.semanticscholar.org/paper/841c1287a6005ac3544c310a770605c00f41ef2e (accessed on 24th October 2025).

49. Tiwari, V., Kumar, A., & Rajkhowa, S. (2021). Green chemistry: Opportunity in drug discovery research (Part 2). *Current Organic Chemistry*. https://www.semanticscholar.org/paper/ac062b80a49b06dc50d7d80d61364ed0651e298c (accessed on 24th October 2025).

50. Trache, D., Tarchoun, A. F., Derradji, M., Hamidon, T. S., Masruchin, N., Brosse, N., & Hussin, M. H. (2020). Nanocellulose: From fundamentals to advanced applications. *Frontiers in Chemistry, 8*. https://doi.org/10.3389/fchem.2020.00392 (accessed on 24th October 2025).

51. Walker, G. M. (2011). 125th anniversary review: Fuel alcohol: Current production and future challenges. https://core.ac.uk/download/228176769.pdf (accessed on 24th October 2025).

52. Winterton, N. (2021). The green solvent: A critical perspective. *Clean Technologies and Environmental Policy, 23*, 2499–2522. https://doi.org/10.1007/s10098-021-02188-8 (accessed on 24th October 2025).

53. Yang, C. (2025). Application of green chemical synthesis technology in fragrance industry. *MATEC Web of Conferences*. https://www.semanticscholar.org/paper/c9c69a1591305a11f0e0efce9a8f85bf9d7de71a (accessed on 24th October 2025).

54. Yang, M., Chen, L., Wang, J., Msigwa, G., Osman, A. I., Fawzy, S., Rooney, D. W., et al. (2022). Circular economy strategies for combating

climate change and other environmental issues. *Environmental Chemistry Letters, 21*, 55–80. https://doi.org/10.1007/s10311-022-01499-6 (accessed on 24th October 2025).

55. Zhao, Y., Pohl, O., Bhatt, A. I., Collis, G. E., Mahon, P. J., Rüther, T., & Hollenkamp, A. F. (2021). A review on battery market trends, second-life reuse, and recycling. *Sustainable Chemistry, 2*, 167–205. https://doi.org/10.3390/suschem2010011 (accessed on 24th October 2025).

CHAPTER 7

GREEN CHEMICAL SYNTHESIS AND DESIGN

Contents

INTRODUCTION

The need for environmentally sustainable procedures in chemical synthesis is increasingly important by the hour as the planet grapples with a range of environmental issues. They range from climate change, reduced resources, and degradation of the environment. Most of the traditional processes of chemicals are energy-intensive and rely on non-renewable energy sources, generating significant amounts of waste and pollution. Thus, green chemistry principles are being taken up more and more as critical tools that can affect not just the field of chemistry, but also broader industrial and agricultural practices. It is universally accepted that continuous pressure on social and environmental systems requires solutions that build sustainability into their very fabric. The intersection of green chemical synthesis and design gives equal weight to not only the success of chemical reactions but to the ecological and social implications of the processes, the beginning of a new trend for more sustainable production and consumption in our charge towards a sustainable future (Wijeratna, 2012). At the forefront is the approach of agro-ecology that integrates ecological science into farm work. This strategy presents a powerful argument for the utilization of the green chemical synthesis for an increase in agricultural productivity at minimal environmental expense. Agro-ecological methods include an array of methods, ranging from recycling biomass to improving soil fertility and reducing the application of chemical pesticides and fertilizers. The emphasis on using domestic resources allows for the creation of agricultural systems that are economically viable and environmentally sustainable. As concerning trends continue—most notably increasing food insecurity and the adverse effects of global climate change—increasingly, more thought is being devoted to the role that green chemical synthesis can contribute. This is within the context of a larger movement towards sustainable agriculture that is context-appropriate and responsive to varied local circumstances (Wijeratna, 2012). Further, this shift to a bio-based economy underscores the significance of green chemical synthesis in establishing new products and materials on the basis of renewable feedstocks. Roadmap Bioraffinage sets out a strategy towards stimulating new biorefining technologies that will assist in meeting this transition in economies such as the Netherlands.' Through exploration of short- and medium-term commercial potential, the roadmap provides a framework for pilot studies and applied research that can spur innovation in sustainable practices. This aligns with the long-term goals of green chemistry, to develop products and processes that avoid waste and reduce the consumption of energy while maximizing efficiency and safety. As the world becomes increasingly green-oriented in its chemical synthesis, there is little doubt that the integration of sustainable principles into the very fabric of chemical design is not a choice but a necessity of a sustainable and just future.

Overview of Green Chemical Synthesis

Green chemical synthesis is a revolutionary method of chemical production with the vision of minimizing pollution to the environment while being economically viable and socially responsive. This follows heightened sensitization to the environmental catastrophes associated with traditional chemical production, which is normally energy requiring and leads to the creation of toxic by-products. In response, green chemistry design principles require developing safer by design and more efficient processes with renewable materials as the first option and with limited use of harmful substances. Through creative synthetic routes following such principles, researchers and industry aim to create a sustainable chemical world without compromising performance or function. Integration of life cycle assessment (LCA) methodology in process design and optimization is crucial for future green synthesis improvement. LCA offers a whole-system approach for evaluating the environment impacts of chemical production from raw material to end-of-life disposal.

Figure 7.1. Green chemical synthesis encompasses a transformative approach to chemical manufacturing, focusing on minimizing environmental impact while ensuring economic viability and social responsibility.

Source: https://globelapharma.com/wp-content/uploads/2024/05/19476ec9-e0b2-45c5-8e57-2119e55f6f12.jpg.

LCA identifies potential areas for improvement by evaluating energy consumption, waste product emissions, and use of resources throughout a process's life cycle. Recent studies indicate that the use of LCA enhances not only environmental impact knowledge but also multi-objective optimization of the process. The combination of LCA methodology and process systems engineering (PSE) provides the potential for thorough investigation tailored to green chemical synthesis problems, ultimately to formulate innovative solutions as coordinated economic targets and environmental accountability (Jacquemin et al., 2011). In addition, technological advances and the availability of new materials are major drivers of the development of green chemical synthesis. Scientists are working on catalysts and reaction conditions to reduce energy inputs and improve yield with less waste production. For example, the availability of bio-based feedstocks has greatly transformed product synthesis with the potential to create sustainable substitutes for fossil fuel-derived products. The latest technologies not only exploit nature processes but also assist in minimizing greenhouse gas emissions associated with traditional chemical syntheses. With industries embracing these green chemical approaches, the shift can lead to enhanced compliance with regulations and enhanced public perception, solidifying the role of green synthesis as an industrial chemistry pillar of the future (Figure 7.1).

7.1. GREEN SYNTHESIS METHODS

For green chemical synthesis, new methodologies must be introduced to minimize the environmental impacts of chemical synthesis processes. Green synthesis methods embrace the utilization of renewable inputs and secure waste minimization, adhering to the sustainability principles that form the foundation of science today. Methods such as water-based synthesis and bio-synthesis of materials, for instance, utilize non-toxic solvents and biological precursors, which may reduce the environmental impact for conventional synthesis. Apart from this, the use of renewable resources in manufacturing allows cascading use, such that the by-products are reused so that resources are optimized. This cross-disciplinary approach to resource utilization aims at designing techniques that not only fulfill the needs of environmental legislations but contribute towards a sustainable future. These innovations can close the divides between industrial processes and environmental responsibility, thereby shaping the paradigm of chemical synthesis in a sustainable direction (Geldermann et al., 2015). The impact of green approaches to synthesis finds relevance in various fields, particularly in the manufacturing of environmental-friendly dyes and their applications in the textile industry. Studies show that green design can be applied to develop dyes that yield hemicyanine dyes not only pleasing in their appearance but also stable and safe in aquatic environments. Computational chemistry used for the prediction of properties such as light stability and fiber affinity allows the process of production to be significantly streamlined, removing tedious and costly procedures traditionally associated with dye manufacturing. This procedure not only enhances the efficiency of discovering new dye structures but also finds application in the purpose of nature conservation. Moreover, utilization of these dyes on products like polyacrylonitrile, cotton, and nylon demonstrates the role such green synthesis procedures play in changing industries that rely greatly on synthetic products in order to shift to environmentally friendly methods without reducing utility or quality (Chen et al., 2019). Peeking into the future, the potential of green synthesis methods is vast in their ability to address environmental problems from traditional chemical manufacturing. While science continues to illuminate the pathways to more efficient use of resources and sustainable production, more and more applications will emerge within various industries. The ongoing development of renewable materials and employing them in the most optimal manner will trigger innovations which can meet consumer behavior halfway and push the responsible shift towards sustainable goods. Interdisciplinary approaches will allow researchers to foster cooperation between materials science, environmental chemistry, and industrial engineering. This complete comprehension will not only promote the green chemical synthesis technologies but also supplement societal changes towards sustainability in consumption and production. Lastly, the use of green synthesis methods will significantly revolutionize industries and usher in an era of environmental responsibility and awareness (Geldermann et al., 2015).

7.1.1. Biocatalysis

Biocatalysis, the utilization of natural catalysts, particularly enzymes, to catalyze chemical reactions, is one of the core pillars of green chemical design and synthesis. It possesses several advantages over conventional chemical processes, including greater selectivity, reduced waste, and milder reaction conditions. By utilizing the unique characteristics of enzymes, biocatalysis efficiently transforms substrates into the desired product through the absence of toxic reagents and harsh conditions usually associated with synthetic organic chemistry. This not only minimizes environmental footprint but also follows the precepts of green chemistry, i.e., efficiency and sustainability. As industries strive to minimize their ecological footprint, biocatalysis offers a highly promising alternative for drug and other chemical manufacture vital to human as well as environmental health and therefore promoting innovation in sustainable development. The pharmaceutical industry, in particular, has increasingly been utilizing biocatalytic reactions, driven by the need for environmentally friendly synthesis procedures. Traditional methods of drug preparation have a tendency to produce toxic side-products and large amounts of dangerous waste. The utilization of biocatalysis, therefore, is in accord with the general green chemistry mission: to promote safer and more environmentally conscious synthesis procedures. By employing biocatalysis in pharmaceutical manufacturing, business enterprises can counterbalance the environment hazards of traditional synthesis, increasing public health gains. The reduced rate of toxin release into ecosystems not only augments drug manufacturing safety but also aligns with global initiatives to promote eco-friendly practices in industry. These advances are a grand revolution within the industry, balancing the demand for powerful pharmaceuticals against protecting the planet. Furthermore, biocatalysis flexibility is not exclusive to the pharmaceutical industry but extends to the development of environmentally friendly consumer products, such as insect repellent. For instance, work on creating green mosquito repellents has seen the emergence of products in the form of natural essential oils derived from crops like citronella and eucalyptus. Such products utilize biocatalytic processes, such as cold pressing and steam distillation, which reduce carbon emissions without compromising efficacy against pests. The synergism of the processes not only emphasizes the potential of biocatalysis for useful applications but also marks the significance of sustainable and responsible development in formulating consumer products. By making ecologically-friendly practice its foremost concern, biocatalysis can make a greater impact in the protection of health and preservation of ecosystems in accordance with green chemistry principles and overall public health and environmental integrity (Figure 7.2).

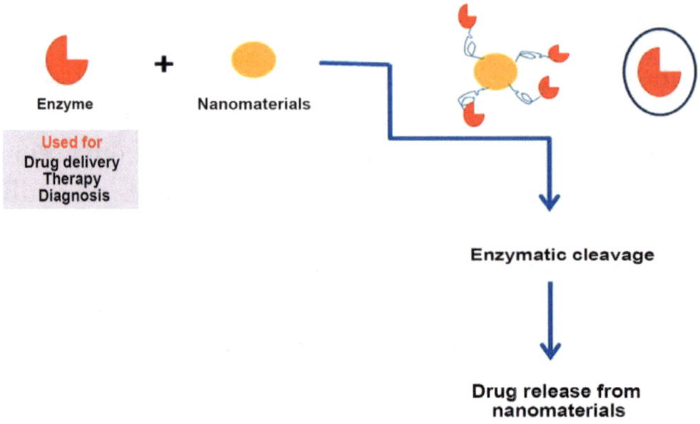

Figure 7.2. Biocatalysis, the use of natural catalysts, primarily enzymes, to facilitate chemical reactions, is a cornerstone of green chemical synthesis and design.

Source: https://www.mdpi.com/catalysts/catalysts-13-01448/article_deploy/html/images/catalysts-13-01448-g006.png.

7.1.2. Solvent-Free Reactions

The application of solvent-free reactions in chemical synthesis is a significant milestone in promoting environmentally friendly methods. Traditional organic synthesis promotes the use of solvents as a significant reagent, which can lead to a range of issues ranging from hazardous waste production to toxicity and emission into the environment. Solvent-free processes, however, decrease the environmental footprint and enhance the reaction efficiency by minimizing the scope of solvent removal processes. These processes can be carried out in solid-state or under solventless conditions, hence leading to a less wasteful and more resource-friendly process of chemical production. The shift towards its use is ideal within the principles of green chemistry, which seeks to keep chemical processes' impact on human health and the environment as low as possible while enhancing innovation in synthetic methodology. At the core of discussion about solvent-free reactions is the application of supercritical carbon dioxide ($scCO_2$) as a replacement reaction medium. Carbon dioxide typically has low chemical reactivity, but when used in supercritical conditions, it presents a way to overcome this limitation.

Figure 7.3. The adoption of solvent-free reactions in chemical synthesis marks a significant advancement in promoting environmentally sustainable practices.

Source: https://www.plusplustutors.com/wp-content/uploads/2025/01/UrAiFOiS-SO-PflirAk4BPw.webp.

Supercritical CO_2 possesses unique characteristics through which there can be varied chemical conversions, and hence it is a viable substitute for solvents that is non-volatile and non-toxic. Of special interest, the integration of catalysis with supercritical CO_2 has shown remarkable benefits such as reduced emission of VOCs, enhanced separation processes, and enhanced product selectivity. The positive potential of CO_2 use in the production of sought chemicals, e.g., dialkyl carbonates from alcohol, points to its likely use as a reducer of toxic chemicals' dependence on conventional use in industrial synthesis (Ballivet-Tkatchenko et al., 2005). In exploring the diversity of solvent-free reaction, gold nanoparticles

(AuNPs) have proven to be particularly valuable catalysts in this regard. Recent research indicates immobilization of AuNPs on functionalized support materials, which rationalizes catalysis in a broad array of contexts, including packed bed continuous-flow reactors. Functionalization of substrates such as silica, alumina, and titania with alkynyl carbamate functionality has proven successful, allowing high gold loading without sacrificing efficient catalytic performance. One major advantage of the method is that it allows for catalyst integrity, which ultimately results in dramatic improvement in reaction conversion rates when utilizing flow systems compared to batch systems (Ballarin et al., 2017). Not only do such developments point to the future possibilities for solvent-free methods enhancing catalytic efficiency, but they also reflect broader trends in green chemistry, where ecologic responsibility and maximum resource optimization are key (Figure 7.3).

7.1.3. Use of Renewable Resources

The move towards green chemical synthesis and design using renewable resources is becoming increasingly crucial with the increasing world energy demands amidst dwindling fossil fuels. The transition underscores the call for new approaches towards energy usage and production. Solar, wind, and biomass are renewable resources that provide viable substitutes for conventional energy sources. By utilizing renewable materials, companies can significantly reduce their carbon footprint and enhance the security of energy. As countries aim to comply with international climate agreements, the integration of renewable resources into chemical synthesis provides the opportunity for cleaner and sustainable processes. Therefore, the identification of the integration and utilization of renewable resources becomes essential for the future advancements of green chemical processes in terms of supplying a foundation for decontaminating the environmental impacts of traditional synthesis methods. One of the prominent processes that best represents the utilization of renewable resources is the Fischer-Tropsch (FT) process, offering the synthesis of liquid fuels from non-conventional carbonaceous materials like biomass. Given the necessity of swift action in tackling climate change, biomass-to-liquid (BTL) processes are shining cases of carbon-neutral liquid fuel production options that address both energy and environmental interests. Classic biomass gasification, however, achieves a mere 25–40% carbon utilization from the feedstock biomass with an inferior economic efficiency. This can be addressed through the addition of sophisticated approaches such as concentrated solar power (CSP) in the BTL process. This hybridization not only maximizes the utilization of carbon but could even be a form of effective storage of solar energy. The combination of CSP with BTL systems ultimately shows how renewable sources have the potential to transform the production of fuel while not weighing down on environmental impacts (James et al., 2013). More so, applications for renewable sources are not limited to liquid fuel production. With the focus on green chemical synthesis in place, it is more vital than ever that knowledge of sustainable design and resource management is kept at the forefront of awareness. By promoting a circular economy approach, industries can maximize recovery and re-use of renewable materials, minimizing waste and dependence on non-renewable sources. This transformation does not just constitute the technological development of chemical synthesis but a change of culture towards sustainable practice. Using renewable resources in chemical reactions invites a modification of the system conducive to caring for the environment and reducing greenhouse emissions. Therefore, the more we become familiar with processes of green synthesis, the more apparent it is that productive use of renewable resources is important for a sustainable future in the area of chemical production (Cukalovic et al., 2009) (Figure 7.4).

Figure 7.4. Renewable resources, such as solar energy, wind, and biomass, present sustainable alternatives for conventional energy sources.

Source: https://earth5r.org/wp-content/uploads/2024/03/6-Types-Of-Renewable-Energy-Sources-Which-Sustainable-Resources-Could-Be-Used-By-Mumbai.jpeg.

7.2. ENVIRONMENTAL IMPACT

The ecological footprint of traditional chemical synthesis pathways has raised intense concerns regarding their sustainability. The traditional approaches tend to use hazardous materials and generate lots of wastes, which have negative impacts on the environment and human beings. The principles of green chemistry, as outlined in Chen (2025), encourage the development of processes that generate fewer wastes and reduce the levels of toxicity of chemical compounds. By employing renewable feedstocks and green solvents, the chemical industry can significantly minimize its footprint. Besides, the inclusion of energy efficient synthesis can make reductions in greenhouse gas emissions, hence saving the world efforts in avoiding climate change. Emphasizing such green approaches serves not only the environmental goals but also the financial burden of waste and resource depletion entailed in traditional methods. Green chemical synthesis new ways, such as the use of catalysis, are essential in minimizing environmental impact. Catalysis enhances reaction efficiency, reducing raw material amounts utilized and minimizing by-products. According to Sun et al. (2025), enhanced photocatalysis and electrocatalysis have opened up new avenues for converting biomass into biofuels, a sustainable alternative to fossil fuels. Such advances in technology enable more effective biomass conversion processes, which are essential for achieving perpetually increasing energy demands without compromising environmental integrity. Systematic investigation of such catalytic reactions not only sheds light on their underlying mechanisms but also allows selection of effective catalysts conducive to certain reactions. The quest for efficient catalysts is an important driving factor in maintaining the utilization of biofuels and mirrors the emphasis on fundamental research for new catalytic strategies. Last, the ecological impact of chemical processes reaches beyond a short-term environmental concern to issues for society and the economy as a whole. Green chemistry encourages an integrated approach to chemical manufacturing that looks ahead on a long-term basis to sustainability. As discussed in Chen (2025), waste and toxicity minimization not only benefit the environment positively but also have positive impacts on public health. Further, as the globe goes towards greener technologies, there is an increasing demand for green products. Implementing green chemical design and synthesis can provide economic benefits in establishing a new market for sustainable products and services. Thus, a concentrated emphasis on green practices in the chemical sector is essential in ushering in an era of innovation that harmonizes economic growth with environmental preservation, engineering a sustainable future for many generations to come.

7.2.1. Reduction of Toxic Byproducts

Minimization of toxic byproducts is a central concern in the design and synthesis of green chemicals. Traditional chemical manufacturing processes have a tendency to generate toxic wastes, which pose a severe risk to the environment and human health. Green synthesis, on the other hand, aims to eliminate or minimize toxic byproducts, thereby enhancing the sustainability of chemical processes. Another key aspect of this approach is employing innocent raw materials and green reaction conditions, which can go a long way in avoiding the creation of unwanted substances. Green synthesis fulfills not just the environmental and ethical requirements but also with increasing regulatory pressure to be in tune with environmental requirements. Thus, it is necessary that researchers and practitioners give priority to methodologies with the aim of reducing toxic byproducts in the synthesis process. Recent advances offer the potential of applying biological materials for reducing toxic byproducts. For example, green synthesis of silver nanoparticles (AgNPs) has been the subject of much attention due to its efficacy as well as low environmental impact.

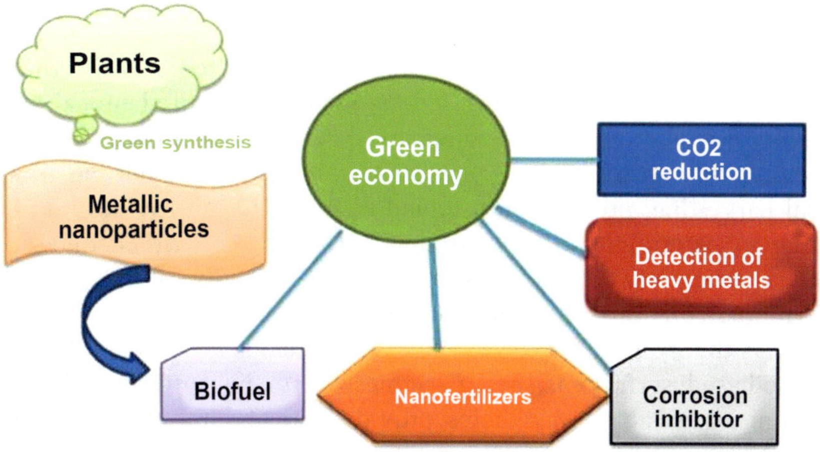

Figure 7.5. Green synthesis not only addresses ethical and ecological concerns but also aligns with the increasing regulatory pressure to comply with environmental standards.
Source: https://www.mdpi.com/suschem/suschem-05-00004/article_deploy/html/images/suschem-05-00004-g001-550.jpg.

Using plant extracts or microbial agents as capping and reducing agents, scientists have produced low-toxicity AgNPs with high yield. Besides eliminating the toxic byproducts typically inherent in conventional synthesis, these novel approaches also engage the unique properties of nanoparticles for applications. AgNPs' green synthesis routes are an illustration of how integrating biological systems with chemical processes can lead to green practices that respect environmental health and safety. Such instances of example show the promise of green chemistry as an antidote to big problems of managing toxic byproducts. Furthermore, reaction conditions optimization in green synthesis is a systematic approach to minimize toxic byproducts in an effective manner. With the use of techniques such as Response Surface Methodology (RSM), researchers have tuned parameters in TiO_2-ZnO nanocomposite synthesis. By studying variables such as initial pH, reaction temperature, and dopant amount, significant improvements can be achieved in the process's efficiency as well as environmental friendliness. In a study, the use of *Vigna unguiculata* L. such extraction not just accelerated the synthesis of TiO_2-ZnO nanocomposites but also ensured colossal reduction in harmful byproducts. Optimization methods of such nature are thus a part of the overall objective of green chemistry, demonstrating that optimized synthesis pathways can lead to useful materials while still meeting environmental stewardship and safety requirements (Figure 7.5).

7.2.2. Life Cycle Assessment

Life Cycle Assessment (LCA) is one of the principal tools to determine the environmental footprint of all phases along the life cycle of chemical products to enable researchers and manufacturers to adopt greener pathways in designing and synthesizing chemicals. Keeping an eye on each step, starting from raw material harvesting, production, use, and disposal, LCA provides a complete approach to determining the environmental fingerprint of chemicals. This method is particularly crucial when designing eco-friendly surfactants, wherein it aids in predicting possible environmental impacts and suggests optimization methods in the synthesis process. With new chemical compounds, such as thiophene-based surfactants, being synthesized to be applied in biosensors and organic transistors, LCA helps provide critical information about their biodegradability and ecotoxicity, and thus guides scientists towards green innovations (Cătălina et al., 2025). The necessity of carrying out LCA extends beyond curiosity on academic grounds and continues to become increasingly important in the shift towards sustainable solutions in the chemical industry. A case would be considering surfactants derived from thiophene, LCA uncovers some chemical processes that can cause significant environmental harm or risks to human health. In characterizing the impact of the entire manufacturing process, LCA allows for informed decisions regarding substitution or redesigning of harmful chemicals with ecologically friendly alternatives. Furthermore, the integration of LCA in the design process ensures that sustainability aspects are incorporated early on, in compliance with green chemistry principles of reducing toxic byproducts and maximizing resource efficiency. In light of growing demand for eco-friendly and innovative chemical products, LCA cannot be substituted in estimating the development of safer materials meeting market and environmental standards (Cătălina et al., 2025). Not only chemical synthesis but also the LCA methodology can be communicated to other sectors, including the construction industry, to present the all-around application of the assessment tool. Since construction activity plays a significant role in global energy usage and carbon dioxide emissions, an understanding of its life cycle impacts is crucial to implement effective green strategies. Progress made recently in the form of machine learning environmental forecasting methods has enhanced the accuracy and applicability of LCA in numerous fields to the extent that it is now feasible to make sustainability choices more consciously. Using powerful predictive models, stakeholders can not only evaluate existing practice but also optimize future design to reduce total environmental impairment. Such synergistic strategies reinforce that LCA is not only a reactive response but also a proactive measure that can promote sustainability in many fields, including chemical design and construction.

7.2.3. Impact on Ecosystems

The interweaving of green chemical synthesis and conservation of ecosystems is becoming more common as businesses strive towards sustainable modes of production and consumption. Green chemistry's emphasis on reducing toxic chemicals in chemical processes essentially attempts to restrict harm to the environment. In support of environmentally friendly reagents and procedures, the branch reduces the harmful impact on plants and animals in infected ecosystems. For instance, the application of biocompatible substances and operations enhances the longevity of natural environments, thereby promoting biodiversity and ecological balance. The Marine Board emphasizes the need for the progression of ecologically friendly practices in marine science, as representative of more universal ecological care, since such practices develop healthier systems with reduced sensitivity to anthropogenic stress (Boyen et al. 2025). Lastly, the use of green chemistry to industrial operations is not a fad but a necessary step towards maintaining our ecosystem integrity. As green chemical synthesis improves continually, it accords very well with agroecology principles, particularly in the management of pests and resources. Agricultural sustainability has shown that approaches such as Agro-Ecological Engineering and Integrated Pest Management (IPM) utilize ecological interactions to enhance the resilience of crops at the expense of reduced chemical inputs. The realization that biodiversity per se is not a determinant of effective pest control has propelled the encouragement of practices that modulate crop and ecological structures to reap functional benefits from biocomplexity (Deguine et al., 2008). This shift can potentially lower reliance on hazardous pesticides, thereby protecting non-target species and agricultural ecosystem services necessary for farming. Here, the amenability of green chemical methods to ecological wholeness becomes a crucial component of sustainable agrofood systems that ultimately translate into biodiversity and ecosystem integrity preservation. Moreover, incorporation of green chemistry concepts into industrial process is crucial in addressing the intricate problems ecosystem is undergoing in contemporary circumstances.

Though industries are adopting sustainable processes, minimization of hazardous waste and eco-detrimental by-products is a significant move towards protection of ecological systems. This transformation, not only conserves local water quality and biodiversity but also promotes broader climatic goals by minimizing greenhouse emissions associated with conventional chemical manufacture. The ongoing optimization of processes, as indicated by advances in pest control and sustainable farming (Deguine et al., 2008), is testimony to the capability of green chemical synthesis to develop paths that are beneficial for restoring ecosystems. Ultimately, the commitment to sustainability will be a primary element to enhance ecosystems while fostering global sustainability objectives, underscoring the ever-vital interconnections between chemistry, ecology, and public health (Figure 7.6).

Figure 7.6. By championing eco-friendly reagents and methodologies, this discipline mitigates harmful impacts on flora and fauna within affected ecosystems.

Source: https://vikalpsangam.org/wp-content/uploads/2025/08/Aatreyee-Dhar-Arunachals-women-healers-plants-cover-1024x576.png.

7.3. ECONOMIC BENEFITS

The financial advantages of green chemistry synthesis are increasingly apparent as companies shift towards environmental-friendly operations. Employing the principles of green chemistry, enterprises can significantly reduce production expenses in terms of waste disposal and energy consumption. For instance, applying novel processes like the reverse Boudouard reaction in combination with traditional processes can lead to cost reduction in the production of green methanol. This process not only decreases the cost of production by an approximate 5% but also enhances the efficiency of the system as a whole through simultaneous production of valuable by-products like biogenic hydrogen. Thus, the use of such integrated systems provides a lucrative economic incentive, making green methanol synthesis even more competitive than fossil-based options, which in turn supports future market trends towards sustainable products (Juan et al., 2025). Other than direct cost saving, green chemical synthesis enhances the competitive edge of a business in the increasingly environmentally conscious market. Businesses that adopt sustainable means tend to have better brand reputation and customer loyalty, which can directly translate to increased sales and market share. Secondly, since the government raises the level of control over environmental activities, businesses that go out of their way to incorporate green technology may evade fines and reap incentives. Its application not only drives innovation but also serves as the basis for building long-term stakeholder relationships that respect sustainability. In accordance with this, embracing green synthesis not only fulfills company social responsibilities but also becomes an economic strategic decision that can determine the viability and attractiveness of a business to investors (Ch et al., 2025). Apart from the environmental benefits, the economic implications of green chemical synthesis also have far-reaching effects on the labor market by promoting the need for qualified personnel in sustainability-driven specialties. As the industry continues to develop, the demand for experts who are competent in green technology, environmental policy, and sustainable manufacturing practices is increasing. This new development not only creates new employment opportunities but also provokes institutions of higher education to modify curricula to address industry demands. Through the development of a workforce trained in sustainable methods, organizations are able to reap higher productivity and innovation. Eventually, the move to green chemical processes is rewarded with a myriad of economic benefits, providing a sound basis for future industrial robustness (Juan et al., 2025; Ch et al., 2025).

7.3.1. Cost-Effectiveness of Green Processes

In recent times, green chemical processes' financial viability has become prominent in talking about sustainable strategies. The shift from conventional to green synthesis techniques usually involves initial expenditures on research and development. But the long-term savings on waste and energy can mitigate these costs. Green synthesis techniques incorporate renewable feedstocks and sustainable resources, thus minimizing reliance on toxic reagents and reducing the cost of hazardous waste disposal. Moreover, the effectiveness provided by such techniques has a tendency to produce higher yields of desired products, enhancing profitability. As stakeholders increasingly prefer sustainability, green processes investment is not just an ethical imperative but also a strategically sound option in competitive markets. Catalyst application derived from plants in green synthesis is a classic example of being cost-effective in carrying out organic transformations. Through the application, for example, of extracts or powders from various plants, scientists have demonstrated achieving significant material and environmental cost savings. Plant catalysts possess the characteristic that they are biodegradable and renewable, factors that translate into reduced procurement and removal cost. This shift to greener catalysts not only lessens the ecological footprint but enhances efficiency in chemical synthesis as well. The findings of such studies prove that when biologically obtained systems are applied as substitutes for conventional methods, operating costs involved with waste and toxic materials decrease, paving the way for industrial applications green in nature (Maniya et al., 2025). This new field of research focuses on the dollar value attached to becoming green. One of the most high-profile demonstrations of the cost-effectiveness of green synthesis is in the creation of nanoparticles, like silver nanoparticles (AgNPs).

Figure 7.7. As stakeholders increasingly prioritize sustainability, investing in green processes becomes not just a moral imperative but also a strategically advantageous decision within competitive markets.

Source: https://miro.medium.com/v2/resize:fit:1400/0*43WJ2mt3qWSuYxLD.png.

Nanoparticles of this kind are of specific interest with their varied applications in medicine and electronics. Traditional synthesis yields toxic byproducts and high energy consumption; however, green methods using biological resources offer an eco-friendly solution. New innovations in the synthesis of AgNPs using green precursors, such as plants and microbes, have proven these reactions not only produce high yields but are also conducted using mild conditions. The cost reduction in synthesis by such environmentally friendly processes is complemented by the reduced toxicity of the resultant products, thus further enhancing their commercial viability to producers and consumers (Sair et al., 2025). Overall, economically viable and environmentally friendly nature of green processes renders them an essential aspect of modern-day chemical synthesis and design (Figure 7.7).

7.3.2. Market Demand for Sustainable Products

Increased awareness of the deterioration of the environment has had a profound effect on demand for green products in the marketplace. Consumers are progressively assigning greater importance to green options in their purchasing behavior, recognizing that their choices serve not only the market but also the well-being of the planets. Consumer culture is shifting across sectors, including cosmetics, food, and cleaning products, where sustainability has become a primary distinguishing factor. As research indicates, businesses are increasingly forced to innovate and design their products based on environmental considerations. This transition compels businesses towards embracing practices that minimize ecological footprints and maximize the use of resources. For instance, the Roadmap Bioraffinage offers strategic advances toward building a bio-based economy focusing on the incorporation of sustainable procurement and production methods as major factors in the solution to evolving consumer demands (Annevelink et al., 2009). Moreover, the intersection of market demand and sustainability is inextricably intertwined with regulatory powers and social pressures. Governments around the world are enacting legislation that will reduce carbon emissions and spur green chemistry principles, further drawing companies into transforming their product lines. Not only does such regulation heighten the visibility of sustainable products in the marketplace, but it also results in consumers calling for certifications that legitimize a product's green claims. While scientific consensus points to the imperative of drastic measures against declining ecosystems, businesses must respond astutely. Green product demand has offered a fertile ground for material and process innovations consistent with sustainability goals. Businesses that act quickly in response to this emerging paradigm can bank on their environmental promises as a source of competitiveness (Gingerich et al., 2015). This nascent tale of consumer demand for sustainable products is an indicator of wider societal consciousness of business responsibility. The popularity of eco-labeling and other forms of green marketing is a sign of heightened consumer interest in sustainability concerns, shifting the terms on which products are conceptualized and marketed. But there is a catch; without tough scrutiny of manufacturers' green claims, customers may inadvertently enable practices that do not lead to genuine environmental benefits. Thus, transparency and accountability become the demand of the hour in the supply chain. In this case, aligning green chemical synthesis and design with market demand is not merely innovation, but conformity to ethical principles that ensure consumer trust building and genuine ecological worth. The test goes beyond compliance to that of being in genuine integration of sustainability through the product life cycle, something that will ultimately lead to long-term market sustainability (Gingerich et al., 2015) (Figure 7.8).

Figure 7.8. Consumers are progressively prioritizing eco-friendly alternatives in their purchasing decisions, recognizing that their choices impact not only the market but also the planets health.

Source: https://indian-retailer.s3.ap-south-1.amazonaws.com/s3fs-public/2023-11/BeFunky-collage%20%2820%29.jpg.

7.3.3. Long-Term Financial Savings

The integration of green practices in the area of sustainable green chemical synthesis has serious long-term financial advantages, particularly for companies seeking to enhance their operational efficiency. By the implementation of techniques such as waste reduction and energy optimization, firms can save costs while simultaneously complying with green legislation. A study on the green supply chain management practices indicates that firms achieve higher cost-effectiveness through these practices, and they are able to save costs for operations and ensure improved profitability. The findings also indicate that firms adhering to the principle of sustainability, besides meeting the requirements of compliance, are favorably placed in the market, and this leads to long-term financial health. Since green supply chain management (GSCM) practices have quantifiable financial benefits, they allow businesses to reinvest cost savings in further innovation and the improvement of their operations, thereby creating a relentless cycle of sustainability and financial stability (Li et al., 2025). Financial savings in the long term are also gained through the utilization of new materials, i.e., mycelium-based green composites (MBCs), which offer cost-effective substitutions for traditional materials. MBC production centers on the preparation and selection of the substrate since their chemical characteristics strongly influence growth and ultimate performance.

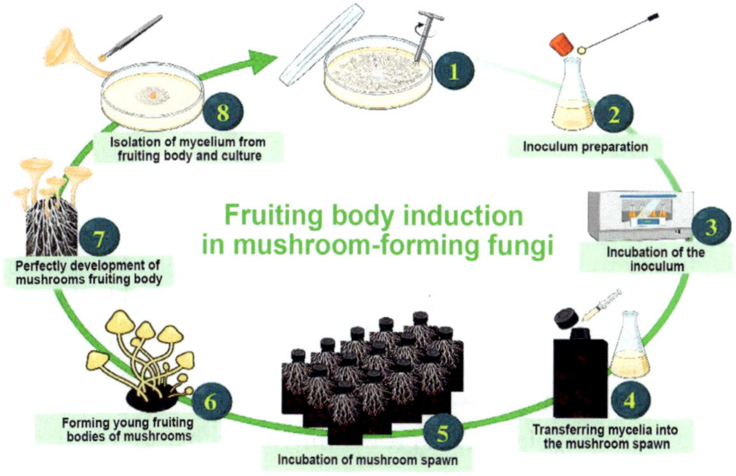

Figure 7.9. Long-term financial savings are also realized through the adoption of innovative materials, such as mycelium-based green composites (MBCs), which offer cost-effective alternatives to traditional materials.

Source: https://pub.mdpi-res.com/biomimetics/biomimetics-09-00337/article_deploy/html/images/biomimetics-09-00337-g005.png?1717397660.

Research has established that while initial investment in developing MBCs may be higher than that of conventional materials, overall returns in the long run such as reduced production costs, less energy requirements, and waste reduction place them as cost-efficient. Also, with the increasing use of MBCs across various industries, the possibility of mass adoption paves the way for extensive cost reduction in supply chains. With an investment in new materials, companies are likely to enjoy significant long-term cost savings while simultaneously fulfilling their sustainability goals, creating a congruent balance between green accountability and financial accountability (Aiduang et al., 2024). Finally, the financial benefits of green chemical synthesis underpin the inherent value of sustainability for long-term fiscal sustainability. The GSCM practices mentioned above illustrate that sustainability is not only an ethical choice but a strategic fiscal decision. According to corporate financial performance models, organizations adopting green measures can enhance revenue growth, customer loyalty, and competitiveness, all of which translate into actual economic benefits. Moreover, with increasing focus on sustainability across the board, companies that are capable of green innovation can differentiate themselves and thus solidify their market share. This proactive strategy not only generates consumer trust but also resonates with evolving consumer preferences for sustainable products. Therefore, the pursuit of long-term cost reduction by means of green chemical synthesis not only aids organizational goals but also contributes to broader environmental sustainability strategies (Li et al., 2025) (Figure 7.9).

7.4. CHALLENGES IN IMPLEMENTATION

The issues of the deployment of green chemical synthesis are complex and most commonly deeply rooted within current industrial settings. Economic viability versus traditional routes is one of the biggest stumbling blocks. Traditional synthetic pathways, involving high yields and lower operational costs, have a tendency to dominate greener routes which might still prove to be loss-making at initial instance. For instance, despite the concepts of CO_2 recycling and valorization gaining popularity, the profitability of such technologies is yet to be confirmed. The existing reactions, e.g., the ones utilized in the Sabatier process, hold the potential for sustainable methane production. However, the necessity for hydrogen delivery and operational efficiencies optimization raises issues of scalability and practicability in real-world applications, raising the need for new business models in financing these green approaches (Gallego et al., 2018). The second challenge emanates from the policy and regulatory frameworks that always lag behind technological progress. Green chemical synthesis transition requires effective and facilitating legislation that stimulates innovation and investment in sustainability. But regulatory frameworks in themselves can be a drawback, particularly if they do not recognize or accommodate new processes and materials on the behalf of green chemistry. The existing framework of a circular economy system, as discovered by most global policies, aims for the reduction of greenhouse gases (GHGs) emissions by utilizing resources more efficiently and controlling waste. Yet, in the absence of open regulatory control and economic incentives, actual implementation of these models can be hindered, thus slowing the transition of green technologies from the lab to industrial volumes (Annevelink et al., 2009). Public awareness and education also present further challenges in the large-scale implementation of green synthesis. The novelty of these approaches mandates a radical shift in not only the academic degree system but also industry training programs to yield a suitably knowledgeable workforce that can innovate and expand sustainable means. Universities and research facilities must focus on incorporating green chemistry principles into their curricula to foster future experts in this field. Furthermore, increased relevance between scientific investigation and the awareness of the general population is essential in order to gain support for green activities. More sophisticated measures of communication are required in order to explicate the benefits of green chemical processes and the necessity of sustainability in order to implement a culture that cares and focuses on environmental stewardship for the chemical industry (Annevelink et al., 2009).

7.4.1. Technological Limitations

The transition to green chemical synthesis and design is tightly bound by technological limitations that hinder the mass use of sustainable processes. The majority of the chemical processes available today emit high volumes of greenhouse gases, and therefore, alternative methods with small environmental footprints must be sought out. However, these new ways of working are often undermined by scalability and even economic viability problems. As an example, the Sabatier reaction for generating methane from CO_2 is promising but limited by the supply of sustainable hydrogen and by energy efficiency. To avoid these limitations, ordered reactors have been proposed that provide enhanced thermal control and enhanced yields. Nonetheless, technology developments required for efficient use of such solutions in industrial processes remain inadequate and hence restrain progress towards more sustainable means of chemical synthesis (Gallego et al., 2018). Moreover, the integration of renewable resources into dominant production systems often encounters great technological hurdles. While cascading utilization of renewable resources is an aspect in which resource efficiency can be enhanced, the frameworks to support such a paradigm are not developed or properly executed. The lack of concrete quantitative models for integrated production and distribution optimization addressing renewable resources indicates a significant knowledge gap in current research. Failure to implement adequate information technology solutions that facilitate value-generating network collaboration results in the potential benefits of renewable resources and the maximized utilization of these being inaccessible. This means there is a long-term disconnect between green chemical design theory and actual application, further supporting the call for further technological innovations (Geldermann et al., 2015). Lastly, despite the growing popularity of green chemistry, there remain substantial technological capability gaps that deter organizations from transition to greener practices. Researchers are continually discovering new products from renewable sources, yet they are rarely used in industries. That is because in most instances, the infrastructure is lacking to support the production of green alternatives alongside traditional chemical processes. Therefore, consumer action remains heavily reliant on traditional practices with little room for adopting sustainable options. Bridging this technology gap is essential for the advancement of green chemical synthesis and design toward facilitating a sustainable future while reaffirming the importance of cross-disciplinary integration in solving these problems effectively (Gallego et al., 2018; Geldermann et al., 2015) (Figure 7.10).

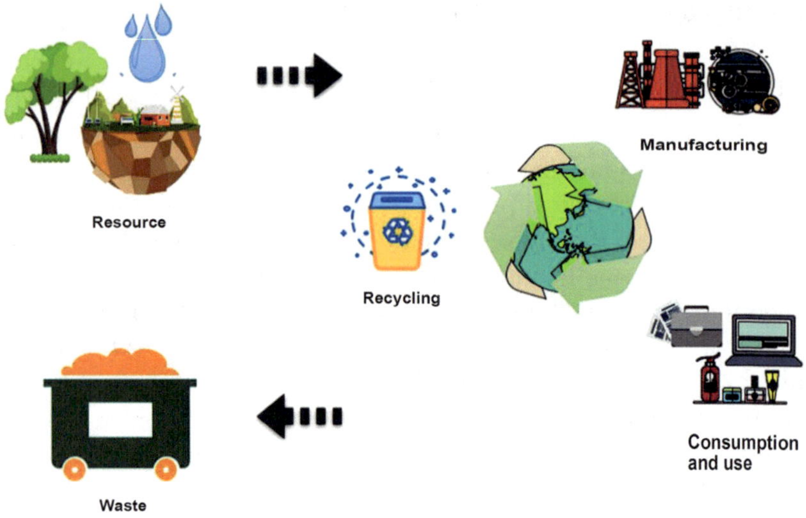

Figure 7.10. Many existing chemical processes generate considerable greenhouse gas emissions, necessitating the exploration of alternative methods that minimize environmental impact.

Source: https://media.springernature.com/lw1200/springer-static/image/art%3A10.1007%2Fs10311-022-01499-6/MediaObjects/10311_2022_1499_Fig2_HTML.png.

7.4.2. Regulatory Barriers

Regulatory hurdles dominate in the field of green chemical synthesis and design, often stifling innovation and adoption of sustainable processes. Regulated hurdles are typically induced by stringent government laws that promote established chemical production practices over new, environmentally friendly alternatives. Regulation mechanisms often fall behind innovation in green chemistry and therefore create space where innovative approaches cannot be effectively commercialized. For instance, the long regulatory approval cycles for new materials and chemicals can discourage industrial application of greener substitutes, with the resultant resorting to traditional practices causing environmental degradation. Therefore, the need for regulatory harmonization becomes apparent as international policy conformity may spur greener alternatives in a non-discriminatory manner, furthering the diffusion of sustainable practices across sectors. Yet another aspect of regulatory barriers is the heterogeneity and complexity of the regulations governing chemical synthesis. These can also differ significantly in countries and regions, making it challenging for firms to comply in a bid to introduce green technologies. As highlighted in the Intergovernmental Panel on Climate Change synthesis report, the issue of conflicting rules can slow not only the exchange of innovative solutions but also the overall momentum towards achieving global sustainability goals (Calvin et al., 2023).

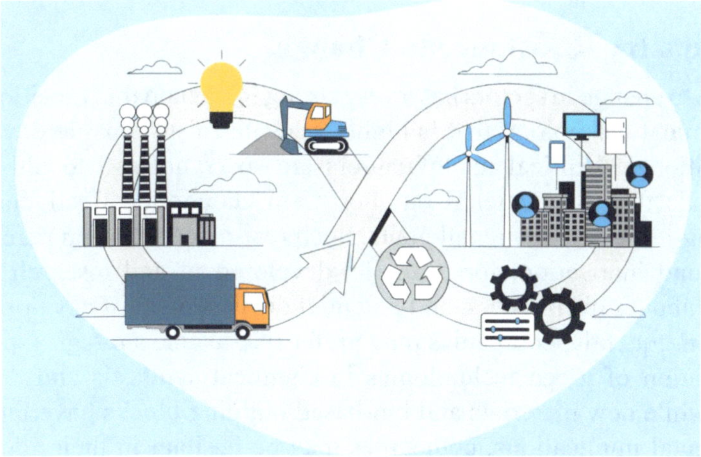

Figure 7.11. As companies seek to adopt sustainable practices, they must navigate not only the technical aspects of regulation but also societal attitudes towards new technologies.

Source: https://www.ismartrecruit.com/upload/blog/main_image/sustainable-business.webp.

The inability of companies to overcome these compliance challenges can put high-stakes investments in research and development into dormancy, effectively resulting in ongoing reliance on conventional chemical processes. Hence, there needs to be a harmonized regulatory strategy towards reform emphasizing streamlined regulation driving innovation adoption in the green chemistry industry. Emphasis on dismantling regulatory barriers also needs to address ethical concerns and public sentiments towards green chemistry. While business goes out of its way to be more sustainable, it must be worried not only about the technicalities of regulation but also about what people think about new technology. Resistance may also result from misinformation or lack of information regarding the safety and effectiveness of green substitutes, adding further to difficulties in complying with regulations. In addition, as outlined in flexible sensor technology research, market barriers to adoption typically comprise nontechnical issues like business, regulation, and ethics (Luo et al., 2023). By involving stakeholders—be they policymakers, industry members, or the broader public—into open debates of the benefits of green chemical synthesis, an informed consensus can help foster an environment where regulatory policies are favorable to creative methodologies. It will be necessary to such efforts as obstacles are broken down and sustainable chemical design is advanced (Figure 7.11).

7.4.3. Industry Resistance to Change

Resistance to change in sectors has a very strong impact on the transition towards green chemical technology that is highly relevant to sustainable development. Most traditional chemical manufacturers are so committed to old processes and methods that are centered on short-term economic effectiveness at the cost of long-term environmental goals. Such resistance relies on perceived risk issues around innovation, for example, development and research expenses, ambiguity about effectiveness, and potential displacement of developed supply chains. Consequently, companies may prefer risk-averse strategies, postponing the application of green technologies in chemical synthesis and design. For instance, while new materials and bio-based building blocks have far-reaching environmental implications, companies may be hesitant in their adoption due to their reliability and financial implications. This is illustrated in the synthesis of unsaturated polyester resin (UPR) from renewable feedstocks illustrated in AbdulRahman et al. (2025). In addition, regulatory systems will reflect and reinforce dominant resistance to change in business markets. The majority of industries are governed by rigid compliance standards that can discourage experimentation with new, cleaner methods. Businesses can perceive the regulatory environment as a second layer of complexity and therefore resist investing in disruptive modifications. Regulatory inertia creates a dependence on legacy processes and materials, which can reenact damage to the environment. In green chemistry synthesis, such inertia needs to be overcome to encourage innovation. For instance, the integration of improvements in materials and biopesticides as demonstrated in recent research demonstrates that regulatory uptake can be eased to the benefit of new ways so that industries can explore greener avenues and answer alarms raised in research like (Zheng et al., 2024).

Lastly, organizational culture will dictate how much an organization will embrace change. Lack of commitment on the part of leaders to sustainability can hinder the promotion of green initiatives for existing chemical processes. Core beliefs of the decision-makers may prioritize shareholder returns over environmental stewardship, and thus it is a common industry hesitation to adopt green chemical synthesis processes. Promoting education and awareness about the long-term returns on sustainable action can do something to change this line of thinking, building an organizational culture that bribes innovation and sustainability. By transcending cultural barriers and top management in sustainable transformation, the industry can enhance its competitive edge and react to worldwide endeavors towards a sustainable world of chemicals, as identified in the progress described in AbdulRahman et al. (2025) and Zheng et al. (2024) (Figure 7.12).

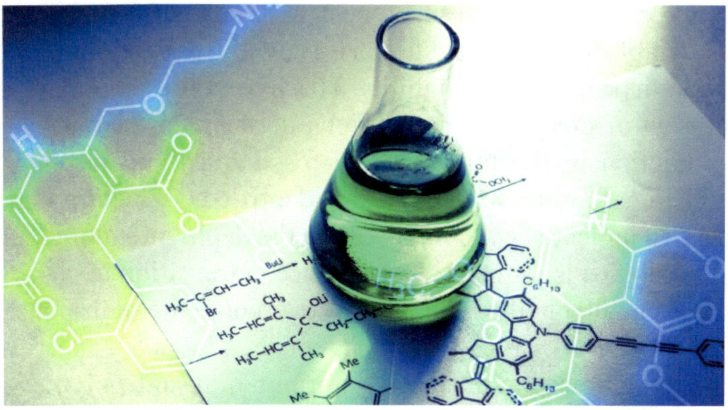

Figure 7.12. In the realm of green chemical synthesis, overcoming this inertia is essential for encouraging innovation.

Source: https://www.innovationnewsnetwork.com/wp-content/uploads/2019/12/ST27-McGillU1-image-%C2%A9-iStock-Garsya.jpg.

7.5. CASE STUDIES

Case study studies within the field of green chemical synthesis and design reports an extensive array of advancements in the use of sustainable approaches. One of the exemplary differentiators is the application of supercritical carbon dioxide ($scCO_2$) as a green solvent, which has been gaining attention from numerous industries. According to research, $scCO_2$ possesses unique properties that render it a good solvent for the extraction of natural products and bioactive compounds with higher efficiency than conventional solvents. Specifically, major industrial giants such as Nestlé have demonstrated the effective application of $scCO_2$

for coffee decaffeination as an indicator of not only process efficacy but also economically viable. These case studies are pivotal in illustrating the ways in which novel solvent systems can be employed to re-design traditional chemical processes into more sustainable forms, as directed by green chemistry paradigms, and contribute to the mitigation of the environmental impact of chemical manufacturing in general. In addition to $scCO_2$ utilization, another significant innovation arrives in the area of utilization of renewable energy sources in chemical manufacturing systems. A revealing case study illuminated by a rigorous operation planning model illustrates the ways in which photovoltaic electricity production and alkaline electrolyzers facilitate effective hydrogen production along with green ammonia synthesis. Through optimal harnessing of interaction between renewable energy volatility and market electricity price, the model indicates the ability of combining renewable energy units and chemical synthesis processes. Results of the case study present economically viable parameters, such as a levelized cost of green ammonia synthesis and high projected profitability. These outcomes elucidate the efficacy of the integration of green energy solutions in chemical synthesis, promoting systematic reduction in the utilization of fossil fuels and moving towards a sustainable chemical future. Moreover, the identified engineering challenges in these case studies indicate how complexity in the deployment of green chemical synthesis technologies is rampant. For instance, a transition to $scCO_2$ and other green solvents is challenged by the need to overcome the high-pressure reactor design, process control, and safety management. The analysis of case studies from industries such as pharmaceutical formulation and polymer processing illustrates how high-end modeling, material innovation, and automation play a significant role in overcoming such challenges. With research into effective deployments in varied applications, industry stakeholders and researchers are able to learn about the directions of sustainable manufacturing. These insights not merely contribute to current practices but also inspire further technological advancements, such that the principles of green chemistry are not merely theoretical but are in fact driving a more circular and sustainable chemical economy in design and production, as seen from studies in the literature (Sajal et al., 2025; Cheng et al., 2024).

7.5.1. Successful Green Synthesis Examples

The advancement in green chemical synthesis has also prompted new techniques in material science, primarily through the synthesis of metal sulfides. Transition metal sulfides, with their multifaceted functional properties, provide a significant alternative to traditional metal oxides. Recent research focuses on enhancing the colloidal stability of the sulfide suspensions, critical not merely for their application in bioimaging but also in other diverse sectors. A key part of this advancement is the development of strong functionalization strategies that are responsive to the less well understood surface chemistry of metal sulfides. This knowledge gap is meant to emphasize the importance of an interdisciplinary experimental and theoretical approach to untangle the complex interactions at play, paving the way for more improved applications of these dynamic materials to a wide range of industries (Dengo et al., 2017). In parallel with advances in inorganic synthesis, the methods outlined in the Roadmap Bioraffinage offer an optimistic prospect for bio-based economies, particularly in the Netherlands. The roadmap provides a series of short- and mid-term, but possible, objectives that include commercialization of biorefinery technologies and establishment of pilot plants. They are operationally-oriented in themselves; they are part of a broader strategic scheme of advancing sustainable operation throughout the economy. Designating four key moonshot strategies, the roadmap brings bold but realistic routes for biological feedstocks' incorporation into conventional manufacturing systems. This combined approach is an exemplar of the policy, research, and innovation interaction needed to drive a successful transition to bio-sustainability, optimizing the efficiency of green chemical synthesis and design (Annevelink et al., 2009). In addition, superior examples of green synthesis are not merely theoretical but occur in reality, demonstrating their efficacy and functionality. For instance, green synthesis of nanoparticles from plant extracts illustrates a novel pathway that avoids the majority of the environmental issues concerned with conventional approaches as well as the production of material rich in catalytic and antibacterial activities. The biogenic route illustrates the ideal way that utilization of natural resources can lead to sustainable solutions in place of hazardous chemical procedures. The wise application of green chemistry principles is not only minimized loss and maximized resource recovery but also allows for the production of materials that are environmentally and health-friendly. Such achievement in green synthesis not only justifies functional efficacy but also inspires a paradigm shift towards more sustainable and responsible chemical construction (Dengo et al., 2017).

7.5.2. Innovations in Green Chemistry

Green chemistry technologies are revolutionizing the design and synthesis paradigm of chemicals, leading to more sustainable practice methods. Central to such advancements is the aspect of catalysis, enhancing efficiency of reactions through lowering the activation energy and maximizing atom economy. Green catalysts have come into prominence even more so since they are designed to generate minimal waste and prevent toxic chemicals in reactions. By facilitating reaction under less severe conditions, such catalysts not only reduce the consumption of energy but also ensure safer production methods. The existing research into various green catalysts bears testament to their versatility and potential, marking a landmark shift in sustainable chemical practice that is in line with environmental sustainability goals (Sakhahari, 2025). Another innovation under green chemistry is the incorporation of biomimicry into chemical synthesis. This approach draws inspiration from natural systems and processes to build new methods that are efficient and sustainable. An example is biomimetic catalysts that replicate enzymatic processes and can lead to reduced energy inputs and higher selectivity in chemical conversions. The promotion of such approaches underscore the importance of learning from nature in order to develop processes that are ecologically sound. This shift toward biomimicry is part of a greater movement to align industrial practices with environmental health, as industries have begun preferring processes that minimize the creation of harmful byproducts and utilize renewable sources (Ch et al., 2025). Furthermore, the development of solvent-free reactions and bio-based materials is one example of how green chemistry innovations are revolutionizing industrial uses. By eliminating harmful solvents from chemical reactions, these methods significantly reduce the environmental footprint of production processes. The synthesis of biodegradable materials and other "green" chemicals illustrates the immense potential of green chemistry to address some of mankind's most significant environmental issues. Green chemistry methods are progressively being adopted within the pharmaceutical industry, where the focus on cleaner synthetic routes has led to lower toxic waste production and improved overall product safety. The economic advantages of adopting such methods, coupled with the requirement for environmental responsibility, place green chemistry technologies in the center of tomorrow's advancements in chemical design and synthesis (Sakhahari, 2025).

7.5.3. Lessons Learned from Failures

In the field of green chemical design and synthesis, failure is typically a motivator toward innovation and improvement. Failures in producing environmentally friendly alternatives to traditional chemical approaches delineate the balance between effectiveness and environmental responsibility. A prime example is found in combating sea biofouling, which has been accomplished conventionally through the use of toxic antifouling compounds. Their initial failure, as evidenced in their detrimental impacts on non-target marine species and the overall oceanic environment, is a valuable lesson that demonstrates the value of sustainable solutions. When scientists turn the spotlight towards less toxic and safer coatings using the properties of the surface to inhibit biofouling, they drive the message home that failure should be utilized to impart better and more responsible approaches to green chemistry (Qiu et al., 2022). In the future, learning from failed sustainable business models highlights the significance of flexibility in the development of green chemical synthesis. Learning from the reasons behind such failures enables scientists and engineers to adjust their methods and come up with more suitable strategies aligned with environmental goals. By systematically evaluating and grouping sustainable business models in various sectors—e.g., energy, healthcare, and agriculture—research has shed light on roads to success previously veiled. Every failure indicates areas where there is limited understanding, inspiring creative minds to venture into new possibilities merging environmental and economic sustainability. This cyclical process is indicative of the background of green chemical design, where learning from past failures is always followed by progressively newer and more sophisticated sustainable alternatives (Nosratabadi et al., 2019). Ultimately, learning from failure in green chemical synthesis not only enhances present practice but also informs research directions in the future. By adopting a culture that sees failure as an essential part of the innovation process, researchers can overcome the nuances of developing environmentally friendly chemical processes. Additionally, this kind of perspective instills a resilience and creativity-driven culture, prompting researchers to experiment with ambitious concepts that are initially impossible. The lessons learned from past failures are a treasure trove of knowledge, informing the continuous conversation of sustainable practices. Continued growth of green chemistry is supported by this candid acknowledgment of past mistakes, as all learning forms the foundation for advancement that prioritizes human health and environmental purity (Nosratabadi et al., 2019; Qiu et al., 2022).

7.6. CONCLUSION

In short, green chemical design and synthesis are not only more than efficiency in chemical manufacturing; they represent a revolutionary approach to address critical environmental challenges. As mentioned throughout this chapter, the incorporation of sustainable methods in chemical synthesis makes it possible for the use of renewable resources, which is greatly needed in our quest for better resource efficiency. The ethos of cascading use, the maxim of maximizing the different uses of renewable materials before their conversion to energy, is the driving philosophy of today's chemical practice. An emphasis on this paradigm can catalyze innovations that not only reduce waste but also promote a circular economy in which materials are cycled through repeatedly, thereby reducing the environmental footprint of chemical processes (Geldermann et al., 2015). Furthermore, the escalating pressures of climatic change and resource depletion challenge the field of chemistry to not only embrace but practically advocate green synthesis protocols. The dire statistics of food insecurity and ecological degradation underline the necessity of radical reform in production systems on the planet. As identified in the ecological agriculture debate, the failures of the traditional models compel us to search for other approaches more aligned with ecological principles. With a vision for green chemical synthesis, we can establish an industry capable of meeting current needs as well as staying responsive to new challenges, thereby ensuring all populations access to desired resources without compromising environmental integrity (Wijeratna, 2012). Finally, this shift towards green chemical synthesis and design is not an exercise in intellectual ingenuity but an imperative need for current and future generations. As the world grapples with the effects of climate change, resource depletion, and environmental distortion, the chemical sector is at a juncture. Green synthesis methods will provide the platform for the creation of environment-friendly modes of production that are economically not suspect. The evolution of the practices has the ability to establish an innovative framework in which chemistry plays a deciding role in formulating sustainable measures that address society's biggest demands. The convergence of the different disciplines is then required to hasten this paradigm shift so that a greener, cleaner future becomes viable (Geldermann et al., 2015; Wijeratna, 2012).

7.6.1. Summary of Key Points

Among green chemistry areas of synthesis and design, biocatalysis is a noteworthy one as a revolutionary area. The recent developments in molecular biology and biotechnology have led to the development of sustainable technology based on enzymatic conversions. These conversions are engineered to counter specific operating conditions, making resultant processes more efficient and compliant with the rules of sustainability. Through rationalization of current enzymes and optimization of novel biocatalytic reactions, the technology has found widespread applications, especially in industrial manufacturing of active pharmaceutical ingredients. The integration of protein engineering with other elements of biocatalysis, including substrate selection and medium design, operates towards maximization of process efficiency and economic viability. This approach not only promotes environmentally friendly processes but also surpasses traditional processes based on performance and stability, favoring biocatalysis as a mainstay of modern-day green chemistry (Roger et al., 2017). The shift towards a circular economy is another critical development in green chemical synthesis and design. Rethinking traditional economic models, this revolution emphasizes the importance of sustainability, with a focus on resource productivity and waste minimization. Big companies, as well as policy-makers, have come to like the economic, social, and environmental benefits packaged in a circular economy. The shift is not, however, straightforward, particularly for small companies that could find it difficult to adopt these new principles. To ease this shift, comprehensive frameworks are required to guide organizations in the creation of effective circular business models. Identification of individual elements such as take-back system and other adoption drivers has seen the implementation of a prolonged business model canvas that provides meaningful ideas and suggestions to companies trying to navigate in this disrupting economic landscape (Lewandowski, 2016). In conclusion, the integration of biocatalysis and circular economy thinking is the essence of green chemical synthesis and design. With the industry progressing further, it becomes increasingly evident that sustainable processes are not only advantageous but absolutely imperative for the future of chemical production. At the focal point of the dual focus on technological innovation and systemic economic restructuring, ongoing research and interdisciplinary collaboration will be crucial. Prying open the quest for sustainability involves perpetuation of innovation in enzymatic processes and robust frameworks that allow for the application of circular systems. This synergistic framework promises to unlock the way to a new paradigm of chemical processes that are not merely efficient but also synergistic with the broader goals of environmental conservation and sustainable development (Roger et al., 2017; Lewandowski, 2016).

7.6.2. Future Directions in Green Chemistry

The future of green chemistry will see revolutionary advancements, particularly in the realm of mechanochemistry. Mechanical force is applied to drive chemical reactions, which is a revolutionary departure from traditional approaches to synthesizing materials. It eliminates the need for toxic solvents, thus addressing one of the major environmental concerns involved with traditional chemical processes. The mechanochemical ability to drive reactions with very little input energy is also consonant with the principles of sustainability and efficiency found in green chemistry. Scaling these processes up for industrial use is to be expected in future advances, which may involve incorporating automation and control technologies to improve reproducibility and efficiency. The avenue of mechanochemistry to help recover and recycle valuable materials makes it a highly promising research area and application. With continued advances in mechanochemistry, it will no doubt take center stage in the future of green chemical synthesis and design, pushing the boundaries of sustainable approaches in the chemical sector (Chetry, 2025). Another possible area of research for green chemistry is exploring plant-based catalysts. These green alternatives leverage the inherent biochemical nature of plant biomass, enabling the sustainable transformation of various organic substances. This shift towards plant catalysts enhances both environmental safety — being biodegradable and non-toxic — and matches the rising demand for renewable feedstocks. The latest advances in this area have reported significant improvement in catalytic activity and selectivity at mild conditions, which defines the prospect of these natural catalysts in minimizing harmful waste. Additionally, recyclability by plant-based catalysts offers a further aspect of sustainability that compensates for the disadvantage of waste traditionally associated with traditional catalytic processes. Future research directions will include optimizing extraction and preparation processes to enhance the usability and effectiveness of these environmentally friendly catalysts in industrial uses (Maniya et al., 2025). Apart from mechanochemistry and plant catalysts, the future of green chemistry must embrace the most recent technological advancements in the form of artificial intelligence and machine learning. These technologies can facilitate the optimization of chemical processes in more efficient design and synthesis processes with sustainability as the focus. Leveraging huge amounts of data, machine learning programs can predict the outcome of reactions, identify safe and green reaction routes, and suggest optimal conditions for green synthesis. This technological adoption not only streamlines research methods but also greener practices more scalable in the chemical sector. The future of green chemistry, therefore, lies not just in developing new materials and processes but in the ability to employ cutting-edge technologies to create predictive models that govern green chemistry practice. Such developments will surely enrich the discipline, enabling greater innovation and environmental stewardship in chemical design and synthesis.

7.6.3. Call to Action for Researchers and Industries

The imperative for action among researchers in the field of green chemical design and synthesis cannot be overstated. With the discipline confronting impending environmental issues and the call for greater resource efficiency growing, the researcher is called upon to utilize innovative modes of working that displace conventional paradigms. Intensification of processes takes center stage in this effort, where intensified equipment and hybrid strategies are created to enhance productivity greatly and reduce energy consumption. The necessity for industrial proofing with combined academic-industry cooperation is evident as seen in the commissioning of the Toulouse plant aimed at accelerating green process engineering uptake. By establishing partnerships with universities, equipment suppliers, and industry leaders, we can establish a robust database of success stories for feeding back into research and utilization in sustainable chemical practice. Industries are important to this transformative agenda as well because they are the primary drivers of chemical production practice. It is the responsibility of industry leaders not merely to adopt green chemical technologies but to also encourage greater cooperation in their industries towards environmental sustainability. Such engagement includes the integration of renewable sources into production and cascading use practices, in which the by-products are effectively recycled into the production process before they are utilized for energy conversion. Interdisciplinarity highlighted in recent studies highlights the value of inter-disciplinary knowledge and practice, focusing on how inter-disciplinary cooperation can enhance the effectiveness of resources. As industries adopt these concepts as a vocation, they will not only improve their environmental profiles but also be in conformity with future regulation and consumer demand for eco-friendly products (Geldermann et al., 2015). In summary, the cry for urgent action is a rallying cry for industries and researchers alike that calls for concerted action with innovation and environmentally friendly approaches in chemical synthesis. As the academic research paradigms shift towards greener approaches, it is the duty of the industry to catch up with such developments through continuous interaction and transparency. The collective effort ought to be towards creating a strong framework for green chemical synthesis that emphasizes continuous improvement and ecological resilience. In issuing this call to action, both the industries and researchers can make a paradigm shift to a greener future, which benefits the environment, society, and economy in the long run. Only by combined efforts shall we be in a position to make a revolutionary impact in the midst of chemical production and design.

REFERENCES

1. AbdulRahman, J. P., Nair, C. P. R., & Mathew, S. (2025). A drive towards green – Unsaturated polyester resins: A critical review of bio-based building blocks and reactive diluents. *Polymers from Renewable Resources, 16*, 123–156. https://www.semanticscholar.org/paper/07700ef209e54e8ee3a2edf46f766cac920375ca (accessed on 24th October 2025).
2. Aiduang, W., Jatuwong, K., Luangharn, T., Jinanukul, P., Thamjaree, W., Teeraphantuvat, T., Waroonkun, T., et al. (2024). A review delving into the factors influencing mycelium-based green composites (MBCs) production and their properties for long-term sustainability targets. *Biomimetics, 9*. https://www.semanticscholar.org/paper/755fb590359538d42a041e6eee90aa63e5d69d40 (accessed on 24th October 2025).
3. Annevelink, E., Broeze, J., van Ree, R., et al. (2009). Opportunities for Dutch biorefineries. https://core.ac.uk/download/pdf/29248632.pdf (accessed on 24th October 2025).
4. Ballarin, B., Barreca, D., Boanini, E., Cassani, D., et al. (2017). Supported gold nanoparticles for alcohols oxidation in continuous-flow heterogeneous systems. https://core.ac.uk/download/84487017.pdf (accessed on 24th October 2025).
5. Ballivet-Tkatchenko, D., Camy, S., & Condoret, J. S. (2005). Carbon dioxide, a solvent and synthon for green chemistry. https://core.ac.uk/download/12039955.pdf (accessed on 24th October 2025).
6. Boyen, C., Børresen, T., Dobson, A., Höfle, G., et al. (2025). Marine biotechnology: A new vision and strategy for Europe. https://core.ac.uk/download/pdf/29241634.pdf (accessed on 24th October 2025).
7. Calvin, K., Dasgupta, D., Krinner, G., Mukherji, A., Thorne, P., Trisos, C. H., Romero, J., et al. (2023). IPCC, 2023: Climate change 2023: Synthesis report. Contribution of Working Groups I, II and III to the Sixth Assessment Report of the Intergovernmental Panel on Climate Change [Core Writing Team, H. Lee and J. Romero (Eds.)]. IPCC, Geneva, Switzerland. https://doi.org/10.59327/ipcc/ar6-9789291691647 (accessed on 24th October 2025).
8. Centeno Gallego, M. Á., Laguna Espitia, O. H., Navarro de Miguel, J. C., Odriozola Gordón, J., et al. (2018). Policies and motivations for the CO_2 valorization through the Sabatier reaction using structured catalysts: A review of the most recent advances. https://core.ac.uk/download/188008301.pdf (accessed on 24th October 2025).
9. Chen, E. (2025). Exploring green chemistry: Sustainable approaches to chemical synthesis. *Revita Latinoamericana de Química*. https://www.

semanticscholar.org/paper/37ef2209f3997399352c45ee568c965c16604a1a (accessed on 24th October 2025).

10. Chen, G., Sun, G., & Tang, S. (2019). Computer-assisted design of environmentally friendly and light-stable fluorescent dyes for textile applications. https://core.ac.uk/download/286361711.pdf (accessed on 24th October 2025).
11. Cheng, L., Ma, L., Cao, Y., Zhao, Z., & Xu, Y. (2024). Operational scheduling model of electricity-hydrogen-chemical systems based on sequential production simulation and its application. *2024 5th International Conference on Smart Grid and Energy Engineering (SGEE)*. https://www.semanticscholar.org/paper/f5ca5cbaff94403d0ec6fa7ed725b2613c5f9b90 (accessed on 24th October 2025).
12. Chetry, A. B. (2025). Mechanochemistry: A new frontier in chemical synthesis. *Journal of Chemical Research, 49*. https://www.semanticscholar.org/paper/b601e758a61cbc16926f466890f1def81ccafbc0 (accessed on 24th October 2025).
13. Choudhury, H. K. (2025). Synthesis of mosquito repellent product through green chemistry. *Journal of Advances in Science and Technology*. https://www.semanticscholar.org/paper/08779d3108baf96c127c86f2afe45cdffed89c02 (accessed on 24th October 2025).
14. Cukalovic, A. (2009). *Fifth International Conference on Renewable Resources and Biorefineries, 10–12 June 2009, Ghent, Belgium*. https://core.ac.uk/download/55849830.pdf (accessed on 24th October 2025).
15. Deguine, J. P., Ferron, P., & Russell, D. (2008). Agroecology and ecological engineering for pest management: Cotton protection as a case study. http://agritrop.cirad.fr/547062/1/document_547062.pdf (accessed on 24th October 2025).
16. Dengo, N., Gross, S., & Vittadini, A. (2017). Functionalization of colloidal transition metal sulphides nanocrystals: A fascinating and challenging playground for the chemist. https://core.ac.uk/download/141699569.pdf (accessed on 24th October 2025).
17. Di Maniya, E., & Benchekroun, Y. (2025). Green synthesis of organic compounds using plant-based catalysts. *International Journal of Emerging Trends in Chemistry (IJETC)*. https://www.semanticscholar.org/paper/23474014b960a4e39f7dc6498519d8cda44c4a41 (accessed on 24th October 2025).
18. Geldermann, J., Kolbe, L. M., Krause, A., Mai, M., et al. (2015). Improved resource efficiency and cascading utilization of renewable materials. https://core.ac.uk/download/151157330.pdf (accessed on 24th October 2025).

19. Gingerich, E. (2015). Eco-friendly marketing: Beyond the label. https://core.ac.uk/download/144553107.pdf (accessed on 24th October 2025).
20. Jacquemin, L., Pontalier, P. Y., & Sablayrolles, C. (2011). Life cycle assessment (LCA) applied to the process industry: A review. https://core.ac.uk/download/12043304.pdf (accessed on 24th October 2025).
21. James, O. O. (2013). Improving carbon and process utilization for biomass-to-liquid fuel operations. https://core.ac.uk/download/16436077.pdf (accessed on 24th October 2025).
22. Lewandowski, M. (2016). Designing the business models for circular economy—Towards the conceptual framework. *Sustainability, 8*, 43. https://doi.org/10.3390/su8010043 (accessed on 24th October 2025).
23. Li, J., Cong, C., Garcia, N. R., Yuan, W., Liang, H., & He, Z. (2025). The impact of supply chain management on corporate financial performance from a green perspective: Basis for sustainable business practices framework. *Journal of Business and Management Studies*. https://www.semanticscholar.org/paper/30ea630ebd0fcb68d8f028be071fb62aa136e8e0 (accessed on 24th October 2025).
24. Luo, Y., Abidian, M. R., Ahn, J. H., Akinwande, D., Andrews, A. M., Antonietti, M., Bao, Z., et al. (2023). Technology roadmap for flexible sensors. *ACS Nano, 17*, 5211–5295. https://doi.org/10.1021/acsnano.2c12606 (accessed on 24th October 2025).
25. Medrano-García, J. D., Chagas, M. T., & Guillén-Gosálbez, G. (2025). Integrating the reverse Boudouard reaction for a more efficient green methanol synthesis from CO_2 and renewable energy. *ACS Sustainable Chemistry & Engineering, 13*, 7088–7097. https://www.semanticscholar.org/paper/1a21a70d01a141f86a49ba8f18ff1ceebec74d16 (accessed on 24th October 2025).
26. Nosratabadi, S., Mosavi, A., Band, S. S., Zavadskas, E. K., Rakotonirainy, A., & Chau, K. W. (2019). Sustainable business models: A review. *Sustainability, 11*, 1663. https://doi.org/10.3390/su11061663 (accessed on 24th October 2025).
27. Ojukwu, U., Arthur, C., Jonathan, W., Ehigiegba, O., Yussuf, M., & Olawuyi, M. (2025). Green chemistry in pharmaceutical synthesis: Implications for sustainable public health. *European Journal of Scientific Research and Reviews*. https://www.semanticscholar.org/paper/3ea7b2d35d9abe52e2b253c0329f127725ff5baf (accessed on 24th October 2025).
28. Qiu, H., Feng, K., Gapeeva, A., Meurisch, K., Kaps, S., Li, X., Yu, L., et al. (2022). Functional polymer materials for modern marine biofouling control. *Progress in Polymer Science, 127*, 101516. https://

doi.org/10.1016/j.progpolymsci.2022.101516 (accessed on 24th October 2025).

29. Sai Laya, C. H., Saroja, K. S. N., Aswini, V. L., Sameera, M., Jayasree, M., Rani, E. U., Kamala, G., et al. (2025). A review of industrial applications of green chemistry. *Journal of Pharma Insights and Research*. https://www.semanticscholar.org/paper/26f8b58af8118b4f6accb051efc5beaaddeeb245 (accessed on 24th October 2025).

30. Sair, Y. Z., Ashraf, H., Ghafoor, M. B. A., Shanza, S., Shabbir, M., Ali, Z., Khan, A., et al. (2025). Eco-innovations in nanoscience: Advances in sustainable synthesis and physico-chemical characterization of silver nanoparticles and their biomedical applications. *Scholars Academic Journal of Biosciences*. https://www.semanticscholar.org/paper/9d5d1b124708881314985ba18dc4d64c7abe405e (accessed on 24th October 2025).

31. Sajal, N. H., & Sutar, R. (2025). Supercritical CO_2 as a green solvent: A comprehensive review of its applications in industrial processes. *American Journal of Advanced Technology and Engineering Solutions*. https://www.semanticscholar.org/paper/d410740935f67b4ddb555ca711b5873aa1a9b868 (accessed on 24th October 2025).

32. Sakhahari, K. V. (2025). Green chemistry and catalysis: Innovations toward a cleaner and sustainable world. *International Journal of Pharmaceutical Research and Applications*. https://www.semanticscholar.org/paper/4a3292276b1c806584dc1e557279ec5e81c0f3e8 (accessed on 24th October 2025).

33. Sheldon, R. A., & Woodley, J. M. (2017). Role of biocatalysis in sustainable chemistry. *Chemical Reviews, 118*, 801–838. https://doi.org/10.1021/acs.chemrev.7b00203 (accessed on 24th October 2025).

34. Stoica, C., Banciu, A., Idriss, H., Lian, J. Z., Pătraşcu, A. M., Cucurachi, S., Richeter, S., et al. (2025). Green design and life cycle assessment of novel thiophene-based surfactants to balance their synthesis performance and environmental impact. *Materials, 18*. https://www.semanticscholar.org/paper/fd74ae2d3176fc270cf0719dc9e869427381b00b (accessed on 24th October 2025).

35. Sun, D., Zhang, Y., Zhou, Y., Nie, Y., Ban, L., Wu, D., Yang, S., et al. (2025). Photocatalytic and electrochemical synthesis of biofuel via efficient valorization of biomass. *Advanced Energy Materials, 15*. https://www.semanticscholar.org/paper/a7db630f2c1c10faa25d3b1c05af89e30821ea07 (accessed on 24th October 2025).

36. Ukamaka, A. C., Sanusi, A. N., Asere, J. B., & Sanusi, H. K. (2025). Machine learning for predicting environmental impact in green buildings: A systematic review. *Asian Journal of Geographical Research.* https://www.semanticscholar.org/paper/c0a2f5d7ed7c3ba85e5ea896d1b7b106194b5fe2 (accessed on 24th October 2025).

37. Wijeratna, A. (2012). *Fed up: Now's the Time to Invest in Agro-Ecology.* https://core.ac.uk/download/71362045.pdf (accessed on 24th October 2025).

38. Yushau, A., Kabo, K. S., & Adabara, S. A. (2025). Response surface methodology optimization and modeling of green synthesis conditions for TiO_2-ZnO nanocomposites using *Vigna unguiculata* L. extract. *European Journal of Chemistry.* https://www.semanticscholar.org/paper/d09a759af86709bbb55a260e884470c352986413 (accessed on 24th October 2025).

39. Zheng, Y., Chen, M., Zhang, R., & Xue, W. (2024). Design, synthesis, antimicrobial activity, and mechanism of novel 3-(2,4-dichlorophenyl)-[1,2,4]triazolo[3,4-b][1,3,4]thiadiazole derivatives. *Pest Management Science.* https://www.semanticscholar.org/paper/aa0d1ed24a93518d17b19598a17c06e3f69afd41 (accessed on 24th October 2025).

CHAPTER 8

WASTE MINIMIZATION AND POLLUTION PREVENTION

Contents

INTRODUCTION

In recent decades, the increased awareness of environmental degradation has heightened the principal focus on waste minimization and pollution prevention methods. These methods aim to limit the generation of waste and discharge of hazardous pollutants into the environment, thereby preventing adverse effects on human health and ecosystems. Industrial processes, often major contributors to environmental pollution, have been under examination for their waste disposal practices. As Lawrence et al. (2005) report, means such as industrial ecology provide frameworks for implementing sustainable practice, which coordinate economic development and environmental responsibility. Through the integration of sustainable practice, not only can the efficiency of operation be increased but even the life cycle of resources can be extended, thereby facilitating a more sustainable sector. The imperative for effective waste control is highlighted by the increasing volume of literature regarding innovative pollution control techniques. The identification of the least risky substance alternatives is crucial in preventing pollution at its origin, as illustrated in Wise (1994). This preventive measure requires cooperation among policymakers, engineers, and industry captains to find integrated solutions in addressing waste reduction. Besides, testing of technologies like bioremediation, stabilization techniques, and new wastewater treatment offers several ways to achieve pollution prevention goals. Industrial organizations can decrease their environmental footprint significantly by employing such technologies while complying with more stringent regulations. Moreover, economic implications of waste minimization and pollution prevention measures cannot be ignored. Firms that adopt waste reduction schemes are likely to achieve cost savings in the long term in terms of greater efficiency and reduced liability for environmental pollution. Successful instances demonstrate how such approaches can lead to innovation and market-based competitive advantage. By making manufacturing processes itself reduce the creation of waste, as extensively argued in et al. (2005), companies can become industry leaders for sustainable production. By embracing waste reduction and pollution prevention as key policies, businesses not only contribute to protection of the environment but also ensure the sustainability of their own position in the increasingly environmentally-conscious market, thereby combining commercial success with environmental stewardship.

Definition of Waste Minimization

An in-depth view of waste minimization is essential in addressing environmental concerns. Waste minimization is a collection of steps that assist in limiting the amount of waste generated at the point of origin and, thus, avoid pollution and optimize the use of resources. It focuses energies on waste generation prior to its occurrence, saving resources and having an enormous difference in decreasing environmental impacts. Effective waste minimization requires a series of approaches, including process modification, use of ecologically friendly materials, and improved working practices. Effective application of waste minimization can allow companies to achieve compliance with environmental regulations at the same time as fostering an organizational culture of sustainability

(Schroeder et al., 2018). Landfill reduction and reducing hazardous waste can allow business firms to improve their sustainability performance and encourage a favorable ecosystem. Another vital element of waste minimization is its link with circular economy practice. Circular economy adopts the concept of maintaining products, materials, and resources in use for as long as they remain useful, thus minimizing wastage creation. This sustainable approach encourages resource recovery and supports eco-friendly production practices. According to recent research, circular economy practices highly match the Sustainable Development Goals (SDGs), particularly responsible production and consumption (SDG 12) (Kalmykova et al., 2017).

Figure 8.1. Effective waste minimization encompasses various strategies, including process modification, the utilization of eco-friendly materials, and enhanced operational practices.

Source: https://res.cloudinary.com/pillarshotels/image/upload/f_auto/web/cms/resources/sustainability/waste-w1800h1360.jpg.

This congruence indicates the broader relevance of waste reduction beyond organizational benefits; it directly aids overall global sustainability objectives. By integrating waste minimization principles into their operational systems, firms are actually participating in the transition towards a circular economy while mitigating environmental concerns at the same time. Moreover, waste minimization efforts have a tendency to produce substantial cost savings for firms. Material wastage minimization can be used directly in reducing the cost of production, as less material wastage translates into lower purchasing costs. By optimizing production activities and resources, companies not only save waste but also optimize their operations efficiency. Additionally, companies that institute waste minimization programs tend to have improved public relations and customer loyalty, as consumers increasingly desire businesses that are green. Lastly, waste minimization support represents a more overall pressure by society on businesses to act responsibly where the environment is concerned (Kalmykova et al., 2017). The inclusion of waste minimization within company strategy is not only for ethical reasons but also gives a competitive advantage through the attraction of an increasing market niche interested in innovation and sustainability (Figure 8.1).

Importance of Pollution Prevention

The importance of pollution prevention cannot be overstated, especially as the global environmental challenges build speed in intensity. Pollution affects not only the local environment but also human health, fiscal viability, and sociological systems. The implementation of effective pollution prevention strategies is paramount to reverse these challenges, thus contributing to a sustainable future. For instance, pollution prevention strategies can significantly minimize the volume of industrial waste that is toxic in nature. This not only relieves pressure on waste management infrastructure but also assists in the achievement of Sustainable Development Goals (SDGs), particularly those related to responsible consumption and production, clean water, and decent economic growth. By prioritizing pollution prevention, policymakers can set up structures that encourage industries to transition towards green practices, incorporate green technology, and foster an environmentally conscious society overall (Schroeder et al., 2018). Beyond the environmental impacts, pollution avoidance is also crucial in public health. The evidence to date has indicated that the healthcare sector is a significant generator of pollution based on intense waste and energy consumption with widespread impacts on community health. Prevention of such pollution is required to enhance health improvements as well as healthcare activities' environmental footprints. Literature points to the fact that a yield strategy to increase output, say, to decrease infection rates must tackle not only traditional practice but also ecological health and clinical needs. In particular, actions that minimize the use of single-use disposables have the potential to halve greenhouse gas emissions. As several healthcare researches have attested, this shift to waste reduction can lead to important improvements in patient safety and sustainable operations (Jodi et al., 2018). Thus, the incorporation of pollution prevention concepts into healthcare policies plays an essential role in promoting a culture of safety and sustainability. Furthermore, the embracement of pollution prevention as a standard practice can bring extensive economic benefits. Through the reduction of wastage and approaching more efficient use of resources, businesses will not only save money but enhance their competitive edge. To illustrate, adopting waste minimization approaches, aside from helping companies comply with environmental regulations, enhances their corporate image and customer loyalty. Value-sensitive customers are choosing companies that are sustainable-driven with greater market share and profitability. It is therefore prudent for companies to examine their business practices and integrate pollution prevention methods as an integral part of strategic planning. This innovation is especially in a period when going green is becoming a deciding factor in purchasing choices, thus uniting pollution prevention with economic prosperity and sustainability in a dynamic market (Schroeder et al., 2018).

8.1. THE CONCEPT OF WASTE MINIMIZATION

The waste minimization philosophy is premised on the fact that efficient management of resources is key to sustainable development. Waste minimization, fundamentally, entails minimizing the generation of waste through deliberate measures and practices by industries. The comprehensive strategy entails process redesigning, the implementation of alternative materials, and improving equipment efficiency. With waste minimization integrated into the manufacturing process, companies can achieve dramatic reductions in environmental impacts while simultaneously lowering raw material and waste disposal expenses. Here, waste management policy needs to align with business goals because non-alignment may result in confusion and inefficiency in application, raising the imperative of certain guidelines that make the vision of zero manufacturing waste possible (Bjelkemyr et al., 2013). Waste minimization as a business process is not only environmentally significant but also economically sensible. It is implied by evidence that waste prevention is far less expensive than treating waste after it is generated, as described in different pollution prevention studies (Bierma et al., 1995). Shifting the paradigm towards proactive waste management fosters a sustainable culture of organizations. With the implementation of waste minimization practices, businesses can improve their competitive advantage while supporting the overall prosperity of the economy. The emphasis placed on prevention and minimization as guiding principles demonstrates an overall society acknowledgment that environmental sustainability and economic balance are not in conflict but, instead, blend harmoniously with one another, advancing innovations to the world and to business sustainability. Furthermore, the complex nature of applying waste minimization principles in various industry sectors requires an individualized approach. It is essential that organizations assess their individualized processes and waste streams to build effective strategies that harmonize with their operation framework. Integrating diverse methodologies—from lean manufacturing practice to innovative recycling programs—can facilitate meaningful waste reduction breakthroughs. These successes, however, call for constant practice and improvement assessment as well as employee engagement at every level. By a common commitment to minimizing waste, businesses are not merely lowering their own environmental footprint but also driving broader change across their sectors, confirming once again that sustainable practices will be paramount to future development and growth (Bjelkemyr et al., 2013) (Figure 8.2).

Figure 8.2. Incorporating waste minimization into business practices is not only an environmental imperative but also an economically sound strategy.

Source: https://earth5r.org/wp-content/uploads/2025/04/Recycling-waste-Management-CSR-ESG-Sustainability-NGO-Earth5R-Mumbai.jpg.

8.1.1. Historical Context and Development

The concept of waste minimization goes back to the age of the industrial revolution when mass production began to create not only economic growth but also extensive environmental degradation. The initial industrialization used to focus on efficiency and profitability at the cost of ecological issues, leading to high waste and pollution. As the number of factories grew, worry about the adverse effects of industry's products on public health and the environment grew as well. The development of waste reduction technologies was made imperative by the growth of industrial processes. Thus, various legislative systems were established in the 20th century due to growing public interest in environmental well-being and sustainability. This historical trajectory represents a shift from a purely economic focus to an integrated approach considering conservation of the environment as a natural part of modern industrial practice, in effect paving the way for the practices of today in preventing pollution and reducing waste. As the 20th century moved towards its later stages, environmental campaigns increased in intensity, and more systematic practice of preventing pollution and reducing waste followed. With the rise of the Circular Economy (CE) in the 1980s and 1990s came a systematic solution to integrating sustainability into production. CE is more than reducing waste; it is about rethinking the utilization of resources along the value chain and, therefore, providing a regenerative solution to economic growth. Moreover, the growing application of case studies linking CE practices with efficient waste management outcomes has encouraged global adoption of these measures by various industries. These practices highlight recovery, recycling, and sustainable consumption as key aspects of modern industrial policy.

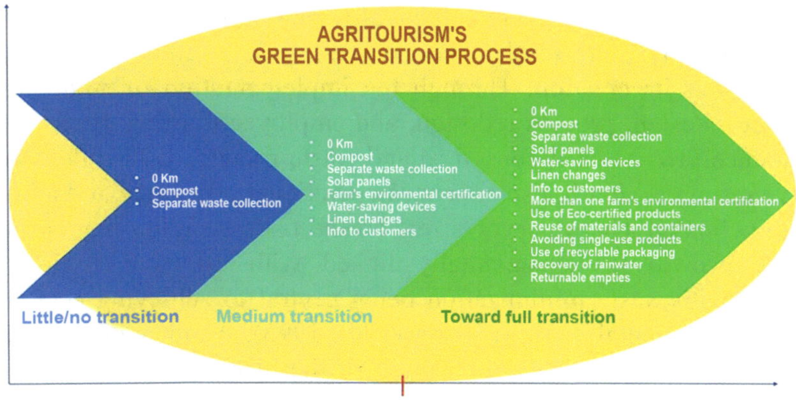

Figure 8.3. The rise of the circular economy (CE) concept in the 1980s and 1990s provided a structured framework for integrating sustainability into production systems.

Source: https://www.frontiersin.org/files/Articles/1174623/fsufs-07-1174623-HTML-r1/image_m/fsufs-07-1174623-g001.jpg.

These developments reflect a deep shift in the global discussion of waste management, demonstrating a dramatic shift toward modernizing strategies that incorporate environmental factors as a fundamental element of economic decisions (Kalmykova et al., 2017). Now, historical experience in minimization guides ongoing technological and policy innovation to reduce environmental prints. Healthcare, on the other hand, presents its own unique challenges due to the dual need of patient safety and waste reduction. Traditionally, the need for clean rooms has resulted in increased single-use equipment, which has created most healthcare waste. However, recent research suggests a potential realignment of such practices, suggesting some middle ground between infection control and pollution prevention. This shift is in alignment with past learning, proving that waste and pollution need not come at the expense of compromise on safety or quality. Instead, a careful understanding of risk classification and waste management can facilitate decision-making on quality. The shifting landscape of waste minimization proves that ancient ways can offer valuable insight to modern-day strategies for reducing pollution to ensure economic practices that are sustainable (Jodi et al., 2018) (Figure 8.3).

8.1.2. Key Principles of Waste Minimization

The key waste minimization principles are those that are taken to minimize the generation of waste at source rather than merely waste management upon creation. Central to this concept is the reality that in the majority of instances, it is economically more prudent not to create waste rather than to manage the waste after its creation. Through the implementation of strategies such as product redesign, process redesign, and employee training, businesses can better reduce raw material consumed and waste produced. This does not just conserve resources, but it also enhances operational efficiency and a healthier economic climate. These concepts rest on the understanding that sustainable economic growth is fundamentally linked with effective environmental protection, with waste minimization being central to achieving economic as well as environmental goals (Bierma et al., 1995). Another distinct element of waste minimization is the application of the hierarchy of waste, ordering waste management practices in a specific sequence: prevention, reduction, reuse, recycling, and disposal. This pyramidal approach encourages firms to favor waste minimization options above recycling and disposal as an initial choice. This proactive approach enables firms to introduce waste minimization techniques into their operations and streamline their production activities to make them more sustainable. For instance, firms are adopting cleaner production practices that streamline operations and minimize waste creation. Implementation of such principles not only ensures environmental compliance but also enhances the image of an organization in the market, a parameter most likely to shape consumer loyalty and decision (Bjelkemyr et al., 2013). Common understanding and communication among stakeholders are also key towards the achievement of waste minimization goals. Internal and external partnerships allow organizations to share knowledge, resources, and best practices in waste management strategies. Employees play central roles to spot areas of opportunity for waste reduction in their routine work. As organizations develop a culture that focuses on waste reduction, employees are motivated to submit innovative ideas and solutions. This collaborative effort results in efficient use of resources and ultimately contributes to an overall vision to promote sustainability in industrial operations. In addition, linking waste minimization activities to corporate social responsibility efforts can highlight a firm's commitment to environmental care, strengthening further its business integrity and market standing (Bierma et al., 1995).

8.1.3. Benefits of Implementing Waste Minimization

Use of waste minimization strategies brings with it many benefits, particularly prevention of pollution. Through reduction of the volume of waste generated in the initial stage, organizations have not only the ability to lower disposal costs, but they also enhance their operational efficiency. These strategies typically entail introduction of cleaner production techniques, which help save significant amounts of resources. For instance, rather than managing waste post-production, companies may re-engineer their operations to lower input materials and energy use. This upfront endeavor, subsequently, helps decrease emissions and mitigate adverse health effects associated with pollution, as suggested by the correlation between lowered industrial pollutant concentration rates and improved public health conditions in the United States (Kumar et al., 2021). Lastly, waste minimization is a friendly green solution that merges economic returns with sustainability. Also, waste minimization promotes innovation and results in the development of new technologies that enhance sustainability. While searching and adopting effective waste reduction strategies, organizations are likely to discover new ways and substances that can be applied in replacement of conventional means. The pilot initiatives of Argonne National Laboratory for the transfer of pollution prevention technologies to the private industry are the best example of how targeted programs can make technological advancements and education (Gatrone et al., 1998). Moreover, companies that implement waste minimization programs can also become more marketable because consumers are increasingly looking for companies that are environmentally friendly. This market transition fosters continuous enhancement and adaptation, creating a culture of sustainability that can lead to long-term economic advantages. This way, the integration of waste minimization not only addresses immediate waste issues but also sets the stage for long-term innovation and growth. Lastly, the integration of waste minimization programs greatly supports corporate social responsibility (CSR) practices, enhancing a company's reputation and stakeholder relationships. As companies actively embark on waste reduction initiatives, they are sending a message to stakeholders—employees, customers, and investors—that they are concerned about the environment and human well-being. Such dedication can yield enhanced employee morale and productivity since employees are proud to be part of an organization that prioritizes sustainability. A good CSR program can also cement customer loyalty among those customers who are becoming more environmentally conscious in their purchases. Furthermore, regulatory agencies have a tendency to favor companies that openly address waste, which can benefit them with added tax benefits and less examination. Thus, long-term savings from waste reduction surpass short-term reductions, and there is strong reason for businesses to adopt such practices into their business models (Figure 8.4).

Figure 8.4. As businesses actively engage in waste reduction initiatives, they signal to stakeholders—employees, consumers, and investors alike—that they are committed to environmental preservation and community well-being.

Source: https://www.mdpi.com/sustainability/sustainability-14-01222/article_deploy/html/images/sustainability-14-01222-g001.png.

8.2. TYPES OF WASTE

It is important to understand the different waste types in order to apply efficient waste minimization and pollution prevention procedures. Generally, waste can be described as following various categories, including municipal solid waste, hazardous waste, industrial waste, and electronic waste. Municipal solid waste applies mainly to day-to-day products discarded by the general public, including food wastes, packaging items, and other domestic wastes. Hazardous waste, on the other hand, refers to materials that are of major danger to public health and the environment, including chemicals, batteries, and medical waste. Industrial waste, which is produced from manufacturing processes, tends to consist of both solid and liquid waste products that have to be handled with caution to avoid their environmental effects. Each type of waste has specific management methods, and thus it is necessary to categorize and manage them in the light of waste minimization strategies. Among the myriad types of waste, e-waste creates distinct problems for waste management strategies. E-waste consists of discarded electronic equipment like computers, mobile phones, and household appliances that have become omnipresent in contemporary living. This kind of waste is particularly difficult to deal with due to its complex composition, with the inclusion of toxic compounds such as lead, mercury, and cadmium. The improper disposal of e-waste not only contaminates the environment, but it also causes severe health risks by polluting soil and water. As highlighted in current research, fresh approaches towards regulating e-waste are critical in stimulating environmentally friendly patterns and reducing the poisonous heritage of abandoned electronics (Stuart et al., 2010). Effective recycling structures and stimulating responsible consumption can be central drivers towards eradicating the effects of e-waste on the environment. Another significant category of waste is hazardous waste, which must be closely controlled and monitored because it has the potential to pose critical environmental and health risks. Hazardous waste results from various industries, including healthcare, manufacturing, and agriculture, and generally includes toxic chemicals, solvents, and biohazardous waste. Proper handling of hazardous waste is necessary as it can lead to soil pollution, air contamination, and harmful health impacts on communities. Hazardous waste reduction measures can involve the adoption of cleaner production strategies, as well as the application of Green Chemistry principles, to guarantee sustainable development at the molecular level (Paul et al., 2009). Through the integration of these strategies into waste policy, industry can effectively reduce the generation of hazardous waste and, in doing so, decrease its environmental impact as well as contribute to overall pollution prevention efforts.

8.2.1. Solid Waste

The handling of solid waste has come to be a source of concern, particularly in the context of growing cities and consumption patterns that are generating increasingly more trash. Landfills, which have been the traditional method of waste disposal, are in themselves an environmental hazard of huge proportions since they can pollute soil and groundwater by leachate and emit methane, a potent greenhouse gas. Secondly, inefficient landfill practices exacerbate public health issues because the proximity to waste plants may lead to respiratory diseases and other ailments in neighboring communities. Hence, policymakers and local communities need to have a comprehensive idea of the generation and management of solid waste in order to plan effective strategies that not only avert pollution but also minimize waste. These issues are tackled by an integration of education, technology, and good regulatory frameworks to offset the growing burden of waste on the environment and health. Mitigating solid waste and preventing pollution are critical by embracing innovative methods in waste management.

Figure 8.5. Landfills, the traditional method of waste disposal, pose significant environmental threats as they can lead to soil and groundwater contamination through leachate and emit methane, a potent greenhouse gas.

Source: https://decodingbiosphere.com/wp-content/uploads/2023/07/pexels-photo-3174350.jpeg.

Minimizing the volume of waste dumped in landfills is possible by moving from traditional ways to waste disposal methods like recycling, composting, and energy from waste programs. Such measures not only divert waste from landfill but also transform it into valuable resources. For example, composting for organic waste can recycle nutrients back to the land for sustainable farming, while recycling of paper, glass, and metals conserves energy and reduces the need for raw materials. Utilizing such processes in conjunction with public education programs can empower people and organizations to participate constructively in waste minimization. Besides, integrating practices such as process modification and source reduction can enhance effectiveness of such schemes by ensuring a holistic approach to waste minimization that benefits the economy and environment. In addressing the issues of solid waste, consideration must be taken of opportunities in pollution prevention and resource recovery. The concept of waste as resource has the ability to turn the notion of solid waste management on its head for individuals, communities, and firms. For instance, the recycling of industrial by-products, such as outlined in recent studies on distillation processes, addresses the potential for recycling value-containing materials out of waste streams. Spent wash distillery, after drying, consists of beneficial nutrients that can be reused as alternative fuel and reduce fossil fuel usage, while providing a green disposal option (Warnakulasooriya et al., 2024). By developing circular economy based on recycling and reusing, all stakeholders are able to not only attain less solid waste but also promote cleaner production process conducive to long-term environmental sustainability. This integrated model highlights the need for coordination between governmental regulations, industry practices, and community participation to assure successful waste reduction and pollution prevention (Figure 8.5).

8.2.2. Liquid Waste

Liquid waste generation is a major issue among different industries, especially those that require heavy usage of water like the fisheries industry. In environments like Muncar Banyuwangi, one of Indonesia's largest fishery business hubs, the production process creates large amounts of liquid waste, which may result in environmental pollution and odors. Such results underscore the importance of implementing waste minimization programs with cleaner production concepts. These principles aim at assessing and refining existing manufacturing practices, thus reducing waste generation at source at once. Companies, through this, will not only maximize their in-process operating efficiencies but also assist in the overall objective of preventing pollution in order to achieve sustainable development objectives as well as regulatory compliance objectives. Effective management methods for liquid waste in industrial contexts stress prevention, recycling, and proper treatment procedures. The first stage involves the determination of key sources of liquid waste during production to enable companies to deploy preventive measures with a focus towards minimizing the volume generated. An illustration includes the adoption of processes that avoid high water consumption during fish cleaning activities aimed at significantly reducing liquid waste volumes. Further, recycling treated wastewater for non-potable use in production facilities offers a viable method of conservation. This also aligns with the first green chemistry principle of preventing waste. By putting these measures in place, not only do industries comply with the environment, but they also promote the sustainability of their operations, achieving tremendous progress in reducing their environmental footprint (Tiwari et al., 2021). Beyond the pragmatic aspects, the effects of poor liquid waste management also extend beyond the short-term problem of pollution to touching public health and local ecosystems. Contaminant leaching from untreated waste to water sources destabilizes aquatic life, disrupts habitats, and risks human communities that rely upon such ecosystems. In light of these challenges, industries are liable to adopt advanced treatments and disposal methods that sufficiently annihilate toxic components in liquid waste. Directed as encapsulated in the green chemistry guidelines, emphasis on the development of safer chemicals and processes is critical. This systematic approach ensures that liquid waste management is included in the overall elimination of toxic outputs, and therefore encourages the waste minimization and pollution prevention agenda in industries (Tiwari et al., 2021) (Figure 8.6).

Figure 8.6. The generation of liquid waste is a significant concern across various industries, particularly those that necessitate extensive water usage, such as the fisheries sector.

Source: https://genesiswatertech.com/wp-content/uploads/2023/09/environmental-impacts-of-industrial-wastewater-discharge.webp.

8.2.3. Hazardous Waste

The issue of hazardous waste is one of the biggest waste management and pollution control challenges. Hazardous waste is waste that has significant or potential hazards to public health or the environment and must be managed with strict handling procedures to minimize risks associated with its storage, treatment, and disposal. Industrial processes ranging from manufacturing to healthcare generate all forms of toxic chemicals, hence leaving hazardous byproducts containing heavy metals, solvents, and chemicals. Manufacturers who are in electroplating processes, among other manufacturing industries, generate several hazardous substances, whose effects can be disastrous if not contained. Thus, the single most important step in avoiding such risks is by effective pollution prevention methods that entwine source reduction as well as waste minimization, which ultimately curtail the amount and toxicity of wastes produced in the very first instance (Dahab et al., 1994). Waste minimization encompasses several methods, which include source reduction, on-site reuse, and recycling, and cumulatively these work towards minimizing the production of hazardous wastes.

Figure 8.7. Industries ranging from manufacturing to healthcare generate a variety of hazardous materials, often resulting in byproducts that contain toxic elements such as heavy metals, solvents, and chemicals.

Source: https://www.chemicals.co.uk/wp-content/uploads/2023/02/chemical-in-protective-suit-hazardous-chemicals.webp.

For instance, process redesign to reduce the use of toxic chemicals and substituting less toxic alternatives can significantly shorten waste at its origin. A number of operational modifications are often reasonably simple to implement and demonstrate striking economic benefits through reduced expense disposal and improved regulatory compliance. Second, the use of cutting-edge technologies in managing hazardous waste—such as state-of-the-art treatment processes—enables industries not only to cut waste generation but also to engage in safe management of waste materials that cannot be removed altogether. By adopting a proactive stance towards waste minimization and pollution control, facilities can improve their environmental record and maximize operational effectiveness, toward a more sustainable industrial situation (Wise, 1994). The importance of proper hazardous waste management is more than regulatory compliance; its impacts reach into community health and environmental quality. Toxic pollutants, when mismanaged, can reach local ecosystems and pose risks to biodiversity and human health through the contamination of water and land. In communities located near industrial plants, the risk in hazardous waste is often amplified by economic and social factors, resulting in disproportionate environmental hazard. Therefore, efficient policies that enhance pollution prevention and waste minimization are critical in the protection of exposed populations against the harmful effects of hazardous waste. Enhanced investment in adequate training of industry personnel and organizational cultural change towards sustainability can lead to significant changes in waste management practices, promoting a healthful environment for everybody (Dahab et al., 1994) (Figure 8.7).

8.3. POLLUTION PREVENTION STRATEGIES

Inclusion of prevention steps against pollution is most important for encouraging sustainable industrial processes. Such steps aim at reducing waste at the source, rather than treatment or disposal following generation of waste. By taking active measures, organizations not only contribute to the conservation of the environment but also achieve economic benefits. As noted in various modules of education, long-term environmental protection is merged with economic growth, emphasizing that prevention of waste is less costly than post-generation control. It urges companies to look at pollution prevention as a competitive strategy, advancing greater economic prosperity in the long run through optimized efficiency in operations and reduced costs of liability (Bierma et al., 1995). Also, the categorization of pollution prevention measures into various frameworks serves to explain how they are applied within industry. As cited in existing literature on manufacturing sustainability, there exists a wide variety of tools and policies that can be used to minimize environmental impacts, particularly industrial waste. The approaches may be categorized based on application area, purpose, organizational agency, lifecycle phase, and waste hierarchy tier. This coordinated approach assists enterprises in identifying means through the complexities of waste prevention strategies, ultimately resulting in convergence towards zero waste production. This entire approach provides collective tools where pollution prevention is not fragmented measures but part of a comprehensive management system (Bjelkemyr et al., 2013). Stakeholder consciousness of pollution prevention strategies similarly boosts their effectiveness. Companies that actively encourage their workforce towards education and training in sustainable methods are more likely to perform well in waste reduction strategies. Staff awareness schemes that explain the benefits of waste reduction, apart from enhanced worker morale, encourage innovation and shared responsibility. With organizational culture that supports sustainability, companies can adopt creative solutions that meet pollution prevention principles. Applying such programs not only portrays corporate social responsibility but also reflects a dedication to environmental stewardship, putting firms in a positive position within an increasingly green-oriented marketplace (Bierma et al., 1995; Bjelkemyr et al., 2013) (Figure 8.8).

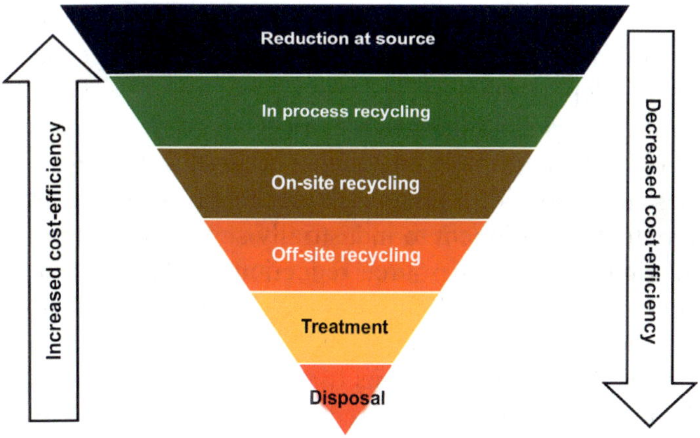

Figure 8.8. The integration of pollution prevention strategies is paramount in fostering sustainable industrial practices.

Source: https://www.researchgate.net/publication/287677398/figure/fig7/AS:590984547151881@1517913104811/Pollution-Prevention-Hierarchy-Cagno-et-al-2005-Gavrilescu-et-al-2008-Schiopu-et.png.

8.3.1. Source Reduction Techniques

Source reduction practices are a highly significant strategy under the broader waste minimization and pollution avoidance program. Directed towards creating less waste at the production source rather than following manufacture, these practices improve sustainability and resource efficiency. Primarily used on manufacturing processes, source reduction refers to designing products to use less material or substituting harmful materials with less dangerous alternatives. For instance, the use of lightweight packaging materials by industries would help reduce the amount of waste to a minimum. Not only is this a cost-cutting strategy for businesses in terms of disposal, but reducing the environmental impact is a strategic objective for global sustainability. Successful implementation relies on collective coordination among manufacturers, regulators, and consumers, which provides a platform whereby reducing waste is a norm in production processes. The principles of Green Chemistry continue to build on the concept of source reduction by focusing on the molecular scale of chemical operations. Industries can design their process through the application of Green Chemistry principles to minimize production of undesirable waste and energy usage. This is achieved through the use of fewer toxic materials and designing the reaction conditions to be optimized, thus leading to a decrease in emissions of pollutants during the course of production. The rigorous analysis of these principles has revealed practices not only meeting environmental objectives but also economical, thereby providing economic advantages and incentives to industries to adopt cleaner technology. Moreover, the Green Chemistry conversation facilitates an interdisciplinarity whereby chemists collaborate with engineers and environmental scientists in developing solution-based methodologies that consider pragmatic concerns in waste minimization, focusing on mutual commitment to industrially sustainable operations (Paul et al., 2009). During research on source reduction mechanisms, it is clear that community participation towards promotion of awareness and attitude change towards generation of wastes is critical. Community programs and education can empower people and organizations to adopt more sustainable practices by informing them about the benefits of reducing wastes. By way of illustration, campaigns promoting the use of reusable bags and containers have been effective in eradicating plastic wastes. Finally, collaboration with local governments and environmental organizations can maximize the effectiveness of these initiatives, so that citizens recognize their role in society. Through networking that supports each other in the implementation and sharing of effective waste reduction strategies, societies are able to work together to minimize waste generation at source. This extent of people's involvement at the grassroots level is crucial in creating an environmentally conscious culture based on pollution prevention and daily sustainability (Tim et al., 2014).

8.3.2. Recycling and Reuse Practices

Effective recycling and reuse techniques are integral parts of an overall strategy for waste reduction and pollution prevention. These techniques not only conserve natural resources through diverting materials from the landfill and back into the economy, but they also eliminate the negative environmental consequences of waste disposal. A circular economy approach, whose target is to ensure constant reuse of resources through recycling and recovery, has a critical role to play in the achievement of the Sustainable Development Goals (SDGs) contained in contemporary policy agendas. Studies have identified recycling operations as particularly valuable for goals in SDG 12 (Responsible Consumption and Production) and SDG 15 (Life on Land), highlighting the potential for sustainable waste management to serve broader environmental objectives (Schroeder et al., 2018). New recycling and reuse processes are also supported by the availability of different tools seeking to facilitate circular economy implementation. For instance, there is a Circular Economy (CE) Strategies Database that classifies various strategies that are capable of joining various components of the value chain.

Figure 8.9. Effective recycling and reuse practices are fundamental components of a comprehensive strategy aimed at waste minimization and pollution prevention.

Source: https://cleanmanagement.com/wp-content/uploads/2021/02/CleanManagementEnvironmentalGroup-seotool-59214-TheImportanceOf-BlogBanner1-1.jpg.

These vary in scope, targeting various products and industries and driving core processes such as recovery and recycling. That there are over 100 cases in the CE Implementation Database is evidence that there are a great number of ready-for-market solutions ready. However, some segments of the value chain, such as production and distribution, continue to be underrepresented in current recycling procedures. Placing emphasis on these components can result in considerable waste management practice improvements (Kalmykova et al., 2017). Besides enhancing resource efficiency, effective recycling and reuse procedures create economic prospects, promoting employment development in green economies. The creation of a circular economy not only confers enormous environmental benefits but also fosters an economic model that is congruent with the decent work ethic principles enshrined in SDG 8 (Decent Work and Economic Growth). By investing in infrastructure that supports recycling schemes, local economies can be boosted and employment created, making them more resistant to economic recession. Cooperation among the various stakeholders, including governments, businesses, and consumers, is essential to unlock the full potential of these practices and overcome obstacles hindering their adoption. Lastly, by giving recycling and reuse priority, we reinforce the interdependency of economic growth and the state of the environment, the determinants of future resilience (Schroeder et al., 2018) (Figure 8.9).

8.3.3. Sustainable Product Design

In sustainable product design, the principles of waste reduction and pollution avoidance are essential to inform manufacturing systems and product life cycles. Central to sustainable design is the dedication to produce products that satisfy customer requirements yet impose minimal environmental footprints. By incorporating techniques such as material efficiency, resource optimization, and lifecycle analysis, designers can reduce waste generated at production and utilization levels. This is also in accordance with the overall objectives of Environmental Management, which seeks to avoid the effects of human activities on natural systems. Adopting sustainable product design practices involves concerted efforts from the industries, coming together of the designers, engineers, and policymakers towards developing innovations that ensure protection of the environment and sustainability (Haloui et al., 2024). To realize effective minimization of waste, circular economy transformation is key. The circular economy model emphasizes the ongoing use of resources by means of recycling, reuse, and remanufacturing principles. Under sustainable product design, manufacturing products that are easily disassembled and composed of recyclable elements is critical. This not only minimizes the contribution of waste to landfill but also minimizes the consumption of virgin materials, which in the majority of cases means more emissions are created in manufacturing. Furthermore, linking consumer incentives and product standards to these principles of design facilitates a smoother transition from linear consumption patterns, ultimately facilitating pollution prevention. The approach is holistic in nature as it focuses on integrating sustainability into product design, which facilitates the coexistence between industrial activity and environmental integrity (Xi et al., 2021). The successful implementation of sustainable product design also relies heavily on consumer education and awareness. As consumers increasingly become eco-sensitive, expecting products that are in harmony with their values, the market is insisting on more disclosure of the environmental impact of products. Educating consumers about the benefits of sustainable products—e.g., more efficient use of energy, reduced waste, and reduced emissions of pollutants—can enhance marketplace acceptance and spur innovation. Overcoming the barriers to sustainability practice demands a multimodal mechanism, one that aims not only at designers and manufacturers but also at consumers themselves through the empowerment of consumers in terms of information about their purchasing behavior. The process of creating an informed customer base can accelerate the move toward sustainable product design, ultimately resulting in significant waste avoidance and pollution prevention on the global scale (Haloui et al., 2024).

8.4. REGULATORY FRAMEWORK

The regulatory framework of pollution prevention and waste minimization plays a fundamental role in developing comprehensive guidelines that industries must follow in order to responsibly operate. Notably, international, federal, and state regulations have the goal of minimizing environmental issues while prompting industries to adopt best practices. For instance, the Paris Agreement identifies the international accord to cap greenhouse gases, which indirectly compels industries to re-examine their waste management processes. By integrating sustainability into regulation, stakeholders are made responsible to reduce waste and implement pollution prevention measures. The regulations impose standards of behavior on companies and reiterate the necessity of compliance, which motivates industries to go beyond compliance towards more sustainable practices with a concern for maintaining environmental health along with economic growth. Moreover, regulatory success heavily rests on incentives available that drive industries to go beyond compliance. The policymakers play a central role in designing programs that encourage companies to adopt cutting-edge technologies as well as environmentally friendly practices. The incentives may be offered in the form of tax rebates, grants, or technical assistance programs meant to facilitate smooth transition to cleaner production. Sembiring et al. (2025) refers to the necessity of clearing out barriers that hinder technological advancements, such as high initial costs and consumer resistance. With ongoing changes in regulatory conditions, these kinds of policies that are inclusive and provide such incentives will play a key role in developing a sustainability culture among industries and making pollution prevention and waste minimization a core part of operational protocols. Finally, coordination between government departments, industries, and academia will be crucial in developing effective regulatory structures that produce sustainability. Collaboration in exploration, development, and implementation of practices to reduce pollution and waste is possible. With collaboration among stakeholders, industries are in a position to gain from advanced technologies like IoT and AI, which enhance operational efficiency and reduce the consumption of resources. Prithiviraj et al. (2025) notes that application of digital tools can significantly facilitate waste management and tracking, thus increasing environmental law compliance. The synergy attained through multi-sector collaborations not only aids in the establishment of effective regulatory systems but also encourages public support and awareness of sustainable efforts. Through combined efforts, the society is able to strive towards a more responsible management of industrial processes to the good of the environment and public health in the long term (Figure 8.10).

Figure 8.10. Policymakers play a pivotal role in creating programs that reward companies for adopting innovative technologies and sustainable practices.

Source: https://www.ey.com/adobe/dynamicmedia/deliver/dm-aid--213a9dc2-24b6-4caa-aea7-958ef71b6b26/ey-ai-system-developers-analyzing-code-on-wall-screen-tv-looking-for-errors-v3.jpg?quality=85&preferwebp=true.

8.4.1. Environmental Protection Agency (EPA) Guidelines

Environmental Protection Agency (EPA) has developed comprehensive guidelines that are the foundation for pollution prevention and waste minimization processes in various industries. The guidelines support an environmental hierarchy that strives to promote source reduction, recycling, treatment, and safe disposal. This systematic framework, if adopted by organizations, results in proper prevention of waste generation and minimizes their impact on the environment. The EPAs commitment to promoting compliance can also be observed through efforts such as the Pollution Prevention Program Implementation Plan, which work in conjunction with waste minimization programs that are prepared at facilities like the Pacific Northwest National Laboratory (PNNL). Such programs reflect the role of federal guidelines in promoting collaborative measures between government agencies and industrial companies, further illustrating the role of regulatory frameworks in facilitating sustainable measures (Engel, 1996). Furthermore, projects funded by the EPA, such as the Waste Minimization Assessment Centers (WMACs), exhibit active initiatives to assist small producers in managing waste generation problems more effectively.

Figure 8.11. The environmental protection agency (EPA) has established comprehensive guidelines that serve as a foundation for waste minimization and pollution prevention practices across various industries.

Source: https://upload.wikimedia.org/wikipedia/commons/a/a7/Zephyr_Dredging.jpg.

Through technical expertise and financing, centers aim to enable underfunded manufacturers with the information and equipment necessary to identify waste streams and develop productive means for minimizing them. All measurements by WMAC teams, for instance, in Kentucky plants, demonstrate the actual savings that can be realized by embracing leading practices, like process improvements in painting and purchases of higher-quality materials. They not only indicate the potential to save much waste but also have substantial cost benefits, fostering a sustainability culture without requiring companies to undergo stringent regulatory requirements. These collaborative frameworks between the EPA and local organizations emphasize the significance of guidelines in the achievement of shared environmental objectives (Fleischman et al., 1995). The EPA guidelines also emphasize tenets of pollution prevention into everyday business operation, particularly as it pertains to hazardous waste management. Proper training and education that is specific to employees are essential components of effective pollution prevention strategies. By integrating environmental stewardship as a part of company culture, businesses are better able to cultivate an enlightened workforce with expertise in waste reduction and management best practices. Such training is complementary to the EPAs critical role of creating a culture of prevention rather than remediation, necessary for long-term sustainability. Moreover, legislation requires organizations to conduct regular reviews of their waste management practices in order to improve and apply adjustments to evolving regulations and advancements in technology. This proactive process not only encourages organizational responsibility but also serves the broader environmental goals established by the EPA, a step towards a greener future (Engel, 1996) (Figure 8.11).

8.4.2. International Regulations and Agreements

The sector of worldwide rules and agreements is important in shaping waste reduction and pollution prevention policy in the world. The agreements tend to set the foundation for comprehension that supports cooperation among states, promoting a unified way of addressing environmental concerns. For instance, the European Union's Directive on Integrated Pollution Prevention and Control is a clear indication of the trend towards integrated pollution control with nations being asked to apply the best practical environmental solutions (BPEO) in the different media. This directive not only enhances the significance of integrated pollution control but also indicates the EU member states' common commitment towards maintaining high environmental standards. Such arrangements show the capabilities of international frameworks to harmonize well various national policies, thereby enhancing their efficacies in environmental pollution reduction and industrial waste generation minimization, as indicated at times of IPC practices debates in Europe (Hersh et al., 2025). The implementation of international guidelines through domestic law is also a testament to environmental health and legislative interlinkages. For example, recent trends in UK legislation on clinical waste management illustrate the challenges of balancing public health policy with environmental policy. The increased focus on single-use, disposable equipment represents recognition of infectious waste threats to health. However, it has created controversy regarding expansion of hazardous waste streams, resulting in increased quantities going for incineration (Pruess et al., 2009). Such dynamics pose the dual challenges before policymakers: ensuring public safety and, at the same time, adhering to green waste procedures. As nations navigate these regulatory fronts, there is also a need to balance the imperatives of public health against sustainable development, adopting a more sustainable waste management strategy. Globally, such agreements have needed to adapt based on shifting challenges of waste and pollution, particularly with respect to contemporary technological innovations. As evolving waste generation patterns around the world dictate, the need for innovative, cross-disciplinary waste management solutions becomes paramount. Future agreements must be driven not just by compliance with regulation, but by the need to develop waste prevention technology and practice research and development. A concerted effort at waste minimization, supported by close international cooperation, would likely deliver more effective prevention and enable sustainable economic activity worldwide. By advancing the arguments in these issues on the international stage, stakeholders are able to establish bases for effective measures that cut pollution while promoting the pillars of waste reduction to achieve the desired environmental effects integrated into various regulatory systems (Hersh et al. 2025).

8.4.3. State and Local Regulations

State and local legislation constitutes a significant component of the overall design of waste minimization and pollution prevention. The legislation is drafted to address specific environmental concerns that arise within a specific jurisdiction that mirrors the distinct ecological and social environment of the residents within the area. Local governments generally make the legislation fit the needs of the residents, having regard to population, industrial processes, and topography. The development of terminology around urban drainage, say, highlights the importance of uncomplicated communication between stakeholders in environmental management. Terminologies appropriate at the local level guarantee that some matters are interpreted uniformly, hence waste minimization is more effective in society (Tim et al., 2014). Finally, context-relevant regulation writing is called for to guarantee a collective approach towards preventing pollution. The importance of state and regional rules also extends to their role towards facilitating compliance across industries and businesses. Regulations have the tendency of necessitating the application of best practices towards effective waste management, thereby pressurizing companies to develop and discover more sustainable alternatives to traditional practices. Such frameworks cause businesses to integrate sustainability into their operational models, embracing environmental, social, and economic dimensions. Literature dictates that integrating sustainability within the framework of a supply chain enables organizations to guarantee long-term existence while embedding efficient waste management (Craig et al., 2008). Local governance is not anymore merely a ruler but an initiator of proactive change, which encourages industries to evolve as per regulations aimed at ensuring waste minimization and preventing pollution at every level. Aside from compliance, state and local laws also tend to catalyze public awareness and community participation in matters of environmental concern. Local programs and education campaigns can draw on the coverage provided by these laws in order to inspire residents to rally behind waste prevention initiatives, and encourage collective responsibility. Establishing regulations to promote public action opens doors for citizens' movements to challenge waste and pollution. Such concerted efforts can reinforce the awareness of sustainable practices among the community and why it is important to conserve local ecosystems. Additionally, as these rules evolve, they can match the emerging environmental concerns of the present day, being applicable to the current social dynamics. The synergy between regulations and community effort thus reinforces not only waste minimization but also an informed public dedicated to sustainable development and anti-pollution measures (Figure 8.12).

Figure 8.12. Local authorities often shape the regulations to meet the needs of their residents, accounting for factors such as population density, industrial activities, and geographic features.

Source: https://publications.wri.org/epubs/824/images/chapter2.jpg.

8.5. INDUSTRIAL APPLICATIONS

Inclusion of waste minimization practices in industrial uses is very important to reduce environmental impacts and maximize efficiency in use of resources. Industry, particularly the manufacturing industry handling basic materials like cement and steel, would significantly benefit if it adopted the principles of circular economy. By redefining traditional production approaches and including waste avoidance and recycling practices, enterprises are not only able to minimize their operating cost but also be capable of being able to make a contribution to emission reduction. 70% of industrial greenhouse gas emissions occur due to the production of these materials, a report says. Therefore, clean production will be necessary in order to become climate neutral. This shift will require policy intervention that is collective and involves stakeholders to make material efficiency the focal point and product standards enhanced for raising recyclability, ultimately leading to a comprehensive waste generation and pollution reduction (Xi et al., 2021). In fisheries, the use of cleaner production techniques is a fundamental approach to waste minimization. Muncar Banyuwangi with a high level of fisheries production activity is faced with serious environmental problems due to waste generated from fish processing processes. Industries can innovate their activities through the implementation of cleaner production techniques and minimize waste at its origin. Fine-tuning the use of water at various stages of production is included in this, as unused water not only equates to wasteful use of resources but also leads to pollution of wastewater.

Further, recycling raw materials and optimizing processing techniques can enhance efficiency and effectiveness and consequently reduce the quantity of waste produced. Cleaner production, therefore, not only addresses short-term environmental concerns but also ensures sustainable industrialization with the ability to bring long-term gains to industry and nearby ecosystems (Widodo, 2016). The incorporation of waste minimization techniques into industrial operations facilitates global sustainability goals via reduced pollution and resource conservation. The transition to a circular economy target not just recycling but the entire life cycle of the products, affirming the belief that waste can be an opportunity for recovering resources. Policymakers play a significant role in going this way by implementing clear policies that encourage the industries to adopt sustainable methods. As seen with the European Union's green ambitions, policy environments can act as drivers of innovation in waste management, driving huge declines in industrial waste and emissions. These efforts highlight the need for inter-sectoral coordination to create a culture where waste is not viewed as a byproduct but a valuable resource whose re-integration into the production cycle awaits. Achievement of these goals will require sustained commitment and alignment of industrial practices with long-term environmental aspirations (Xi et al., 2021).

8.5.1. Case Studies of Successful Waste Minimization

Through analysis of successful waste minimization techniques, industry examples highlight the ground-breaking nature of implementing sustainable practices. Industrial ecology principles are an exemplary case in point, as they emphasize interdependence between industrial operation and environmental health. The approach offers a chance to reconceive waste as an asset, thus enabling reuse and recycling in production processes. For instance, closed-loop processes where waste materials are reused and recycled as inputs for other processes have been employed by chemical companies to reduce overall waste generation. Besides minimizing environmental impacts, this also saves money in terms of disposal costs and enhanced resource efficiency, a major tenet of waste minimization as described in Lawrence et al. (2005). But another powerful illustration can be found in the clothing industry, which has long been severely criticized for its environmental contribution. Leaders in the industry have heavily invested in cleaner production technologies that aim to reduce waste and toxic release at the source, rather than controlling such issues with end-of-pipe technology. The integration of wastewater treatment and pollution prevention methods has facilitated companies to manage their resources efficiently and lower total waste output. For instance, a major textile firm implemented better methods of dyeing using fewer chemicals and less wastewater output. This precautionary measure not only ensures compliance with stringent environmental laws but also enhances the company's image, attracting environmentally conscious customers and investors, as offered in Al-Yousfi (2004). These achievements in waste minimization do not have any bounds; even small and medium enterprises (SMEs) have also been able to implement the same. A good example of this is of a local food processing company that did a detailed waste audit and subsequently redesigned its operations to minimize food waste. With the help of solutions like wastage monitoring and staff training on sustainable processes, the company was able to significantly reduce its wastefulness. In addition, by partnering with neighborhood composting centers, it diverted organic refuse from landfills to useful applications in the agricultural sector. This strategic combination not only saves resources but also promotes bonding between people and supports community relationships, showing that waste reduction can promote greater social and ecological advantages. Overall, the diverse case studies best illustrate the efficacy of innovative methods of achieving substantial waste minimization and thereby supporting the broader goals of pollution reduction and sustainability for industries today.

8.5.2. Best Practices in Manufacturing

The integration of green practices at the manufacturing phase is essential to minimize waste and prevent pollution. The producers are increasingly recognizing the necessity of integrating green chemistry and energy-saving technologies into the manufacturing process. The practices not only reduce environmental footprint but also enhance business efficiency. For instance, by adopting circular economy and sustainable sourcing methods, manufacturers can ensure raw materials are responsibly sourced and wastage is minimized throughout the product life cycle. This shift is also in most instances encouraged by regulatory systems that, even if at times perceived as a drawback, generally promote more sustainable industry practice. Businesses utilizing these innovations usually gain cost reductions that enhance their competitiveness in the market, supporting the argument that environmental factors may supplement economic performance (Muthoni, 2025). An important component of best manufacturing practices is to address the tremendous pollution industrial processes incur. The sector has long been a major source of air and water pollution, primarily due to the emission of harmful emissions like greenhouse gases and particulate matter. As industry development increases, so does the demand for strong measures against pollution, particularly in the large cities where vehicle exhausts exacerbate air pollution. In order to solve this problem, businesses are encouraged to make use of cleaner technology that reduces emissions at the source, for instance, by using advanced filtration systems and innovative waste management practices. By applying pollution prevention, manufacturers can turn towards sustainability, not only fulfilling regulatory obligations but making a positive impact on public health and environmental integrity as well. Furthermore, innovation and ongoing improvement are included in the proper implementation of best practices in manufacturing. Organizations are increasingly involved in research and development to discover means that further decrease wastage and maximize sustainable production. This drive for innovation leads to the use of best practices such as lean manufacturing, which decreases wasteful practices and increases productivity by seeking out and eliminating them. By putting money into employee training, and incorporating employee feedback into making operational adjustments, manufacturers can foster a culture of sustainability and efficiency. Also, sharing effective case studies across the industry helps to establish standards and encourage widespread implementation of best practices. As producers become more sustainable, the emphasis remains on integrating pollution avoidance mechanisms into their operational processes as part of their core procedures, thereby setting up a more sustainable production environment (Muthoni, 2025) (Figure 8.13).

Figure 8.13. The manufacturing sector has long been a major contributor to air and water pollution, primarily due to the emission of harmful pollutants like greenhouse gases and particulate matter.

Source: https://cdn.britannica.com/51/155351-050-0A512566/Air-pollution-factory-smokestacks.jpg.

8.5.3. Role of Technology in Waste Reduction

The advent of technology has significantly altered the waste reduction scenario, with new solutions coming to the age-old problem of how to create and manage waste. Advances in technology allow for the development of leaner production systems that minimize material waste at the point of production. For instance, technologies such as computer-aided design (CAD) and simulation software allow manufacturers to optimize their production lines, thereby minimizing the use of energy and unnecessary materials. Additionally, technologies such as the Internet of Things (IoT) create networked spaces where information offered in real time directs waste management operations, enabling organizations to monitor the generation of waste and identify trends. Such a shift towards data-driven decision-making predisposes a more sustainable production culture, which subsequently enhances pollution avoidance and overall waste reduction in industries. Also, technology has a part to play within the recycling and waste treatment sector, transforming the way wastes are handled. Novel recycling technology, such as advanced sorting machines with artificial intelligence and machine learning, enhances the efficiency of separating recyclables from waste streams. By improving these processes, technology not only increases the quantity of materials actually being recycled but also reduces the cost of waste management. In addition, developments in clean technology have enabled the conversion of waste into energy and promotion of a paradigm of circular economy where resources and materials are being reused continuously. This integration with the CE values indicates how technology can serve as a driver of the attainment of different SDGs, particularly those of responsible consumption and production as highlighted by Schroeder et al. (2018). Last but not least, applying technology in waste reduction strategies ensures economic and environmental benefits. Firms adopting these technological innovations are able to incur lower operational expenses by being able to consume (and therefore, minimize) resources and optimize waste management strategies. Additionally, the congruence of such practices with the natural-resource-based view (NRBV) implies that organizations are able to construct competitive advantage through harnessing sustainable resources and capabilities, thereby further solidifying their environmental stewardship. With continued research development, it is important that organizations learn about the interties between technology, waste management, and sustainable enterprise solutions, whereby investments in technology not only bear a financial effect but also contribute to solving world sustainability objectives as provided for in Stuart et al. (2010) (Figure 8.14).

Figure 8.14. Innovative recycling technologies, such as advanced sorting systems equipped with AI and machine learning, enhance the efficiency of separating recyclable materials from waste streams.

Source: https://miro.medium.com/v2/resize:fit:1400/1*lTC06CFrq96VJNZUk4Y-XA.png.

8.6. WASTE MINIMIZATION IN HOUSEHOLDS

Household waste minimization concept is a vital element in creating sustainable communities. Through activities that focus on waste reduction, families can significantly minimize their impact. Public awareness programs aimed at enlightening homeowners about proper waste segregation and reduction methods are the main programs behind such efforts. Small actions, such as shopping bag use and bulk purchasing, not only reduce packaging waste but also establish a culture of sustainability. They have tangible dividend advantages, such as decreased landfill burdens and enhanced resource efficiency, which sit very well within broader waste management policy critical to avert pollution. The integration of learning materials into community outreach can be the catalyst for such a cultural shift, whereby waste reduction becomes the norm rather than the exception for families. Moreover, household waste diversification goes beyond participation; it tends to demand changes in local policy frameworks. As households become wiser with regard to waste management, they can nudge urban policies toward the promotion of recycling and composting initiatives. Functional recycling programs that are created to meet domestic needs can increase the level of participation, which in turn, in research, has been observed to contribute to environmental and economic benefits. The interaction between household cooperation and municipal incentive policies points to the importance of coordinated policy frameworks and grassroots movements. Indeed, research indicates that neighborhoods with effective waste management systems exhibit better declining waste generation levels, which bear testament to the effectiveness of holistic approaches. Out of a symbiotic relationship between households and the government, there can emerge a long-term model for waste reduction that will ultimately translate into pollution prevention efforts. Also, innovation in technology holds out new possibilities for household waste reduction. Improved waste monitoring software and networks of smart homes provide citizens with the opportunity to monitor usage and waste precisely. These technologies allow individuals to make empowered decisions about their consumption patterns and their management of waste. Behavioral science research indicates that, where households are offered information about their production of waste, they are more likely to implement reduction strategies. These studies point towards the need for technology-supported solutions augmenting traditional waste prevention strategies. Through integrating these technologies into everyday practices, households can contribute significantly to achieving broader environmental objectives. This technology-person behavior interaction dynamic is a significant component of a systems approach to pollution prevention and waste reduction (Bjelkemyr et al., 2013; Mazzanti et al. 2025) (Figure 8.15).

Figure 8.15. As households fine-tune their waste management practices, they can influence municipal policies to support recycling and composting initiatives.

Source: https://www.deskera.com/blog/content/images/2022/09/zibik-iR4mClgg-zEU-unsplash.jpg.

8.6.1. Strategies for Reducing Household Waste

Practical actions towards reducing household wastes are essential in addressing the rising challenge with environmental degradation and natural resource exhaustion. The first rudimentary approach is adopting the three Rs: Reduce, Reuse, and Recycle. Households have the capacity to reduce waste substantially by promoting low consumption, using products with minimal packaging, and buying durable products instead of disposable ones. This action not only assists in curbing the accumulation of waste but also encourages a transition towards green consumption practices. Besides, education plays an important role in the awareness-raising process concerning the implications of waste. By involving people in awareness-raising campaigns, families are better placed to understand how their consumption behavior affects the environment and why they need to embrace greener lifestyle practices, as embodied in the substantial legislative efforts of the European Union towards better waste management (Rymarchuk et al., 2024). Other than the three Rs, composting is also a good method in managing organic waste generated in homes. By composting lawn clippings and food waste, households can help minimize the volume of waste sent to landfills by a significant margin while, at the same time, generating nutrient-enriched soil to be utilized for gardening. Composting not only minimizes food waste but also educates people about the life cycle of organic matter, generating higher awareness of food sustainability. Many cities and municipalities are also launching pilot composting projects that invite participation through conducting workshops and educational content, emphasizing the shared benefits of reducing household waste (Mikavica et al., 2024). With individuals transitioning from traditional waste disposal methods to composting, local practice is brought in line with environmental goals at large, capturing the spirit of wastes minimization. Another significant strategy involves applying community involvement and policy to aid waste reduction. Local authorities can initiate programs such as zero waste programs, swap meet among communities, and bulk purchasing cooperatives to reduce household waste efficiently. These programs not only present solutions in the real world but also foster community among the participants who share a common goal of reducing waste. On an even larger scale, regulatory measures such as prohibiting single-use plastics and incentives to business for more sustainable practices consolidate efforts at reducing waste on a grander scale. Through the provision of a setting where sustainable decisions are made available and promoted, communities can play a big role in supporting national waste minimization efforts as well as pollution reduction, as seen through the European Unions integrated strategy towards enhancing waste management regimes (Rymarchuk et al., 2024). This combination of community action, together with personal responsibility, is the key to bringing about sustained changes in waste management procedures.

8.6.2. Community Involvement and Education

In viewing the connection between education and community participation, the position of educated citizens in fostering waste minimization and prevention of pollution is made clear. The active participation of communities in education programs on environmental stewardship increases conformity with sustainable practices. Education programs that raise awareness on waste disposal and pollutions' adverse effects inculcate responsibility among the residents. For example, children exposed to recycling principles in schools tend to take these ideas home with them, so to speak, thereby instigating a ripple effect. This community-based teaching approach fosters awareness of immediate challenges, such as poor sanitation and waste management, which are key factors in informal settlements. In settings such as those that have been considered in the Zambian government's efforts at cholera prevention, education in the community holds promise to prevent pollution-related health hazards greatly, improving overall public health outcomes among at-risk populations (Chakanyuka et al., 2025). Moreover, the inclusion of local stakeholders like NGOs and community leaders is the key when creating tailored educational interventions directed at certain environmental issues. Such a collaboration ensures that the content is applicable to culture and resonates with community members, thus enhancing engagement and interaction. For example, projects incorporating local knowledge and indigenous approaches to environmental education not only enhance residents but also promote community cohesiveness. By offering workshops, community cleanups, and awareness campaigns, such collaborations foster increased awareness of waste minimization techniques and pollution prevention measures. It builds a dynamic culture of sustainability, where communities have control over their ecological well-being. This is consistent with the bigger sustainability goals, as seen in mainstreaming environmental education in terms of mass events like the Olympics, whereby participation by communities dictates the events ecological footprint (Brani, 2025).

Figure 8.16. Communities that actively engage in educational initiatives about environmental stewardship demonstrate a higher adherence to sustainable practices.

Source: https://one-more-tree.org/wp-content/uploads/2024/04/2149040313.jpg.

To facilitate maximum effectiveness of community education and outreach, continuous program improvement and evaluation must be implemented. Feedback loops need to be established to gauge the effectiveness of education programs and improve them based on the community needs as well as shifts in the external environment. This regular check permits a continuously changing learning environment that provides opportunities for participants to participate actively with respect to evolving issues of waste and pollution. Moreover, forming partnerships with education facilities can render educational resources and training materials more valuable in terms of quality so that they may be able to transfer optimal waste management practices. Investing in such initiatives ultimately yields a more educated populace prepared to battle pollution at the local level. By emphasizing community learning and participation, not only is waste reduction in the short term covered but also a foundation for long-term ecological sustainability and community resilience in the face of the ongoing environmental challenges (Brani, 2025; Chakanyuka et al., 2025) (Figure 8.16).

8.6.3. Impact of Consumer Choices

Consumer decisions play a central role in shaping the environmental impact, particularly in the reduction of waste and pollution control. When consumers make decisions to adopt sustainable products and services, their aggregate purchasing power can induce market transformation. Consumer decisions are capable of influencing the mode of production pursued by businesses, with numerous firms proceeding in environmentally sustainable production when demand for sustainable products rises. Therefore, integrating concepts such as the circular economy becomes the cornerstone, which begets a recycling and reuse culture and which aligns with broader environmental goals. By actively deciding to purchase goods that operate based on the concept of sustainability, consumers not just usher in less waste but also compel businesses to implement greener practices, thus instilling a cascade effect that facilitates overall well-being for nature (Schroeder et al., 2018). In addition, consumer decisions have a reach that is more significant than just buying; they are capable of encouraging legislative and corporate overhauls intended to avoid pollution. As consumers become more focused on knowing how products will affect the environment, brands are forced to account for what they do on behalf of sustainability. This transparency promotes competition among businesses to obtain increased levels of environmental responsibility. An educated consumer population will lead to a massive growth in environmentally conscious businesses and practices, highlighting how individual choices are connected to the broader systemic change. How far consumers can influence the adoption of Circular Economy practices also speaks to their power to contribute to the fulfillment of Sustainable Development Goals. For instance, consumer preferences form the values that feed into responsible production and consumption directly, reiterating the importance of value-driven decisions in the market (Reike et al., 2017). Besides retail decisions, activism and consumer education can heavily utilize the potential of individual decisions to curtail waste and pollution. Through the use of grassroots organizations and online forums, consumers can enforce an awareness campaign towards the environmental consequences of unchecked consumption. The advocacy can instigate changes in what society values and what is considered important and acceptable concerning waste, which forces the consumer and the producer alike to switch to sustainable alternatives. Participatory initiatives such as community recycling schemes also empower consumers by providing them with actual means through which they can contribute to avoiding pollution at the grassroots level. Through the development of a culture of sustainability and shared responsibility, consumer choice facilitates tremendous change, in addition to more extensive environmental policy and ultimately to a more sustainable future (Schroeder et al., 2018).

8.7. CONCLUSION

Lastly, the implementation of waste minimization and pollution prevention programs is essential in achieving sustainable urban sustainability and mitigating the negative effects of waste and pollution to human health and the environment. The achievement of the Zero-waste City Pilot (ZWCP) policy demonstrates that massive waste management programs could assist in substantially reducing the amount of solid waste and carbon emissions. The empirical findings indicate that the ZWCP policy enhances green technology innovation and enhances the legislation of the environment, both of which have a two-way impact that decreases pollution as well as carbon. In addition, the results observed indicate that the efficiency of the policy is particularly crucial under certain city settings, further solidifying the importance of tailor-made strategies in implementing waste reduction policy as stated by Chai et al. (2025). The principal polluters—industrial activities, motor vehicle exhausts, and unregulated agricultural output—highlight the need for concerted and extensive pollution control action. Understanding the causatives of pollution, such as over-reliance on fossil fuels and inapt handling of waste, enables policy makers and people to develop effective prevention and mitigation responses. Research indicates that the levels of pollution continue to rise as a consequence of urbanization and industrialization, thus necessitating the implementation of innovative approaches to environmental sustainability. The development of specific policies and the creation of public awareness about the polluting sources is crucial in the implementation of pragmatic waste management practices based on the sustainability principles. Lastly, the realization of a zero-waste future demands concerted effort with the integration of government policies, industry practice, and civic engagement. Through cross-regional collaboration and best practice sharing, people have managed to manage resources effectively with the least amount of waste and negligible environment degradation. The success of such policy enactments as the ZWCP not only presents an example of reducing pollution but also environmental health and economic prosperity interdependence. Consequently, long-term devotion and creativity in waste management and avoidance of pollution strategies become paramount while creating resilient cities that will be in a position to support future generations. This four-pronged approach not only benefits regional ecology but is also instrumental in supporting global environmental goals, thus creating the foundation towards a greener tomorrow, as noted by Chai et al. (2025).

8.7.1. Summary of Key Points

The alliance of technology and green living has become increasingly applicable, particularly in waste minimization and avoidance of pollution. As technologies go digital, they have a significant impact on carbon imprints. For instance, the growth in mobile phone usage—projected at reaching 70% of the global population by 2023—is matched by increased waste from manufacturing due to the complexity of products in smaller sizes. Besides, the International Energy Agency (IEA) reports that video streaming is associated with high hourly emissions, reflecting the need for effective management of digital usage of resources. However, encouraging environmental sustainability through digital innovation is possible, as technologies can realize a 20% reduction in energy, material, and mobility requirements by 2050, thereby depicting the need for integrating digital transformation within waste management systems to enhance eco-efficiency (Oo et al., 2023). Meanwhile, emerging waste management strategies are transforming, such as being witnessed through the utilization of biological resources for sustainable approaches. The silkworm pupae are a milestone case with high nutritional value that not only ensures added benefits for human and animal welfare but also serves to avoid excessive waste levels. Typically considered as by-products, the incorporation of silkworm pupae in food products can avoid high levels of pollution discharges from traditional disposal procedures. Further, mulberry leaves that constitute the silkworms' primary diet show their potential for carbon sequestration by absorbing poisonous toxins from the environment. Such findings highlight the dual utility of such findings in halting waste and conserving sustainability to form a palpable link between environmental healthiness and economic sustainability in the context of insect-based feeds (Hăbeanu et al., 2023). The overall thrust towards minimizing wastes and preventing pollution gives evidence of the fact that successful methods need to address technological improvements as well as biological interventions. The digital revolution allows for optimization of efficiency and resource usage, whereas alternate feedstocks such as silkworm pupae provide environmental advantages and lesser ecological footprints, which demonstrate the diversified nature of sustainability. Addressing these matters must formulate a comprehensive view into our waste generation and pollution strategy, pointing out that the benefits of integration can yield significant ecological and economic gains. With our advancing knowledge, it is necessary for the stakeholders to examine these emerging ways critically and encourage their increased application to form an eco-friendly future (Oo et al., 2023; Hăbeanu et al., 2023).

8.7.2. Future Directions for Waste Minimization

As the societies of the world continue to grapple with the issues of waste pollution, the future of waste minimization is all the more a matter of urgency. Application of new technologies for the management of waste is an excellent step forward. Higher-order recycling techniques, such as biochemical recycling and chemical recycling, can convert plastics previously deemed too difficult to recycle into valuable raw materials, thereby easing the burden on landfills and the environment. In addition, intelligent waste management system design employs the Internet of Things (IoT) to optimize waste collection routes and scheduling to reduce greenhouse gas emissions resulting from transporting the waste. Such technologies are highly in line with public health requirements, particularly considering that wasted products contribute significantly to ocean pollution that endangers human health as highlighted by Philip et al. (2020). By leveraging technological innovations, societies are able to transition towards sustainable waste management that optimizes urban living and produces healthier results. The use of a Circular Economy (CE) paradigm also enables systemic change in waste reduction processes. The central concept of this paradigm is to keep the value of resources within the economy for as long as possible, and this discourages resource depletion as well as reduces the burden of pollution. This requires a paradigm shift in the direction of waste-to-resource operations—where resources are continuously recycled and reused in numerous value chains. Developing databases listing numerous CE strategies, as seen in Kalmykova et al. (2017), is a valuable tool for organizations to adopt sustainable practices. Moreover, stakeholder engagement at every level—from producers to consumers—ensures that waste reduction becomes everyone's responsibility. Emphasizing such collaboration encourages not only waste reduction but also contributes to broader sustainability objectives, reinforcing interconnectedness of environmental stewardship. Finally, education and public engagement are integral components of future waste reduction. Awareness-raising about environmental consequences from waste generation fosters consumer responsibility and active involvement in recycling and composting programs. Education initiatives can be highly effective in schools, where kids learn early on the importance of sustainability, which translates to long-term behavioral change. Further, holding individuals accountable for local waste management lessens reliance on central waste systems and therefore generates pride and accountability towards local environments. Outreach programs can also inform people about pollutions impact on health, for instance, marine pollution as stressed by Philip et al. (2020), and push people to act against such hazards at a grassroots level. Placing knowledge and public involvement first, communities will ensure a culture for sustainability that will underpin future waste minimization and pollution avoidance efforts.

8.7.3. Call to Action for Individuals and Organizations

The increasing feeling of urgency about waste reduction and prevention of pollution calls for a powerful call to action by people and organizations both. Collective action has to see to it that a culture of sustainability and conscious consumption is developed. People can do a lot through simple but effective lifestyle changes, such as reduced use of single-use plastics, participation in local clean-up activities, and promoting recycling schemes in their areas. These little acts can make larger movements, getting other people to act too. By being an active participant in these actions, not just do people lower their carbon footprint, but also live a shared consciousness that supports sustainability. Organizations are the only thing that can bring sustainable acts into the system, though.

Through the adoption of circular economy (CE) principles, businesses can minimize their waste creation and enhance resource conservation, remaining connected with the Sustainable Development Goals (SDGs) (Schroeder et al., 2018). Adoption of CE practices does not just prevent pollution but also offers economic benefits, enhancing innovation and new market opportunities. For instance, companies can shift from linear business models to sustainable models with an aim towards minimizing waste and efficient utilization of resources. Such should be complemented with commitments not only by the leadership but by all employees, ushering in a collective approach towards achieving sustainability goals. In addition, the interplay of individual actions and corporate responsibility underscores the coordination imperatives among different sectors of the industry. As society becomes increasingly demanding of corporate social responsibility, organizations are being called on to put their missions in alignment with society's broader goals, for example, waste minimization and prevention of pollution (Roland et al., 2009). Being socially responsible could stand to gain significantly with enhanced brand loyalty and stakeholder trust. By creating alliances with the community and encouraging employee participation in sustainability initiatives, organizations are able to create a strong narrative of their responsibility towards the environment. Through such combined efforts, not just do they avoid wastage and pollution but also create a culture where efforts towards sustainability become an integral part of everyday living.

8.7.4. Final Thoughts on Environmental Responsibility

When thinking about environmental stewardship in the context of pollution prevention and waste minimization, it is obvious that novel materials such as biodegradable plastics have possibilities along with limitations. Research has increasingly focused on the impacts of long-lasting plastics in our ecosystem, prompting the investigation of alternatives such as poly(lactic acid). However, as pointed out, such substances may not decompose well under all states of the environment, such as seawater, thereby implying that the development of alternative substances per se is not sufficient in the absence of a critical understanding of their performance within the actual environment (Haider et al., 2018). This underlines the necessity of not only enhancing material science but also ensuring that such innovations advance actual environmental sustainability and that this requires an emphasis on field testing under actual conditions. In addition, the shift from linear to circular economy is a central strategy for enhancing environmental accountability. In such a situation, businesses have to reconsider product and recycling design to develop closed-loop systems that eliminate wastage (Bocken et al., 2016). By implementing approaches that prioritize slowing, closing, and narrowing the resource loops, businesses can drastically reduce their environmental footprint. Such a shift in paradigm not only encourages sustainable practices but also welcomes shared consumer responsibility over purchasing and product life cycle decisions. Lastly, integrating these cycle principles into company operations is a forward-thinking philosophy to being green that can lead to significant waste and pollution reductions. Overall, the way to environmental responsibility for waste reduction and pollution control is complex and encompassing involving full consideration of material innovation, business model transformation, and consumerism. As outlined in recent studies, effective functioning of biodegradable products and circular economic processes will necessitate ongoing dialogue among policymakers, companies, and scientists (Haider et al., 2018; Bocken et al., 2016). Through collaboration and an ongoing awareness of our environmental impact, society can establish a platform for sustainable action that not only protects our natural systems but also ensures prudent stewardship for future generations. As group efforts continue to increase, it is important to stay committed to advocating for environmental responsibility at every level.

REFERENCES

1. Al-Yousfi, A. (2004). Cleaner production for sustainable industrial development: Concept and applications. *Practice Periodical of Hazardous, Toxic, and Radioactive Waste Management, 8*, 265–273. https://www.semanticscholar.org/paper/6d88d8ea03c21ba8d71fef1f12175901cfe41b30 (accessed on 24th October 2025).
2. Ambec, S., Cohen, M. A., Elgie, S., & Lanoie, P. (2013). The Porter Hypothesis at 20: Can environmental regulation enhance innovation and competitiveness? *Review of Environmental Economics and Policy, 7*, 2–22. https://doi.org/10.1093/reep/res016 (accessed on 24th October 2025).
3. Anastas, P. T., & Eghbali, N. (2009). Green chemistry: Principles and practice. *Chemical Society Reviews, 39*, 301–312. https://doi.org/10.1039/b918763b (accessed on 24th October 2025).
4. Arifalo, S. F. (2024). Waste management strategies adopted by poultry farmers to prevent environmental pollution in Ondo State, Nigeria. *Ondo University Journal of Agriculture and Social Sciences*. https://www.semanticscholar.org/paper/e65560c00762488c427ec67278126d7f27764d8e (accessed on 24th October 2025).
5. Bénabou, R., & Tirole, J. (2009). Individual and corporate social responsibility. *Economica, 77*, 1–19. https://doi.org/10.1111/j.1468-0335.2009.00843.x (accessed on 24th October 2025).
6. Bierma, T. J., & Waterstraat, F. L. (1995). Pollution prevention and business management. Curricula for schools of business and public health. Volume 1: Modules 1–3. https://core.ac.uk/download/4812417.pdf (accessed on 24th October 2025).
7. Bjelkemyr, M., Jönsson, C., Kurdve, M., Shahbazi, S., et al. (2013). Industrial waste management within manufacturing: a comparative study of tools, policies, visions and concepts. https://core.ac.uk/download/42143949.pdf (accessed on 24th October 2025).
8. Bocken, N., de Pauw, I., Bakker, C., & van der Grinten, B. (2016). Product design and business model strategies for a circular economy. *Journal of Industrial and Production Engineering, 33*, 308–320. https://doi.org/10.1080/21681015.2016.1172124 (accessed on 24th October 2025).
9. Braniște, G. (2025). Contemporary environmental challenges impacting the Olympic legacy. *"Sports. Olimpysm. Health," international scientific congress*. https://www.semanticscholar.org/paper/f5ea36d07bfd5f90f4f93e63b4afcb81d0b35313 (accessed on 24th October 2025).

10. Budrevičiūtė, A., Damiati, S., Sabir, D. K., Önder, K., Schuller-Goetzburg, P., Plakys, G., Katileviciute, A., et al. (2020). Management and prevention strategies for non-communicable diseases (NCDs) and their risk factors. *Frontiers in Public Health, 8*. https://doi.org/10.3389/fpubh.2020.574111 (accessed on 24th October 2025).

11. Carter, C. R., & Rogers, D. S. (2008). A framework of sustainable supply chain management: moving toward new theory. *International Journal of Physical Distribution & Logistics Management, 38*, 360–387. https://doi.org/10.1108/09600030810882816 (accessed on 24th October 2025).

12. Chai, Z., Zhu, X., Sun, Y., & Chen, X. (2025). Evaluating pollution reduction and carbon mitigation in China's zero-waste cities. *Sustainability*. https://www.semanticscholar.org/paper/b16244dcea963317d73c444378c8b94ebffffd97 (accessed on 24th October 2025).

13. Chakanyuka, T., Siamabele, B., Makungu, M., Nsama, P., & Kalonje, V. (2025). Interventions by Zambian Government to prevent the resurgence of cholera in unplanned settlements: Lessons from Misisi Compound in Lusaka. *International Journal of Advanced Multidisciplinary Research and Studies*. https://www.semanticscholar.org/paper/631f04864a0a2ff30c66448543525f6c6f807bb7 (accessed on 24th October 2025).

14. Dahab, M., Montag, D., & Parr, J. (1994). Pollution prevention and waste minimization at a galvanizing and electroplating facility. *Water Science and Technology, 30*, 243–250. https://www.semanticscholar.org/paper/75e01734a8458665a036a324e15d3a9cb668c001 (accessed on 24th October 2025).

15. Engel, J. (1996). Pollution prevention program implementation plan. https://www.semanticscholar.org/paper/c3f066507d46c2a09c35771b4232bf9719add09b (accessed on 24th October 2025).

16. Fleischman, M., Harris, J., Handmaker, A., & Looby, G. (1995). Waste minimization assessment for a manufacturer of iron castings and fabricated sheet metal parts. https://www.semanticscholar.org/paper/579a12e1a8c65a4a7b0ced11e5e64125b4da33ed (accessed on 24th October 2025).

17. Fletcher, T. D., Shuster, W. D., Hunt, W. F., Ashley, R., Butler, D., Arthur, S., Trowsdale, S., et al. (2014). SUDS, LID, BMPs, WSUD and more–The evolution and application of terminology surrounding urban drainage. *Urban Water Journal, 12*, 525–542. https://doi.org/10.1080/1573062x.2014.916314 (accessed on 24th October 2025).

18. Gatrone, R., McHenry, J., Myron, H., & Thout, J. R. (1998). Waste minimization and pollution prevention technology transfer: the Airlie

House Projects. https://www.semanticscholar.org/paper/02ea8fb02415993c8e9c8e5ff5c3d4e23ae6f078 (accessed on 24th October 2025).

19. Hăbeanu, M., Gheorghe, A., & Mihalcea, T. (2023). Nutritional value of silkworm pupae (*Bombyx mori*) with emphases on fatty acids profile and their potential applications for humans and animals. *Insects, 14*. https://www.semanticscholar.org/paper/cc9eca5b1c99cc37a028146a59b7035dc2f09fa1 (accessed on 24th October 2025).
20. Haider, T., Völker, C., Kramm, J., Landfester, K., & Wurm, F. R. (2018). Plastics of the future? The impact of biodegradable polymers on the environment and on society. *Angewandte Chemie International Edition, 58*, 50–62. https://doi.org/10.1002/anie.201805766 (accessed on 24th October 2025).
21. Haloui, I., & Yang, L. (2024). Integrating environmental management and industrial ecology: approaches, tools, and models for sustainable development. *International Journal of Advanced Research*. https://www.semanticscholar.org/paper/2e07130b9c71397329395e9aa872d990c73d073e (accessed on 24th October 2025).
22. Hart, S. L., & Dowell, G. (2010). Invited editorial: A natural-resource-based view of the firm. *Journal of Management, 37*, 1464–1479. https://doi.org/10.1177/0149206310390219 (accessed on 24th October 2025).
23. Hersh, R. (2025). A review of integrated pollution control efforts in selected countries. https://core.ac.uk/download/pdf/9307942.pdf (accessed on 24th October 2025).
24. Kalmykova, Y., Sadagopan, M., & Rosado, L. (2017). Circular economy–From review of theories and practices to development of implementation tools. *Resources Conservation and Recycling, 135*, 190–201. https://doi.org/10.1016/j.resconrec.2017.10.034 (accessed on 24th October 2025).
25. Kitanovski, V., Lutovska, M., Trifunov, Z., Demiri, S., Popovska, O., & Kjosevski, S. (2025). Incorporating nexus ecolabels into household level food waste mitigation strategies: Environmental perspectives from life cycle assessment. *Journal on Processing and Energy in Agriculture*. https://www.semanticscholar.org/paper/6c8a0905be478ab611f0c800f64670b47d30d5ea (accessed on 24th October 2025).
26. Konar, B. (2025). Biomedical waste treatment: Technologies, challenges, and management strategies. *International Journal of Scientific Research in Engineering and Management*. https://www.semanticscholar.org/paper/0d6d8a2d738d1afdca5cec9e31ad0c58e06d8df0 (accessed on 24th October 2025).

27. Kumar, A. V., & Madiraju, S. V. H. (2021). Role of pollution prevention activities to improve public health and wellbeing. *Proceedings of the 3rd International Electronic Conference on Environmental Research and Public Health–Public Health Issues in the Context of the COVID-19 Pandemic*. https://www.semanticscholar.org/paper/46a2097c15f4d58ed689aee21d617f94f8229180 (accessed on 24th October 2025).

28. Landrigan, P. J., Stegeman, J. J., Fleming, L. E., Allemand, D., Anderson, D. M., Backer, L. C., Brücker-Davis, F., et al. (2020). Human health and ocean pollution. *Annals of Global Health, 86*, 151–151. https://doi.org/10.5334/aogh.2831 (accessed on 24th October 2025).

29. Le Blanc, D. (2015). Towards integration at last? The Sustainable Development Goals as a network of targets. *Sustainable Development, 23*, 176–187. https://doi.org/10.1002/sd.1582 (accessed on 24th October 2025).

30. Mazzanti, M., & Zoboli, R. (2025). Waste generation, incineration and landfill diversion. De-coupling trends, socio-economic drivers and policy effectiveness in the EU. https://core.ac.uk/download/pdf/6300636.pdf (accessed on 24th October 2025).

31. Mikavica, I., Ranđelović, D., & Mutić, J. (2024). Microplastic sources, reduction and remediation: current state and future trends. *Metallurgical and Materials Data*. https://www.semanticscholar.org/paper/61d004b3839f261097e0ecaa0a0a349064bdc8a4 (accessed on 24th October 2025).

32. Muthoni, K. L. (2025). Sustainable practices in pharmaceutical manufacturing. *Newport International Journal of Public Health and Pharmacy*. https://www.semanticscholar.org/paper/310b95c646915d31d3ebb522d084fe7d0635335d (accessed on 24th October 2025).

33. Oo, K., Jonah, K., & Thin, M. M. Z. (2023). A systematic review of the pros and cons of digital pollution and its impact on the environment. *Journal of Sustainability and Environmental Management*. https://www.semanticscholar.org/paper/4e6cd94d31a052ab835611a15364086fc812cb37 (accessed on 24th October 2025).

34. Prithiviraj, N., & Kumar, J. K. (2025). Environmental safety in industrial waste management: Innovative solutions for hazardous material handling. *International Research Journal on Advanced Engineering and Management (IRJAEM)*. https://www.semanticscholar.org/paper/c635ca58a6a932e4ea5c730e9efacddc87ddb119 (accessed on 24th October 2025).

35. Pruess, A., Kaiser, B., Sunstein, C., Sunstein, C., Waldby, C., Kahneman, D., Hutchins, D. C. J., et al. (2009). The precautions of clinical waste:

disposable medical sharps in the United Kingdom. https://core.ac.uk/download/47067.pdf (accessed on 24th October 2025).

36. Reike, D., Vermeulen, W. J. V., & Witjes, S. (2017). The circular economy: New or refurbished as CE 3.0? –Exploring controversies in the conceptualization of the circular economy through a focus on history and resource value retention options. *Resources, Conservation and Recycling, 135*, 246–264. https://doi.org/10.1016/j.resconrec.2017.08.027 (accessed on 24th October 2025).
37. Rissman, J., Bataille, C., Masanet, E., Aden, N., Morrow, W. R., Zhou, N., Elliott, N., et al. (2020). Technologies and policies to decarbonize global industry: review and assessment of mitigation drivers through 2070. *Applied Energy, 266*, 114848. https://doi.org/10.1016/j.apenergy.2020.114848 (accessed on 24th October 2025).
38. Rymarchuk, R., & Mykhaniv, V. (2024). European Union legislation regarding waste processing technology. *Scientific and Informational Bulletin of Ivano-Frankivsk University of Law named after King Danylo Halytskyi*. https://www.semanticscholar.org/paper/1ea3d006e5c9d10dc421243a135609b3e2157e6a (accessed on 24th October 2025).
39. Schanes, K., Dobernig, K., & Gözet, B. (2018). Food waste matters - A systematic review of household food waste practices and their policy implications. *Journal of Cleaner Production, 182*, 978–991. https://doi.org/10.1016/j.jclepro.2018.02.030 (accessed on 24th October 2025).
40. Schroeder, P., Anggraeni, K., & Weber, U. (2018). The relevance of circular economy practices to the Sustainable Development Goals. *Journal of Industrial Ecology, 23*, 77–95. https://doi.org/10.1111/jiec.12732 (accessed on 24th October 2025).
41. Sembiring, N., Khalif, F. A., Herlim, C., Tarigan, E. B., & Puspitasari, M. (2025). Redefining food processing: Smart innovations for efficiency and eco-friendly production. *BIO Web of Conferences*. https://www.semanticscholar.org/paper/fa2f3caab094ccc6b81f759cdadb40ead6b9ea0a (accessed on 24th October 2025).
42. Sherman, J. D., & Hopf, H. W. (2018). Balancing infection control and environmental protection as a matter of patient safety: the case of laryngoscope handles. *Anesthesia & Analgesia, 127*, 576–579. https://doi.org/10.1213/ane.0000000000002759 (accessed on 24th October 2025).
43. Surroca, J., Tribó, J. A., & Waddock, S. (2009). Corporate responsibility and financial performance: the role of intangible resources. *Strategic Management Journal, 31*, 463–490. https://doi.org/10.1002/smj.820 (accessed on 24th October 2025).

44. Tiwari, V., Kumar, A., & Rajkhowa, S. (2021). Green chemistry: Opportunity in drug discovery research (Part 2). *Current Organic Chemistry*. https://www.semanticscholar.org/paper/ac062b80a49b06dc50d7d80d61364ed0651e298c (accessed on 24th October 2025).
45. Ubaidillah, Z., Al Husna, C. H., Salsabhila, E. M., Putri, S. A., & Dwi, P. P. (2025). Empowering communities through health education: Raising awareness on microplastic pollution and prevention in Malang. *ABDIMAS: Jurnal Pengabdian Masyarakat*. https://www.semanticscholar.org/paper/ba8f7486415b0013eecac7c751eb54225fbd760e (accessed on 24th October 2025).
46. Vaghani, B. M. (2024). Integrating quality at source into supplier management: A pathway to cost efficiency and regulatory compliance. *International Journal of Engineering and Computer Science*. https://www.semanticscholar.org/paper/0f2fea9e7d51c0b0ea083c3610711c4db349a512 (accessed on 24th October 2025).
47. Wang, L. K., Hung, Y., Lo, H., & Yapijakis, C. (2005). *Waste Treatment in the Process Industries*. https://www.semanticscholar.org/paper/0e73721537e40b9bd4d219fea4dc6da69f07abf2 (accessed on 24th October 2025).
48. Warnakulasooriya, G. A. P. K. G. G., Amarasinghe, A. M. P. C., Lankathilake, L. J. M. U., & Wijesekara, E. R. J. M. D. D. P. (2024). Distillery spent wash as an alternative fuel in boilers and potash recovery from that ash remaining in boilers. *Proceedings of International Symposium on Earth Resources Management and Environment*. https://www.semanticscholar.org/paper/08292513be48c660fab055d3f229ca7bd826d926 (accessed on 24th October 2025).
49. Widodo, L. (2016). Peluang penerapan produksi bersih pada kawasan industri perikanan Muncar Kabupaten Banyuwangi Jawa Timur. *Volume, 11*, 95–103. https://www.semanticscholar.org/paper/5aefbb7738ec5c353b5a61a4655e3ce067666685f (accessed on 24th October 2025).
50. Wise, D. (1994). Process engineering for pollution control and waste minimization. https://www.semanticscholar.org/paper/1d9c6a9e49f1a04c79ab3c2a2f96e01fb85da343 (accessed on 24th October 2025).
51. Xi, S., Lettow, F., & Neuhoff, K. (2021). Climate neutrality requires coordinated measures for high quality recycling. *Volume, 11*, 187–195. https://www.semanticscholar.org/paper/9ccc1191d87bed6811a332cffe348927a30cf030 (accessed on 24th October 2025).

CHAPTER 9

TOXICOLOGY, SAFETY, AND ENVIRONMENTAL IMPACT

Contents

INTRODUCTION

The field of green chemistry plays a central role in minimizing the negative effects of conventional chemical methods on human health and the environment. Green chemistry is essentially the building of chemical processes and products aimed at minimizing toxicity and waste and maximizing sustainability. This new approach aims at safer alternatives, lower toxicity chemicals, and the efficient use of materials, hence fulfilling the urgent need for environmentally friendly practices throughout the manufacturing of chemicals. Prioritizing safety and diminishing toxicity, green chemistry is not only in congruence with regulations but also assists in fostering confidence among the public towards chemical application. Hence, the concepts of green chemistry form a robust foundation for re-examining traditional methodologies and for guiding acceptable and sustainable development processes industry-wide, leading to safer products and an improved environment. The development of such metrics as the E factor and atom economy has far evolved the green chemistry science to be measured quantitatively with chemical processes. These measurements provide a foundation for waste formation and resource use assessments, thereby promoting the development of more sustainable and greener processes. The E factor, which is waste per unit product, allows chemists to identify where waste reduction will occur by streamlining reaction pathways and resource utilization. In addition, atom economy, a measure of how effective a reaction is in terms of incorporation of starting materials into final products, makes it possible to create processes that produce less waste byproducts. However, even as such mass-based measures are helpful, they must be supported by other measures, such as life cycle assessments, to provide an overall estimate of environmental impact across the entire product life cycle (Roger, 2017). As a green chemistry challenge and opportunity, the use of nanomaterials (NMs) on a perspective of toxicity and environmental impact is what this special issue aims to discover. The novel physical properties of engineered as well as natural nanoparticles (NPs) yield enhanced performance for every application, but their toxic potential needs to be studied in depth. To determine associated risks, one needs to examine the nanoscale properties of NPs and how they interact in biological systems. With increased production and application of NMs, risk assessment techniques and legislation enter the scenario in their application in a safe manner. Toxicological profiling comparison between synthetic and natural nanomaterials can be applied to drive safer design principles, which in turn directs towards the general goals of green chemistry. This evaluation not only highlights the importance of toxicology and safety in the development of novel materials but also highlights the need for innovative approaches that prioritize environmental sustainability above all else (Jeevanandam et al., 2018).

Overview of Toxicology and Safety in Chemical Processes

The toxicology discipline is crucial to the evaluation of chemical process safety, particularly in green chemistry. The research of chemical and biota interactions is crucial to identify potential hazards and risk assessment compared to chemical exposure. Toxicological data analysis informs regulatory standards and safety protocols, and it results in the development and implementation of safer alternatives. In green chemistry, the focus is not just to reduce toxicity but to enhance the sustainability of chemical processes as well. For instance, the use of less toxic chemicals during the manufacturing process minimizes the risk of exposure to toxicity by the workers and the environment to ensure overall safety in chemical manufacturing. Accordingly, the use of toxicological design during the design phase of chemical products and processes ensures a preventive measure for safety and environmental issues, which underscores the significance of toxicology in the practice of responsible chemistry. In additional detail regarding the complexity inherent in toxicological studies of chemical processes, the role of contaminants' effect on the ecosystem, especially those caused by industrial processes, should be considered. For example, pollution resulting from abandoned mines has been cited to result in massive ecological destruction, affecting aquatic organisms as well as the overall health of river ecosystems (Jarvis et al., 2009).

Figure 9.1. The use of less hazardous substances in production processes minimizes the risk of toxic exposure to workers and the environment, promoting overall safety in chemical manufacturing.

Source: https://blog.storemasta.com.au/hubfs/BWE%20images/BWE%20HERO%20IMAGE/Staff%20in%20PPE%20carrying%20a%20toxic%20chemical%20container-1.jpg.

The presence of heavy metals such as cadmium, lead, and copper has been found to inhibit aquatic biodiversity because species respond differently based on numerous factors, such as competing ions availability and pH. These findings underscore the importance of continued monitoring and assessment of water quality to ensure the safety and resilience of ecosystems, with challenges for environmental regulation and policy. In the end, more knowledge about chemical toxicity's impact on biological systems is essential in building solutions minimizing risks while maintaining environmental integrity. In urban environments, chemical pollutant control and their toxicological characteristics become more complicated due to multidimensional source processes. Urban runoff transports a combination of pollutants like heavy metals and polycyclic aromatic hydrocarbons (PAHs), which impact sediment quality and aquatic life significantly (Beasley et al., 2001). Studies have shown that such pollutants can be capable of altering the ecological integrity of water bodies by reducing species diversity, hence influencing the foundations of aquatic food webs. Despite the stark dangers posed by urban runoff, there still remains a knowledge gap with regard to the dynamics and interactions of such pollutants in different environmental conditions. This is a challenge to stakeholders who aim to formulate effective regulatory measures. Therefore, it is crucial to encourage intensive and process-based field measurements that intensively investigate the interaction between chemical pollutants and sediment chemistry in an attempt to encourage an informed strategy towards the safety and management of urban waterways according to the principles of green chemistry (Figure 9.1).

9.1. UNDERSTANDING TOXICOLOGY IN CHEMISTRY

Toxicology as a chemical discipline is a vital field that focuses on the adverse effects of chemical substances on the living body and the ecosystem in general. It is vital in knowing how some compounds respond in living systems, which helps to determine potential risk involved in exposure to toxic substances. There is more emphasis in green chemistry on more environmentally friendly chemical manufacture with the reduction or elimination of harmful chemicals use and emissions. Toxicological information use in the design of chemical processes enables scientists to develop safer chemicals with a less damaging environmental impact but still ensuring human health safety. Such integration not only promotes chemical manufacturing processes that are innovative but also represents regulatory efforts that seek to safeguard ecosystems from pollutants and harmful factors, exemplifying the importance of toxicology as a part of chemistry practice and training. The ecological impact of exposure to chemicals is profound and typically complex, particularly considering the interaction of metals combined with each other and their combined effect. An example is closed-down mines, which are such a problem as regards leaching toxic metals into nearby bodies of water to contaminate them chemically and seriously disrupt aquatic life. A report identified the lack of environmental quality standards (EQS) for metals, including cadmium and lead in rivers with severe mining activity over the years. Results have indicated that aquatic organisms' diversity decreases with increasing metal concentration, in turn bringing the long-term persistence of such ecosystems into question (Jarvis et al., 2009). Understanding such toxicological interactions provides a better assessment of environmental safety as well as synergistic metal effects, which are pivotal in designing useful pollution remediation technologies in chemical operations. Besides, the intricacies presented by metal toxicity in aquatic environments demand covert approaches towards monitoring and control. Current water quality monitoring methods are usually lacking in the detection of episodically high metal levels and lack the capability to separate the toxicological effects of metals from other causes of environmental stress. The development of models relating chemical data to ecological responses can facilitate more efficient regulation of pollutants. For instance, knowledge of the function of pH, dissolved organic carbon, and other factors of water chemistry enhances the metal bioavailability and toxicity predictive capacity (Arnold et al., 2001). Improved approaches, including biodiagnostic indices as well as total ecological assessment, are required to balance the risks of chemical exposures. In conclusion, enhancing toxicological systems with green chemistry not only enhances public health outcomes but also promotes responsible environmental stewardship (Figure 9.2).

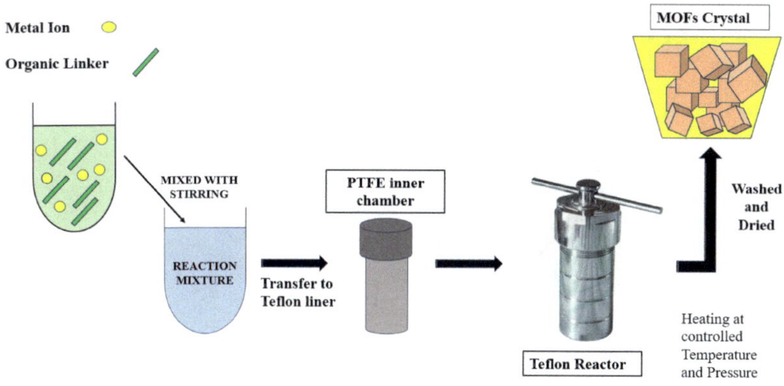

Figure 9.2. The ecological ramifications of chemical exposure are profound and often complex, particularly when considering the mixture of metals and their interactions.

Source: https://www.eeer.org/upload/thumbnails/eer-2023-636f1.gif.

9.1.1. Definition of Toxicology and Its Relevance to Chemistry

Toxicology is defined as the scientific discipline that investigates the character, action, and detection of dangerous substances, particularly chemical substances that can adversely impact human health and the environment. Its core involves the investigation of exposure risks of various chemicals, including pharmaceuticals, industrial chemicals, and agrochemicals. This field has a natural affinity with chemistry as it is based on an understanding of chemical structures, properties, and chemical reactions. By its complete evaluation of toxic substances, toxicology gives information that can mitigate deleterious effects on living organisms and the environment. The increasing understanding of these interactions has raised awareness of the value of toxicology in recent chemical research, particularly in the field of green chemistry, which focuses on safety and sustainability in the environment during the design and production of chemicals.

Figure 9.3. Green toxicology is emerging as a progressive avenue that intertwines toxicological assessments with cutting-edge technologies like organs-on-chips and predictive modeling using artificial intelligence.

Source: https://pubs.acs.org/cms/10.1021/acs.chemrestox.0c00260/asset/images/large/tx0c00260_0008.jpeg.

The use of toxicology in chemistry comes into clear perspective in the framework of green chemistry, which promotes the development of products and processes that lower toxicity and minimize effects on the environment. Because traditional approaches to establishing chemical safety can be reliant on animal testing and may be time-consuming, new options are necessary. Green toxicology is being designed as a novel route that integrates toxicological assessments with emerging technologies like organs-on-chips and predictive modeling using artificial intelligence. This convergence of disciplines not only enhances the validity and efficiency of toxicity evaluations but also enables key concepts of green chemistry to promote safer, alternative paradigm testing. These developments are capable of generating an ethically enhanced practice in chemical effect assessment to complement policies aimed at protecting both public health and the environment. Amid growing concern about chemical safety, especially of chemicals like glyphosate and its analogs, toxicology progressively becomes essential. Recent research has put the complexity of glyphosates' toxicological profile in the limelight, with potential impacts on health previously under estimated. The imperative for exhaustive toxicological examination is underscored by evidence of glyphosate residues being left in foodstuffs and in water sources, challenging the regulation standards. Not only does this suggest a deficiency in the fundamental safety studies that have been historically utilized, but it also points towards new methods that consider long-term effects on both human health and environmental integrity. Incorporating toxicological evaluations into green chemistry platforms is required, as it ensures more informed decision-making for the approval and regulation of chemicals with potential harmful effects on safety and environmental sustainability (Maertens et al., 2024; Cuhra et al., 2016) (Figure 9.3).

9.1.2. Types of Toxic Substances and Their Effects on Health

The separation of toxic agents plays an important role in analyzing their health effects and lowering exposure-linked risks. For the most part, toxins are separated into heavy metals, organic agents, and biological agents. Heavy metals such as lead, cadmium, and mercury are some examples of heavily documented harmful effects on human health primarily through bioaccumulation within living systems. Lead, for example, causes neurological developmental problems in children, whereas exposure to cadmium has been associated with damage to the kidneys and osteoporosis. Identification of the routes through which these metals enter the body, usually via contaminated water or foodstuffs, is fundamental in developing effective public health intervention. Hence, the identification of the sources and types of harmful chemicals allows for certain actions to limit exposure and protect community health in accordance with green chemistry guidelines for the reduction of hazardous chemical usage in industrial processes. Organic compounds, such as persistent organic pollutants (POPs), are another very significant category of harmful chemicals with extensive health implications. These chemicals, such as pesticides, polychlorinated biphenyls (PCBs), and dioxins, are environmentally resistant to degradation and will persist in ecosystems, ultimately affecting human health via the food chain. POPs have been associated with many adverse health effects, such as endocrine disruption, developmental problems, and cancer. Understanding the way in which these chemicals interact within biologic systems informs risk assessment and regulatory policy. Hence, efforts at minimizing emission of POPs into the environment align with the concepts of green chemistry, which encourages the design of chemical products and processes that reduce or eliminate the formation of dangerous substances (Jarvis et al., 2009). Microbial toxins, including viruses, bacteria, and fungi, are also a key category of toxins that affect human health. Food or water sources containing contaminants can lead to the transmission of such pathogenic agents, inducing diseases manifesting in various intensities of severity. Aquatic pathogens caused by contamination with metals, for instance, can debilitate aquatic organisms and eventually human health by consumption of water or fish. Environmental body water quality criteria must consider such biological risks along with chemical pollutants so that comprehensive health protection strategies can be formulated. The overlap of chemical and biological poisons indicates holism in toxicology, safety, and environmental policy with an emphasis on innovative solutions in green chemistry to address such cross-cutting challenges (Jarvis et al., 2009) (Figure 9.4).

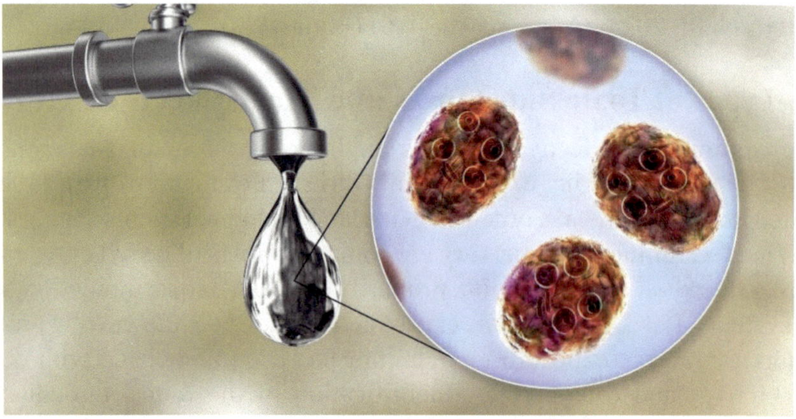

Figure 9.4. Contaminated water supplies or food sources can lead to the spread of these harmful agents, resulting in diseases that manifest in varying degrees of severity.

Source:https://carewater.solutions/wp-content/uploads/2023/01/%D8%A7%D9%84%D8%A3%D9%85%D8%B1%D8%A7%D8%B6-%D8%A7%D9%84%D9%85%D9%86%D9%82%D9%88%D9%84%D8%A9-%D8%B9%D8%A8%D8%B1-%D8%A7%D9%84%D9%85%D9%8A%D8%A7%D9%87.jpg.

9.1.3. Methods for Assessing Chemical Toxicity

Toxicity evaluation of chemicals is critical in ensuring public security and environmental protection. Traditional methods, such as animal testing, have been employed for centuries to determine the toxic effect of chemicals. Such methods, however, have created ethical concerns and are both time and cost-consuming. Therefore, there has been a huge emphasis on alternative methods, namely in silico evaluations, which use computer-based models to provide predictions regarding the effect of toxicity. Tools like ProTox 3.0 are the best example of this paradigm shift with integration of molecular similarity and machine learning to forecast various endpoints of toxicity like acute toxicity and organ toxicity. This tool not only has a user-friendly approach but also provides critical outputs like confidence scores and toxicity profiles, making it a tool of the trade for toxicologists and risk assessors (Banerjee et al., 2024).

Their synergy provides a streamlined, moral route for measuring chemical safety in green chemistry applications. Aside from in silico methods, the growing use of nanomaterials presents new toxicity issues. Nanoparticles (NPs) possess unique properties different from their bulk counterparts, which bring about enhanced biological activity and potential toxicological effects. The toxicity of NPs depends on a number of factors, including size, shape, and surface charge, all of which determine how NPs interact with biomolecules. For effective evaluation of NP toxicity, these physicochemical parameters and how they influence the biological response must be considered. Review articles have stressed the need for a clarification of NP toxicity mechanisms, including cellular uptake and ROS generation, to enhance safer nanomaterials for biomedical applications (Abbasi et al., 2023).

Adequate understanding is essential to developing greener chemistry strategies and minimizing adverse environmental impacts sourced from such new materials. Lastly, a holistic strategy for chemical toxicity assessment involves combining traditional and innovative methods for strengthening conclusions about safety profiles. A combination of *in vitro* assays, in silico predictions, and high-throughput screening can provide a better overall picture of chemical behavior under varying environments. Combination of these diverse approaches enables one to catch potential negative effects at the early phase of chemical design, reducing the extent of animal tests and making regulatory systems more effective. Moreover, the more data become available due to these diverse methods, the more precise becomes toxicity modeling in its predictions, enhancing the safety of chemical design. The incorporation of rigorous toxicological assessments in green chemistry not only guarantees public health security but also aids global sustainability by making it easier to develop safer, environmentally friendly alternatives in industrial applications.

9.2. SAFETY PROTOCOLS IN GREEN CHEMISTRY

Green chemistry safety procedures are key to minimizing toxicological risks and fostering environmental sustainability. Green chemistry is in line with the concept of decreased toxicity and dependence on the environment by encouraging inherently safer chemicals and processes. Safety procedures are strong where risk evaluation and designing safer alternative chemicals are involved, which not only restrict the production of hazardous waste but also reduce occupational exposure. These protocols are enforced through stringent regulations that facilitate the controlled use and operation of chemicals throughout their life cycle, from manufacturing to disposal. Implementing such safety standards is a collaborative effort between researchers, manufacturers, and government regulators in defining and preventing potential risks associated with chemical activities and products, ultimately creating a safer, more sustainable world. Moreover, solving the challenges of nanotechnology in green chemistry demands a strict safety protocol.

With the development of nanotechnology offering new materials and applications, including those in agriculture and food packaging, the demand for proper safety protocols becomes very essential. Nanomaterials may have novel attributes and behavior that can present unexpected toxicological hazards, and thus, certain guidelines are needed to regulate their use. For instance, the integration of bio-synthesized and bio-inspired nanomaterials can give sustainable options, but innovations must be accompanied by strict risk assessments to ensure that they are safe for consumers and the environment. The development of governing frameworks for nanotechnology, as according to He et al. (2018), underscores the need for further research and public awareness toward making safe practices and increasing acceptance of such new materials. Finally, there should be enforcement of safety practices that involve continuous education and training for personnel working in the green chemistry industry. Continuing education courses on current developments in new standards in safety and emerging practices are critical to equip researchers and practitioners with information to address toxicological complexities in their research and practice. As further highlighted by Muratov et al. (2020), understanding the interrelation of quantitative structure–activity relationship (QSAR) modeling and safety protocols can allow scientists to make accurate predictions of new chemicals toxicological effects. Through fostering a safety culture in education, then, not only does this enhance the scientific communitie's ability to innovate responsibly but also provides confidence to the public regarding the usage of safer, cleaner chemical technologies. By prioritizing safety at every level, green chemistry can give confidence that its environmental benefits are realized without compromising human health or natural balance (Figure 9.5).

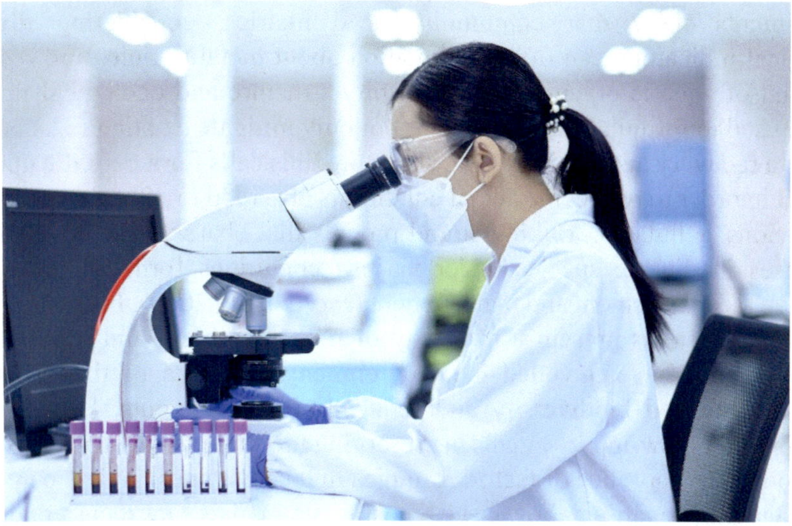

Figure 9.5. Nanomaterials can exhibit unique properties and behaviors that pose unforeseen toxicological risks, necessitating specific guidelines that govern their use.

Source: https://www.azonano.com/images/Article_Images/ImageForArticle_6792_17235348285454645.jpg.

9.2.1. Risk Assessment and Management Strategies

In relation to risk assessment and management in green chemistry, one has to evaluate the environmental impact of harmful chemicals like heavy metals involved in industrial activities like mining. Extensive risk assessment considers both acute and chronic exposure, and this can play a wide-ranging impact in aquatic environments. As an example, studies have indicated that abandoned mines are a major source of chemical pollutants, leading to water quality degradation and aquatic biota health compromise. This illustrates the significance of effective management strategies that are capable of quantifying the threats posed by metals like cadmium, lead, and zinc, as highlighted in prevailing environmental quality standards (EQS). A major component of such plans is the development of innovative techniques founded on empirical data and biological indicators for determining the ecological state, thereby enabling more effective evaluations of environmental impact and potential recovery pathways for affected ecosystems (Jarvis et al., 2009). Further, incorporation of novel monitoring processes in risk management schemes enhances the ability to assess the multilateral impact of toxicants in various environmental contexts.

Experiments show that community-based metrics, considering different biological indicators, can offer information about metal toxicity and recovery potential of aquatic organisms. Community structure and ecological function can be used to formulate strategies that not only mitigate existing risks but also promote recovery and resilience. Furthermore, this calls for a periodic updating of EQS in order to reflect the actual conditions observed in affected ecosystems of previous pollutants. Making sound models that characterize physiological responses in diverse species under various conditions forms a key part of effective risk assessment and management (Jarvis et al., 2009). In farming, low-input farm systems present a new way of risk assessment and management. Recent studies have unveiled the environmental benefits of these systems, such as enhanced biodiversity and reduced chemical inputs. However, the contribution of low-input systems in preserving food safety and quality is not yet questioned. The absence of factorial studies in today's times hinders perceptive understanding regarding how each constituent in these systems contributes towards nutritional implications. Through adoption of an environmentally and health-effect consideration risk assessment system, stakeholders can make choices in favor of sustainable agriculture. This integration of environmental toxicology and agribusiness principles provides the potential for developing integrated management strategies to counteract losses in biodiversity while ensuring food production safety, and thus enhance a healthier ecosystem.

9.2.2. Safety Measures in Laboratory and Industrial Settings

Inside the laboratory and in the industrial setting, the application of strict safety measures is the most essential to protect individuals and the environment. Safety measures entail a set of steps, including the use of personal protective equipment (PPE), such as gloves, goggles, and lab coats, as the initial protective shield against harmful material exposure. Accurate training on the dumping and handling of harmful substances is also essential. Employees should be educated on the potential risks that encircle chemicals handled by them and how to respond in the event of accidents, spills, or exposure. Education forms a culture of safety, reflecting that knowing and anticipating risks is crucial in minimizing risks. Hence, by adopting these standards not only is individual health ensured but working conditions as a whole are made secure. In the area of green chemistry, safety guidelines also furnish a paradigm by which to evaluate the environmental impact of chemical procedures. Through emphasis on sustainable procedures, professionals are compelled to balance competing materials and methods with regard to their compatibility with environmental safety considerations. Substitution of hazardous solvents with less toxic alternatives represents one such move towards the reduction of potential health risks and promoting ecological responsibility. Apart from substitute approaches, applications of technologies like nanotechnology also help enhance safety in agricultural applications by limiting the environmental impact of pesticides and fertilizers. The accurate application of nanoparticles, which are under consideration for their potential to promote plant health, should then be complemented with sufficient safety tests to avoid the negative implications of their production and application, highlighting the nexus between laboratory safety and broader environmental conditions (Vânia et al., 2021). In addition, continued research highlights the need for integrating safety measures into the emerging fields of nanotechnology and biotechnology. As these regions expand, they do so with accompanying unique risks and tremendous possibilities for green chemistry breakthroughs. For instance, while nanoparticles have demonstrated significant promise in optimizing agriculture methods—like optimized nutrient usage and disease resistance—there are still environmental issues and toxicity concerns. Therefore, a comprehensive risk assessment and definition of regulatory requirements are necessary in guiding studies and application. This dual approach will ensure that innovations will benefit food security and agricultural system sustainability, in essence following the principles of green chemistry. As industry stakeholders embrace this mandate, they protect not only public health but also pave the way for progress that holds respect for sustainability and environmental stewardship (Mittal et al., 2020) (Figure 9.6).

Figure 9.6. Within laboratory and industrial settings, the implementation of stringent safety measures is paramount to protect personnel and the environment.

Source: https://image.kindle-tech.com/images/articles/281/b43f7768e22f05b-50277b363e62c36ad650ab01e3e324.jpg.

9.2.3. Regulatory Frameworks Governing Chemical Safety

The regulatory framework for handling chemical safety is critical to safeguarding public health as well as the environment. The Toxic Substances Control Act (TSCA) in the United States plays a central role in controlling the introduction and use of chemicals by demanding safety evaluation before new chemicals are introduced into the marketplace. This legislative framework demands that manufacturers provide information on the toxicity and environmental characteristics of their chemical products. Loopholes in and the paucity of funds within regulatory agencies, however, often get in their way, and those substances continue to enter by documented danger. Federal and state law also mesh in many ways, differing significantly and even causing confusion regarding compliance. An integrated review of such as such frameworks is therefore crucial since it points to areas that could compromise safety and public health programs. Moreover, industry impact on rule-making is one of the sources of growing concern in chemical safety governance.

Associations such as the American Chemistry Council have been implicated in donating millions into lobbying activities with an aim of shaping regulations that promote chemical production interests. As Greenhaw et al. (2013) point out, the Office of Advocacy, under the Small Business Administration, has come under attack for prioritizing corporate interests over public health by formulating their comments and recommendations based on industry talking points rather than on science-based analysis. Not only does this continue to undermine the integrity of regulatory frameworks, but it raises questions about whose interests regulatory agencies truly serve. As the public increasingly learns about the potential harm that comes from chemical exposure, urgency must be applied to transparency and accountability when these frameworks are created and enacted. To further drive green chemistry initiatives, regulatory frameworks must shift to prioritize more sustainability and safety in chemical manufacturing.

Green chemistry promotes the design of chemical products and processes that eliminate or avoid the use of hazardous materials, and this is almost perfectly in line with regulatory aims of decreasing toxicity and environmental impact. Policymakers and funders can follow the following approaches: integrating principles of green chemistry into current regulation, and thereby promoting innovation while maintaining safety practices. Active cooperation among scientists, regulatory agencies, and funders is the greatest requirement to address the problem of hazardous chemicals in the environment. By ensuring an active culture of chemical safety that opens doors to green chemistry, the stakeholders are able to come together towards the future where public health and environmental quality are ensured amid development.

9.3. ENVIRONMENTAL IMPACT OF TRADITIONAL CHEMISTRY

The impact of traditional chemistry on the environment is a critical consideration, particularly with the result of chemical reactions reflected in various environmental issues. The conventional methods are often laden with hazardous chemicals, whose byproducts lead to extensive contamination of air, land, and bodies of water. Industrial procedures in traditional chemistry tend to generate enormous volumes of waste that can inundate waste management facilities and contribute to ecosystems' degradation. This is exacerbated by a lack of institutions for adequate assessment or tackling of these impacts. The discharge of hazardous chemicals into the environment can have far-reaching consequences on human health and biodiversity, highlighting the need for an urgent transition toward more environmentally friendly chemistry practices (Hanno et al., 2018).

Figure 9.7. Manufacturing practices in traditional chemistry frequently generate large quantities of waste, which can overwhelm waste management systems and contribute to the degradation of natural habitats.

Source: https://watermanaustralia.com/wp-content/uploads/2023/12/mwtp-31.jpg.

In addition, the environmental footprint of traditional chemistries can also be quantified by other aspects such as the E factor, which is used to quantify waste per unit of product harvested. With heightened emphasis on sustainability, the shortcomings of traditional metrics have been exposed to demonstrate that they are not sufficient to meet extensive environmental issues. Atom economy and process mass intensity (PMI) have been in fashion but are still insufficient to expose the complete environmental cost of chemical processes. Life cycle assessment (LCA) is becoming increasingly important to study not only waste generated but also lifecycle impacts of chemical production, from resource extraction to end-of-life disposal. This systems method is significant in opening the door to greener chemical practices (Roger, 2017). Transitioning from traditional to green chemistry entails the redesign of chemical production, with a strong focus on sustainability and minimizing harm. The development of green chemistry aims not only to reduce the environmental impact of chemical processes but also to make safer chemicals, which are less toxic and less environmentally persistent. Installation of renewable feedstocks and process development aimed at maximization of atom economy and minimization of waste is of utmost significance during such transition. Green chemistrys principles necessitate a transition towards the circular economy, promoting resource efficiency and waste reduction as fundamental principles. This advancement is crucial in the battle against the hazardous ecological issues caused by traditional chemistry, towards an improved and safer future (Hanno et al., 2018) (Figure 9.7).

9.3.1. Pollution and Waste Generated by Conventional Chemical Processes

Within the ambit of traditional chemical processes, pollution and waste are colossal environmental issues that have a direct impact on human health and environmental equilibrium. The traditional manufacturing processes tend to release harmful chemicals, which remain in the environment and enter the food chain. These harmful chemicals carry numerous risks, ranging from interfering with endocrine systems, affecting the brain functions, to causing reproductive health complications among populations. Anthropogenic chemical contamination has been discovered to represent one of the most critical risks to human beings, but proof and perception of its scale remain disunited (Naidu et al., 2021). Ignorance not just hinders good policy-making, but also supports perilous practices which compromise ecosystem integrity and human security. The agricultural sector further demonstrates the decimating impacts of conventional chemical processes that breed wastage and contamination. Reliance on synthetic fertilizers and pesticides has been linked to widespread loss of crops, not just impacting productivity but also the health of surrounding ecosystems. For instance, almost a third of global crops are threatened by a variety of reasons, including chemical runoff that contaminates soil and water supplies. This reliance on harmful chemicals necessitates the identification of eco-friendly alternatives, such as nanotechnology, which has proven to enhance agricultural resilience and food security. Through the application of new solutions, the sector can reverse the adverse impact associated with conventional methods and encourage a healthier relationship with nature (Mittal et al., 2020).

Notably, there is an urgent imperative for a global shift towards cleaner methods in chemical processes to minimize the impact of negative effects of pollution and waste. While there are existing measures of pollution control in place, their implementation is prone to being below that required for radicle improvements in human health and environmental defense (Naidu et al., 2021). The integration of green chemistry principles ensures a path to achieve sustainable production processes, minimizing harmful emissions and waste generation. By directing emphasis towards the invention of less harmful chemical substitutes, educating the public, and investment in research for new technologies, the community can minimize the detrimental consequences of conventional procedures. This paradigm shift is necessary not only for current ecological equilibrium but also to guarantee the preservation of future generations' health and happiness in a world with continuously increasing pollution (Figure 9.8).

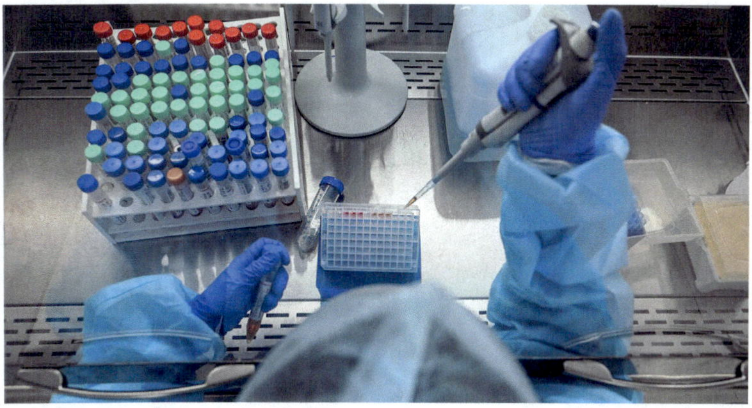

Figure 9.8. By prioritizing the development of safer chemical alternatives, fostering public understanding, and allocating resources for innovative research, society can address the detrimental impacts of conventional processes.

Source: https://carnegieendowment.org/static/media/images/GettyImages-1228545815edit-1.jpg.

9.3.2. Long-Term Ecological Consequences of Toxic Chemicals

The long-term effects of toxic chemicals on the environment are a worrisome issue in environmental science. A notable example is ocean pollution, a critical issue that originates primarily from land sources. The pollution leads to the release of an array of toxic materials, including heavy metals and microplastics, into the ocean. Once these chemicals enter the ocean, they can disrupt marine life, leading to lower biodiversity. Besides, the introduction of such pollutants affects not just marine animals and plants but also the physical and chemical properties of the water, having a cascading effect on the whole system. Habitat destruction, such as on coral reefs, contributes to the issue, calling for urgent action to address such environmental issues through exposure to toxic chemicals, with the additional problem of climate change (Philip et al., 2020). The environmental impact of toxic chemical pollution extends beyond physical harm; it also has strong impacts on food webs and public health. For instance, harmful substances like methylmercury and polychlorinated biphenyls (PCBs) have been recorded to bioaccumulate in aquatic organisms, forming a poisonous pathway that not only endangers wildlife but also the people involved in seafood consumption. Continuous contact with these toxins can lead to extensive health inequalities, particularly among vulnerable populations.

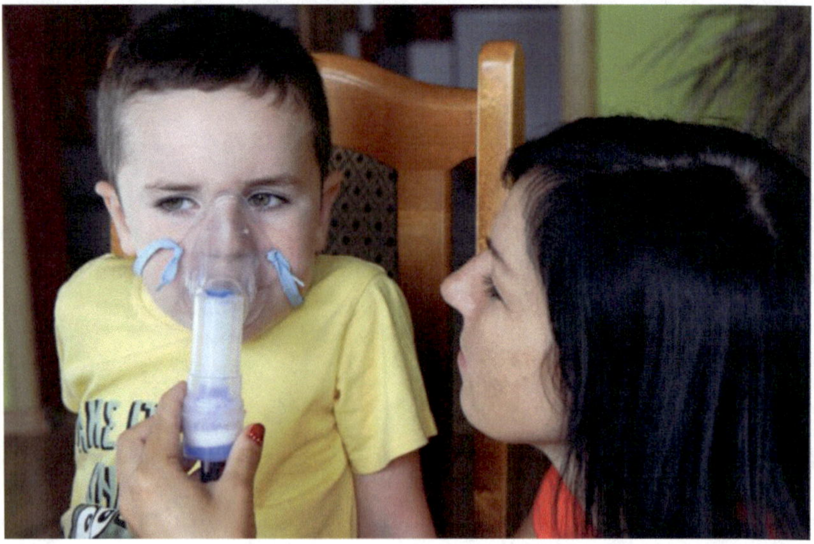

Figure 9.9. Chronic exposure to these contaminants can lead to significant health discrepancies, particularly in vulnerable populations.

Source: https://www.epa.gov/sites/default/files/styles/medium/public/2020-10/air_child_inhaler2.jpg?itok=_SRlNSPt.

Long-term health consequences, ranging from neurological impairments in children to high cardiovascular disease risks in adults, are the direct consequence of this bioaccumulation, which is a demonstration of the impact toxic chemicals have on ecological processes on very long-time scales. As such, the necessity for curbing and dealing with these pollutants stands paramount to the conservation of both ecological health and human well-being (Philip et al., 2020). Furthermore, the connectivity of ecosystems also emphasizes the necessity to grasp the combined effects of toxic chemicals. And as the pollution keeps mounting by the minute, so does the frequency of poisonous algal blooms (HABs) triggered by nutrient leaching due to excessive farm procedures. Not only do these blooms decrease the oxygen in water bodies but also create harmful toxins for marine organisms and human health. Their impact is felt across an ecosystem with a reduced species count and altered predator-prey interactions. Solutions to these problems require a holistic approach to chemical toxicity and comprehensive environmental management. Green chemistry-driven methods can assist us in developing safer alternatives and solutions to lesser ecological harm through addressing the need for a less toxic chemical presence in the environment (Figure 9.9).

9.3.3. Case Studies of Environmental Disasters Linked to Chemical Use

Environmental catastrophes brought about by the use of chemicals have extensive implications for the ecosystem and general health. A good example is the Bhopal, India, disaster where a 1984 gas leak released methyl isocyanate, a highly toxic substance used in the manufacture of pesticides, into the atmosphere. Thousands of individuals were exposed to it. The disaster took immediate lives and long-term health complications among many of the survivors, illustrating how exposure to chemicals can bring about irreversible harm. In such a case, the necessity of rigorous risk assessment and management by the regulatory bodies becomes apparent. When communities are grappling with the effects of such accidents, it is important that the role of rigorous safety protocols based on green chemistry principles to prevent such disasters in the future be realized. One such example in this regard is the environmental degradation resulting from the chemical runoff in agriculture. Various pesticides and fertilizers, if not properly controlled, can contaminate water bodies with the potential for large-scale ecological disruption. The process of eutrophication in water bodies, which is characterized by over-enrichment of nutrients and the consequent algal growths, is a direct consequence of such runoff. These conditions not only impact aquatic life but also lead to threats for human communities relying on these water bodies. Chesapeake Bay case is the classic struggle between farm output and the protection of the environment. Counterbalancing its effects requires adopting greener methods and promoting more sustainable methods of farming that cut down on chemical dependency, therefore fulfilling the principles of green chemistry. The Love Canal disaster illustrates the long-term effects of bad chemical waste disposal. Residentials near Love Canal in New York State began developing serious conditions related to toxic waste buried in the neighborhood in the 1970s. Not only did the industrial chemical disposal lead to contamination of the quality of life among the residents, but it also spread a nationwide environmental health epidemic. The situation triggered public awareness and regulation reforms concerning the handling of hazardous wastes. The development of environmental toxicology, as discussed further in a Witness Seminar examining such events, highlights the critical necessity of ongoing vigilance in evaluating chemical safety and toxicity. With a greater comprehension of environmental dangers, these lessons must be incorporated into green chemistry protocols to protect both human wellness and ecological balance (Solecki et al., 1999). Further, the discussion on this historical context highlights the continued necessity to construct stable foundations for environmental protection, considering the historical evolution of attitudes toward chemical risk (Christie et al., 2004).

9.4. GREEN CHEMISTRY PRACTICES TO MITIGATE TOXICITY

Green chemistry is the evolution of chemical processes and products with the basic objective to lower the toxicity of chemical processes and products, an objective coordinated with improving human health and environmental safety. Through the innovative effort to produce safer alternative chemicals, green chemistry methodologies highlight that the use of harmful substances should be avoided in the design step of chemical synthesis. For instance, electronics manufacturers have adopted new methods that do not involve the use of poisonous substances such as brominated and chlorinated compounds, which are known to create hazardous waste. This approach not only meets strict international standards but also creates a culture of sustainability for industries that traditionally employed hazardous chemicals, showing a significant shift toward safer manufacturing and reduced ecological injury (McPherson et al., 2009). Moreover, the inclusion of public health considerations in green chemistry use optimizes their ability to reduce toxicity. The evaluation of emergent aquaculture production technologies illustrates the potential for an integrative approach balancing both environmental and public health implications. Research on tilapia culture illustrates the necessity of comprehensive public health impact assessments, essential to understanding the implications of novel aquaculture systems for ecosystems and communities. With the use of both qualitative and quantitative methods, data can be acquired about workplace hazards along with environmental hazards associated with such processes with a view to ascertaining safer alternatives in harmony with green chemistry principles. Such audits illustrate the interconnections between safety, public health, and the environment, cementing the need for collaboration in carrying out green chemical processes (Azim et al., 2012). Ultimately, the push for green chemistry is encouraged by the realization that reducing toxicity is not an academic exercise but a practical necessity for sustainable development. Constructing effective green chemistry practices requires ongoing research, stakeholder communication, and a strong commitment to innovation. With sectors increasingly embracing green practices, the role of education in creating an informed workforce that values safety and environmental integrity becomes more significant. Equipping emerging scientists and engineers with the green chemistry concepts can help deliver innovations that combat the deleterious effects of toxic chemicals, making the quest for economic development without compromising public health or ecological balance. Hence, green chemistry's promise to transform industries while fostering health and wellbeing continues to be a critical area of study and action (Figure 9.10).

Figure 9.10. Establishing effective green chemistry practices requires ongoing research, collaboration among stakeholders, and an unwavering commitment to innovation.

Source: https://s7d9.scene7.com/is/image/mmmspinco/solv-web-header-image-how-we-innovate-entp:M--large-tablet?ts=1745342183700&dpr=on,2.625.

9.4.1. Use of Renewable Resources and Sustainable Materials

The incorporation of sustainable sources and renewable resources in the application of green chemistry is crucial to address the environmental concerns that traditional chemical practices raise. The exploitation of fossil fuels not only causes degradation of resources but also exerts a massive toxicological impact through the by-products of processing and extraction. Making the transition to energy from renewable sources, such as biomass, wind, or solar power, has the potential to minimize the adverse effects inherent in conventional chemical production. For example, using feedstocks from agricultural residues or other renewable sources can eliminate the carbon and particulate matter footprints of production, eventually resulting in a cleaner chemical sector. With the use of these alternatives, not only is the ecological equilibrium of ecosystems guaranteed, but human health risks associated with pollution and chemical exposure are also prevented, forming a central pillar of an integrated approach towards environmental responsibility in chemistry. Additionally, sustainability of materials used in chemical processes is paramount in the prevention of chemical pollution. Synthetic compound impacts extend beyond their immediate purpose; the life cycle of these types of material—from production to disposal—far surpasses environmental toxicity. Green materials, such as bioplastics and biodegradable polymers, have a far-reaching capacity to alleviate long-term chemical disposal and contamination risks, which invariably express themselves in air, land, and water. Cabernard et al. (2021) emphasizes the need to adopt sustainable methods of making plastics, with the reality that while increasing global demand exists, so too does their environmental footprint. It is a call for using materials which not only effectively carry out their purpose but also will decompose naturally and leave minimal long-term pollution, therefore placing emphasis on the focal role of sustainable material choice in green chemistry. Finally, the initiative to utilize renewable sources and green materials also indicates the global response to poisonous chemical pollution and its mass health impact. Naidu et al. (2021) points out the fractured understanding of the threats of chemical pollution and notes the imperfection of an integrated approach to reducing anthropogenic chemical hazards as a priority. By prioritizing renewable and sustainable materials, industries and scientists can redefine the chemical map, prioritizing materials that are healthy for human well-being and the environment. This redefinition is critical not just to reduce existing pollution levels but also to prevent future threats from toxic chemicals in the environment. Integrating sustainable principles into chemical research and production is congruent with public health objectives, reflecting a path towards a safer and more sustainable chemical sector supportive of environmental protection and human health.

9.4.2. Development of Less Hazardous Chemical Processes

Development of safer chemical processes is a major achievement in the field of green chemistry that demands sustainable processes that expose humans to minimal risks of environmental and health harm. The move is occasioned by increased awareness of the adverse effects associated with conventional chemical processes, which release toxic byproducts and are involved in the use of risky substances. Of particular focus here is the transition to safer substitutes that still fulfill very stringent environmental standards without being any less effective. For instance, some electronics companies have re-engineered their manufacturing processes to rule out the application of brominated and chlorinated chemicals, thereby reducing toxic waste and improving worker safety (McPherson et al., 2009). This joint endeavor emphasizes the significance of the intersection of innovation, regulatory compliance, and corporate social responsibility towards achieving safer chemical processes. In driving the shift to safer processes, decision analysis is a crucial method in their evaluation and selection.

Figure 9.11. Utilizing decision analysis within alternatives assessment allows stakeholders—from regulatory bodies to businesses—to make informed choices regarding the adoption of safer chemicals.

Source: https://tks-hpc.h5mag.com/tks-hpc/hpc_today_3_2024/regulation_-_substances_on_the_spotlight/51809/regulation_2.960_0_1.jpg.

This framework for analysis integrates various forms of analysis to prioritize substitute chemicals according to their safety, efficacy, and environmental sustainability. Using decision analysis for alternatives assessment helps the stakeholders—regulatory bodies to firms—make informed decisions regarding using safer chemicals. The benefits of this analytical process have been highlighted in a recent workshop, which brought scientists and experts across disciplines together to talk about the potential role that decision science could play in enhancing alternatives analysis (Blake et al., 2016). By making a systematic comparison of risks and benefits of different chemical alternatives, the stakeholders are in a better position to place their decisions in context, and in doing so, they develop a sustainability and safety culture within the chemical industry. In integrating alternatives analysis into decision-making processes, business organizations are able to transform fundamentally the manner in which they handle chemical safety and environmental responsibility. As corporate responsibility takes hold, businesses are discovering that sustainable practices are not only a regulatory mandate but also a market imperative. The contribution of decision science to this transformation is central, as it provides methodical methods for assessing and formulating more secure chemical alternatives. Suggestions of the aforementioned workshop emphasize transdisciplinary co-operation, sequential tool development, and educational extension (Blake et al., 2016). By co-operative culture between disciplines and engagement with multiple stakeholders, less hazardous chemical process development can be significantly hastened and ultimately lead to safer products and a reduced environmental footprint in chemistry (Figure 9.11).

9.4.3. Innovations in Biodegradable and Non-Toxic Alternatives

The evolution of the biodegradable and non-toxic alternatives in green chemistry is a major paradigm shift in the environmental practice of sustainability and safety. Advances in this field are mainly for designing materials and processes with minimized ecological effects and enhanced safety for human health. For instance, biodegradable solvents manufactured from renewable source materials not only have reduced toxicity but also peak performance in most chemical processes. According to research, the alternative solvents must meet strict standards such as stability, non-volatility, and synthesis using environmentally friendly methods in order to be truly regarded as green alternatives. This strategy aligns with the principles of green chemistry, which emphasize sustainability in processes and new alternatives to conventionally used, toxic chemicals (Radošević et al., 2016). Some of the notable advances in this field are supercritical fluids and ionic liquids that can potentially function as ecologically friendly solvents that can replace traditional organic solvents. Supercritical carbon dioxide ($scCO_2$), in particular, has been recognized for its unique characteristics to enable efficient extraction processes with the unwanted environmental implications of typical solvents. Studies highlight the characteristics of $scCO_2$ as non-toxic and capable of dissolving a vast array of compounds, thus showcasing its multi-purpose applicability in industries. Yet, ionic liquids like imidazolium salts have also been proposed as alternative solvents, albeit with caution regarding their manufacture from petrochemical feedstocks, disputing their overall green character. Thus, while these solvents are promising, their relative toxicity and environmental picture must be appraised in a balanced way in determining their long-term sustainability (Samorì, 2010). The studies on alternative solvents do not only satisfy the need for clean chemical processes but also lay the foundation for developments which have the capability to enhance environmental protection further. Recent developments involve the development of furan-based ionic liquids from renewable feedstocks that have reduced toxicity profiles compared to traditional solvents. Using such materials allows scientists to circumvent the toxicological challenges of chemicals from petrochemical sources, hence being in tandem with green principles. Further, the use of switchable polarity solvents to extract valuable resources from biomass is a novel way of increasing the efficiency of extraction in addition to decreasing associated environmental impacts. With these technologies in ascendance, ongoing research is required to establish their viability and efficacy to ensure that the shift towards cleaner alternatives is effective and long-lasting (Samorì, 2010).

9.5. ASSESSING THE SAFETY OF GREEN CHEMISTRY PRODUCTS

Green chemistry products' safety assessment entails a new strategy that reconciles toxicology and environmental principles. Traditional methods adopt the use of animal testing and trailing chemical safety evaluations, which could result in devastating human and environmental health impacts. Compared to these, green toxicology is an active strategy toward chemical safety, whereby it focuses on not being toxic at the initial stages of product formulation. By applying advanced techniques such as organs-on-chips and AI-powered predictive models, researchers are able to assess risks, exposures, and hazards more effectively. Not only does this transition allow for quicker identification of safe alternatives, it also reduces dependency on animal testing, aligning with moral standards and enhancing public trust in chemical safety procedures (Maertens et al., 2024). To assess the sustainability of green chemistry processes, several metrics have come to be seen as indispensable tools. The E factor and atom economy are two of the oldest green chemistry metrics, giving numerical indications of waste amounts and resource use in chemical processes. These mass-based metrics, though invaluable, lack the completeness needed to accurately represent the environmental footprint of green chemistry products. To support these measurements, life cycle analyses (LCA) are increasingly being used, allowing the overall picture of a product's environmental footprint from conception to demise. Additionally, metrics such as process mass intensity (PMI) and reaction mass efficiency (RME) also allow for greater insight into sustainability. These criteria collectively provide a multidimensional model for the assessment of green chemistry products' safety and environmental impacts, leading to researchers and industries towards more eco-friendly options (Roger,

2017). Ultimately, the transition towards safer and greener chemical products demands the collaboration of scientists, policymakers, and industries. Green toxicology progress is aligned with initiatives by the government, such as the European Green Deal, to drive sustainable innovation in the chemical industry. By encouraging policies favoring the implementation of green practices, stakeholders can influence the creation of safer substitutes for chemicals with reduced adverse effects on human health and the environment. This alignment of safety controls with environmental objectives can make it easier to shift towards a paradigm of developing and regulating chemical products. As green chemistry keeps progressing, embracing such holistic strategies will become essential to making chemical safety the principle that guides a future where functionality and profitability are also taken into account (Maertens et al., 2024) (Figure 9.12).

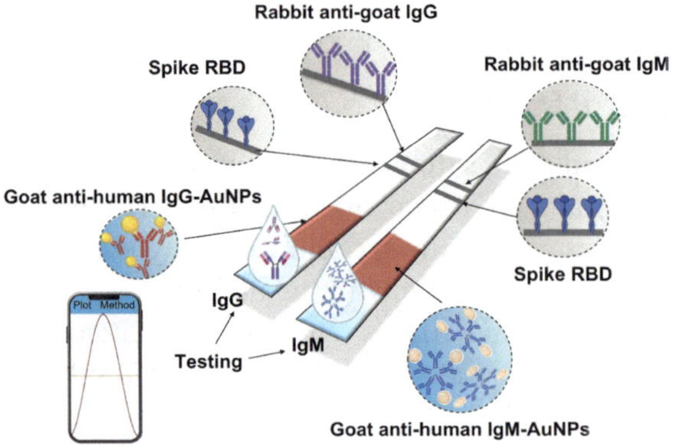

Figure 9.12. By utilizing advanced techniques such as organs-on-chips and AI-driven predictive models, researchers can assess hazards, exposures, and risks in a more efficient manner.

Source: https://www.ahajournals.org/cms/10.1161/CIRCRESAHA.122.321877/asset/ef110a89-4431-4746-bd5f-52ce09df9a72/assets/graphic/circresaha.122.321877.fig04.jpg.

9.5.1. Evaluation of the Safety Profiles of Green Chemicals

Green chemical safety profiles are evaluated most importantly to confirm that alternatives to traditional toxic chemicals do not pose unforeseen risks to human health and the environment. Green chemistry as a science focused on the design of chemical processes in which less energy and less hazardous waste is consumed, promotes safer material use. However, the transition to greener technologies requires rigorous testing methodologies to measure the toxicity and environmental degradation of chemicals. For instance, the pattern of environmental degradation of green chemicals and their interaction with biological systems could inform regulatory policies and consumer safety protocols. With increasing businesses going green, there needs to be stringent toxicological tests that uncover potential adverse effects and establish safety profiles comparable to or higher than those of the conventional counterparts. An additional significant component in determining the safety profiles of green chemicals involves embracing new technologies and methods, such as quantitative structure-activity relationship (QSAR) modeling. This computer resource has an important role to play in the toxicological properties of new synthetic compounds based on their molecular structure, enabling scientists to assess the risk of new green chemicals before extensive use. It is essential that QSAR modeling is supplemented with experimental data to improve the validity of predictions and inform safer green chemical design. In addition, as the field evolves, the coming studies must focus on refining these models to accommodate more varieties of chemical classes and biological activities for complete safety assessments.

These advances are crucial in eliminating public fear about the environmental consequence of adopting new chemical methods. Further, the use of nanoparticles in agricultural systems through green chemistry warrants careful consideration of their safety. Nanoparticles have proven the potential to increase crop yields and pest resistance with less environmental footprint than traditional agriculture. It is, however, urgently necessary to evaluate their longer-term effects on water, soil, and ecosystems. Sustaining the safety of nanoparticle products entails complete information about their interactions with crops as well as the neighboring ecosystems, as highlighted in recent research on agrotechnological innovations. Through employing an interdisciplinary approach and emphasizing the importance of safety testing in the production of such green chemicals, the agricultural sector is capable of aligning with sustainability goals while avoiding unforeseen hazards (Mittal et al., 2020; Muratov et al., 2020).

9.5.2. Comparative Analysis of Green vs. Traditional Chemical Products

Comparison of green chemistry products with their traditional counterparts reveals immense differences in safety, toxicology, and environmental impact. Traditional chemical products, for the most part based on non-renewable resources, utilize hazardous materials that are largely responsible for risks to human health and ecosystems. Such chemicals tend to possess lasting and bioaccumulative nature, creating long-term environmental contamination and accumulation toxicity in food chains. On the other hand, green chemistry promotes safer alternatives that are focused on resource conservation and less toxicity. Green chemical processes with renewable feedstocks can minimize toxic byproducts and maximize biodegradability, hence decreasing their environmental footprint. Such seminal steps stress the necessity of revisiting conventional chemical methods in light of sustainable processes that respect current health and environmental tenets. Considering the environmental issues, typical chemicals such as per- and poly-fluoroalkyl substances (PFAS) evoke critical questions regarding their persistence and bioaccumulation. Current ecotoxicological discussions highlight a sub-set of PFAS with terrifying characteristics that undermine human health and environmental integrity. While these findings are well-documented for certain PFAS such as perfluorooctane sulfonate, there is considerable dearth of data regarding the larger class of PFAS compounds.

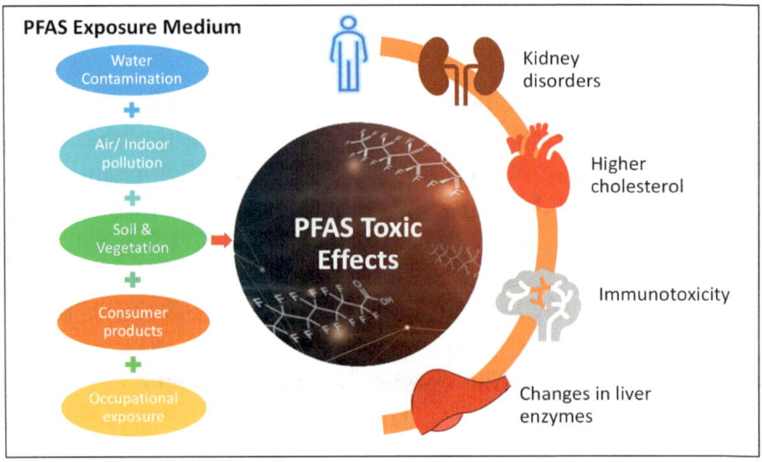

Figure 9.13. Examining the ecological implications, traditional chemicals such as per- and poly-fluoroalkyl substances (PFAS) raise critical concerns regarding their persistence and potential bioaccumulation.

Source: https://pub.mdpi-res.com/pollutants/pollutants-04-00009/article_deploy/html/images/pollutants-04-00009-g002.png?1709730545.

These lacunae cast light on the need for rigorous monitoring and predictive modeling to completely understand their effect. Yet, even in comparison to green chemicals, which are designed keeping environmental considerations in mind, they have less adverse consequences and more transparent risk profiles. This fundamental divergence highlights the pressing necessity to move towards green alternatives in addressing the many environmental problems generated by traditional chemical commodities. Quantitative structure-activity relationship

(QSAR) modeling is a main contributor to the quantitation of the toxicity and safety of both green and traditional products. QSAR models provide the possibility of thorough understanding of chemical structure-biological activity correlations, thereby enabling the prediction of possible toxicities without exhaustive experimental evaluation. Comparative evaluation not only enhances ability to assess risk but also enables safer alternatives for development by emphasizing safer chemical structures thereby align with green chemistry principles. In addition, adopting new methods will increase our ability to bridge current knowledge gaps about toxicological profiles of a variety of chemicals, such as less researched PFAS. As indicated in recent debate in the field, integrating modern methodologies, such as in silico and *in vitro*, has the ability to provide significant input in regulatory settings formulating chemical safety and efficacy. Constructing these relationships well informs a more nuanced understanding of the relative merits of green in comparison to traditional chemistry (Gerald et al., 2020; Muratov et al., 2020) (Figure 9.13).

9.5.3. Role of Life Cycle Assessment in Safety Evaluation

Application of life cycle assessment (LCA) within the framework of green chemistry is key in evaluating the safety and environmental impact of chemical products. LCA fully takes into account the environment and potential impacts of a product's entire life cycle from raw material extraction, right through to production, use, and disposal. By taking into account the holistic approach of chemical processes, LCA allows researchers and business stakeholders to identify and stop harmful impacts on human health and the environment. As industries transition to more sustainable processes, understanding the life cycle of chemical substances will offer a comprehensive evaluation that takes into account the value of innovation without overlooking possible adverse effects, particularly from new chemicals such as surfactants (Cătălina et al., 2025). This approach thus promotes a more informed decision-making process regarding chemical safety and environmental stewardship. The application of LCA reiterates the importance of sustainability in the quest for novel chemistries. For instance, as environmentally friendly surfactants become more and more requested by consumers, LCA provides important data on their environmental footprint. Studies indicate that despite the promising biodegradability observed in recently synthesized surfactants from thiophene, their lifecycle impact can reveal extreme environmental and health consequences, particularly for their polymers (Cătălina et al., 2025). By identifying some of the production lifecycle steps that result in such unfavorable outcomes, stakeholders are able to engineer chemical syntheses that lead to even more sustainable substitutes. These specific advancements lead to the generation of greener chemical products that are aligned with green chemistry principles and ultimately a sustainable circular economy. Moreover, LCA is crucial in comparisons between different synthesis pathways utilized in the production of chemicals. An example is when comparing alternative reaction routes, i.e., enzymatic and petrochemical routes for manufacturing bio-based monomers, LCA explains the contrasting aspects of these. Despite enzymatic processes being able to show reduced overall environmental impacts, undesirable implications like greater electricity consumption must also be considered. Such analyses facilitate strategic process efficiency and sustainability enhancements, informing how LCA makes it possible for paramount changes to be identified for enhancing safety performance. Thus, LCA is not only a tool for assessment but also a template for continuous improvement in safety evaluation throughout the whole discipline of green chemistry.

9.6. CASE STUDIES IN GREEN CHEMISTRY APPLICATIONS

The pharmaceutical industry is frequently conflicted with the environment due to its high contribution towards pollution as a result of using toxic chemicals and waste generation. A major case study of the use of green chemistry illustrates how adopting eco-friendly measures can help reverse these issues. Focusing on Green Chemistry principles such as atom economy and catalysis, producers are capable of simplifying chemical processes while minimizing waste and energy by a significant percentage. One such initiative highlighted the transition from traditional solvents to greener alternatives, effectively preventing hazardous emissions. This strategic shift not only complies with evolving regulatory standards but also holds the potential for an increasing consumer demand for sustainable products. Ultimately, these case studies imply that the application of green chemistry has the potential to render pharmaceutical production more environmentally friendly, directing towards the intersection of toxicology and environmental defense within the industry (Achieng, 2025). In the case of electric vehicles (EVs), the application of nanotechnology is a radical step towards green mobility. From case studies of the development of lithium-ion batteries and the more recent solid-state counterparts, researchers discovered significant environmental advantages and safety aspects associated with these new technologies. Nanomaterials improve the efficiency of energy storage and are capable of leading to the development of lighter vehicles, hence achieving improved overall performance with less carbon emission. Apart from this, studies also continue to address the importance of utilizing sustainable production processes and functional recycling systems for these materials. Proper management of toxicity and safety when applying nanotechnology is important so that greener transportation does not introduce new hazards. With examples of the latest developments, case studies demonstrate how nanotechnology is at the core of creating green approaches in the motor industry (Esanakula et al., 2025). Apart from certain industries, the intersection of green chemistry and safety is further beyond industries. Analysis across various case studies identifies that the principles of Green Chemistry can be applied as equally across industries, introducing a general embracement of reducing environmental footprint. Employing eco-friendly materials and processes aids in reducing toxic byproducts and subsequently introducing public health and safety. For instance, solvent-free reaction industries report fewer cases of toxic exposure among employees and surrounding communities. By avoiding toxic chemicals and opting for greener alternatives, organizations meet regulatory needs as well as ensure the well-being of the environment. Overall, such examples reflect the significance of green chemistry in fostering safety, control over toxicity, and stewardship over the environment (Achieng, 2025; Esanakula et al., 2025).

9.6.1. Successful Implementation of Green Chemistry in Industry

The successful use of green chemistry to industry is very promising for reducing toxicological risks and environmental impacts of traditional chemical processes. By optimizing chemical product and process design for minimizing the creation and consumption of harmful chemicals, industries can become more sustainable. For example, the use of solvent-free synthesis and renewable feedstocks can greatly lower toxicity levels in the manufacturing processes. Such alterations not only enhance consumer and worker safety but also reduce the environmental effects linked with chemical waste and emissions. As industries innovate and implement green chemistry techniques, they are aligned with global sustainability targets, thereby reducing healthcare and environmental costs associated with chemical pollutants (Philip et al., 2020). For the realization of the transition to green chemistry, robust training and education programs must be established within the sector. This is crucial in establishing a pool of staff with information concerning the principles of sustainable practices that ensure higher innovation in solutions in green chemistry. Investing in training programs ensures that chemists, engineers, and manufacturing staff gain necessary skills to produce fewer harmful chemical processes and products. Furthermore, interdisciplinary cooperation among universities and industrial actors can accelerate knowledge and technology transfer from research to practice (Muratov et al., 2020). Not only will this cooperation provide professionals with the ability to apply theoretical concepts into practice, but it also enhances sustainability culture in industrial practices.

Lastly, the largest challenge is scaling up green chemistry principles to other sectors. Sectors must inculcate a profound cultural dedication to sustainability reflected in their business systems and strategic goals. This requires not only a willingness but also regulative support and economic motivation to incorporate green chemistry into production systems. Globally, monitoring systems and facilitating the transition toward greener practices will be key to tracking progress and driving innovation. With policymakers and stakeholders in the industry taking their seats to tackle the present challenges of pollution and environmental deterioration, green chemistry-led multidisciplinary strategy offers a viable path toward the realization of cleaner, healthier industrial operations promoting both ecological harmony and human well-being (Philip et al., 2020).

9.6.2. Research Initiatives Focused on Reducing Chemical Toxicity

Research projects aimed at reducing chemical toxicity are an essential field of scientific research under the broad framework of green chemistry. The essence of it is the obligation to develop alternatives that not only reduce human health hazards but are also eco-friendly. Several multidisciplinary research programs are ongoing that explore the design of safer chemicals, new processes that restrict harmful compounds, and complete chemical lifecycle analysis. Through these programs, scientists hope to replace dangerous substances with those that are inherently safer, thereby rendering chemical manufacturing procedures safer. Such proactive approach eventually boosts consumer protection as well as ecological integrity, ensuring that chemical technology development occurs hand in hand with natural conservation. A vital element of such research work is the systematic evaluation of existing chemicals, particularly those with established toxicological profiles. Such is evident in research studies investigating the environmental impact of materials like spent lithium-ion batteries. These studies painstakingly record the potential sources and routes of pollution from these chemicals, thereby providing vital information for policy-making and regulatory decisions. Not only is the aim to identify the immediate effects of such chemicals on the environment but also to predict long-term environmental impacts. The integration of these findings into chemical product regulation and design plays a key role in mitigating their adverse effects, as consistent with the overall goal of increasing sustainability in the industry. In addition, technological innovation has fueled research efforts toward safe storage and use of alternative energy sources, correlating them directly with reducing chemical toxicity. For instance, the scientific challenges concerning hydrogen storage in porous media are a cutting-edge area that can promote more secure energy alternatives. The studies conducted here are not just concerned with the efficiency and productivity of hydrogen as a clean source of energy but also with its eco-friendliness compared to traditional fossil fuels. Through these channels, scientists are paving the way for an economically viable energy economy that will decrease reliance on perilous chemical procedures, which will ultimately translate into improved public safety and greater ecological balance. Together, these endeavors are integral parts of a multi-faceted strategy directed at improving toxicological issues while bringing about a greener tomorrow through ethical chemistry practices.

9.6.3. Collaborative Efforts Between Academia and Industry for Safer Practices

The interaction between the industry and academia is key in facilitating safer practice within the areas of toxicology and environmental footprints of green chemistry. The complementarity of each sector combined, working to advance goals, utilizes each other's strength; academia stimulates innovation based on research and the industry with practical applicability and compliance within legislations. For instance, initiatives that merge research in schools with an industrial focus result in developing safer chemical processes taking into account human health and a sustainable environment. These partnerships can pave the way to new practices, for instance, the synergy of Computer-Aided Drug Design (CADD), which facilitates drug discovery by making predictions about toxicological outcomes at early stages, ultimately resulting in reduced-risk compounds in drug use (Sarfaraz et al., 2023). Also, collaborative research with academia fosters the sharing culture that is required to foster innovation in safer chemistry standards. Both partners can share long-term discussions on regulatory requirements, ethical manufacturing, and sustainability initiatives through collaborative research. For example, Avantiums research into biobased polymers illustrates how research from academia can inform industrial processes for replacing fossil-based materials. Through close collaborations with scientists, Avantium navigates the complexities of product development, life cycle analysis (LCA), and recyclability in ways that environmental values are inherent in emerging material innovation (de Jong et al., 2022). Such collaborative frameworks not only enhance product safety but also promote a shared interest in public health and ecological integrity. Finally, continued interaction between the industry and academia offers opportunities for capacity building and training in safe chemical practice. Industries both ensure future employees are well educated on toxicology, regulation, and environmental sustainability through active student and researcher involvement in practical project work. This proactive approach guarantees new professionals are sufficiently prepared to face safer chemical practice challenges. Moreover, joint projects as they evolve have a tendency to find cost-efficient solutions that can be economically scaled up, to meet pressing global needs for more sustainable materials and safer consumer goods. Built into the synergy of these partnerships is the forward-thinking frame of mind necessary to propel safety and environmental responsibility in the case of green chemistry in leveraging the accelerated pace of technological advancements.

9.7. CONCLUSION

In looking at the intersection of toxicology, safety, and environmental impact in the realm of green chemistry, one finds that amazing strides may be made to minimize ecological disruption and maximize human health. Central to that effort is a recognition for the way metal impurities, particularly in the guise of defunct mines, pose a twofold danger of environmental devastation and biological hazard. Existing environmental quality criteria may be ineffective in the protection of ecosystems since studies reveal the dominance of metals such as cadmium, lead, and zinc in water bodies poses a significant threat to aquatic organisms, likely to result in lower biodiversity (Jarvis et al., 2009). This highlights the need for more sensitive measures that account for long-term changes in ecologies and community adaptations, which fall short in traditional measures of evaluation. Moreover, the role of metal bioavailability in food safety contributes to the intricacy of toxicity comprehension. Research has discovered that nickel, as an essential micronutrient, can be a health concern when found at high levels in food items like legumes and nuts (Babaahmadifooladi et al., 2020). Distinguishing between the overall quantity of nickel and the bioaccessible portion that might affect human health is significant. In the absence of comprehensive bioaccessibility testing, estimated dietary intake may not be an accurate reflection of current levels of exposure, leading to spurious alarm or tolerance of food system safety standards. This calls for more stringent methods that reflect the complexity of metal toxicity concerns so that both ecological and human safety receive appropriate consideration when establishing environmental regulations. Lastly, addressing the issues resulted from toxicological risks in green chemistry must include a strategy that adopts comprehensive environmental studies and improved safety protocols. The serious gaps discovered in current legislative frameworks must be addressed through persistent research that informs policy changes, particularly for metals obtained from industrial waste. By refining our understanding of the impact metals have on the ecosystem as well as food safety, we can be assets towards the development of a greener and more sustainable green chemistry model that ensures both environmental health as well as public well-being. Combining ecological resilience with responsible chemical practice will be essential as we move toward a future green chemistry that breaks forth without compromising the safety of our natural resources and food supplies.

9.7.1. Summary of the Importance of Toxicology and Safety in Green Chemistry

The integration of principles of toxicity and safety in green chemistry is crucial to ensure environmentally friendly production practices. The understanding of toxicological impacts enables chemists to foretell and prevent detrimental effects associated with chemical reactions and products. By testing the chemical compounds' toxicity, scientists can develop safer alternatives with maximum efficiency and minimize damages to human health and the environment. For instance, mining effects on water quality emphasize why one needs to know the impact individual metals such as cadmium and zinc pose to aquatic life. The awareness enables chemists to create products that do not store harmful impurities in the environment, ultimately resulting in healthier environments and safer communities (Jarvis et al., 2009).

Figure 9.14. By evaluating the toxicity of chemical substances, researchers can design safer alternatives that maintain high efficacy while minimizing harm to human health and ecosystems.

Source: https://cdn-ilcjceb.nitrocdn.com/VtFsciNiiemJiXTnmldbhHuQqIuDi-jYB/assets/images/optimized/rev-b5b2fd0/www.ecoonline.com/wp-content/uploads/2024/12/How20to20Avoid20PFAS20in20Workplace20_20Feature20Image-7.jpg.

Another significant move is incorporating safety measures with toxicological testing in chemical development's life cycle. That is it. The green process safety is not only able to safeguard workers who work with potentially harmful materials, but also ensure products that are rendered from the process are safe for use and the environment. Urban areas have contaminants like zinc runoff posing a significant threat to aquatic life, which are an essential component of the food chain. By emphasizing safety and environmental integrity, green chemistry can generate chemicals that are both sustainable and efficient, and promote biodiversity simultaneously, as well as reduce the impact of industrial processes on the environment (Beasley et al., 2001). In addition, the incorporation of toxicology, safety, and environmental impact promotes the application of a single approach in green chemistry. This type of method not only involves designing less toxic materials but also addresses the broader context of chemical use and disposal. Chemists can predict biological response and design site-specific strategies to offset metal toxicity in aquatic environments by taking into account the interactive effects of pollutants, such as metal mixtures in sediment. O.n-going R&D in these areas can help frame guidelines and regulatory standards that facilitate the production of green chemicals, among others, hence promoting sustainable practices in many industries (Jarvis et al., 2009; Beasley et al., 2001) (Figure 9.14).

9.7.2. Future Directions for Research and Practice in Green Chemistry

In planning ways forward for green chemistry practice and research, it is necessary to prioritize interdisciplinary approaches incorporating safety, toxicology, and environmental effects assessments. Existing research aligns with adding regulatory frameworks and technological innovation as being the vital components in addressing the issues of industrial pollution. There is an evident necessity for systematic literature review since it can help to elucidate gaps in existing methodologies for resource-efficient practice. Additionally, as industries pick up pace to adopt sustainability, exploring ongoing state-of-the-art technology innovations in waste management technologies, i.e., nanotechnology and bioremediation, will be a necessity. By emphasizing an integrative approach, future studies can help facilitate effective strategies that assist in overcoming infrastructural as well as resource-based challenges, which remain to threaten the adequate control of industrial waste and pollutants, thereby guaranteeing human health as well as environmental integrity (Singh et al., 2023). The future of green chemistry also involves a solid understanding of ocean pollution and its extensive implications on human health and ecosystems. As industrial processes become more and more the source of marine pollution, the field of green chemistry to design sustainable means of waste prevention and reduction of pollution is of greatest significance. This necessitates focus on novel processes and materials that minimize the release of toxic byproducts and release of toxic components into marine ecosystems. There needs to be an integrated research agenda to assess and minimize the effects of toxic polluters, which have disproportionate effects on vulnerable populations. Firm environmental surveillance and collaborative efforts from policymakers and industry actors will also empower scientists to develop focused interventions, driving greener conduct that protects aquatic ecosystems (Philip et al., 2020). By advocating for the kind of solutions that address industrial and marine pollution, green chemistry can play a major role in curbing the wider effects of human actions. Lastly, the transition to a circular economy provides the key platform for advancing green chemistry initiatives. Future research has to embrace sustainable design principles with a focus on the lifecycle of chemicals and materials in order to reduce waste and environmental burden. By stimulating an innovative culture where material recyclability and the manufacture of biodegradable products are high priorities, researchers can facilitate a more sustainable economy. Additionally, interdisciplinary collaboration between chemists, ecologists, and social scientists will enhance the understanding of chemical processes as well as the societal effects of their applications. It will make it easier to identify regulatory compliance measures and consumer training effectively, thereby enhancing the acceptance of green chemistry processes among various industries. Eventually, it will enable an integrated strategy that will facilitate the deep changes needed to promote toxicologically acceptable and sustainable chemical practices in the ongoing development of industrial processes (Singh et al., 2023; Philip et al., 2020).

9.7.3. Call to Action for Adopting Green Chemistry Principles Globally

The strong imperative for global call to action on the use of green chemistry principles is triggered by the baying environmental crisis fueled by conventional chemical practice. As manufacturing industries around the world face mounting pressure to minimize their ecological footprints, green chemistry is a route to a sustainable future. Through a concentration on the thirteen guiding principles of green chemistry, professionals are able to design processes that are safety-focused, reduction-wasteful, and enhanced for sustainability. These values are the foundation for not just enhancing the effectiveness of products but also significantly decreasing toxicological impacts on human and environmental well-being. Moreover, establishing interdisciplinary collaboration between academia, business, and policymakers has the potential to make the transition towards greener processes meaningful and feasible, thus developing a culture of sustainability that resonates in international markets (Singh et al., 2023). Effective implementation of green chemistry principles requires case studies describing their actual applications in industry. For instance, in agriculture, the use of bio-based fertilizers and pesticides from renewable feedstocks has been seen to not only eliminate chemical toxicity but also increase crop yields. This approach addresses the devastating health risks that emerge due to poisonous mycotoxins prevalent in crops like maize that primarily affect regions like Tanzania (Rosentrater et al., 2015). By employing green chemistry concepts, industries can innovate with caution against public health risks through toxin impurity minimization during food production. In-depth discussions and accounts of effective greening of methodologies will be essential to demonstrating the feasibility and merits of such systems, elicit much more encouragement towards mass adoption. Stakeholder involvement at all levels is at the heart of driving green chemistry forward on a global basis. Schools play a central role in developing an appreciation of environmental issues and innovative solutions among the next generation of scientists and engineers. Curriculum design with green chemistry principles can enable a generation of leaders to address urgent environmental issues. Moreover, government drives and incentives can induce industrial growth of greener processes by making research and development expenses available to find sustainable alternatives. Here, concerted policies that encourage inter-sectoral cooperation will be instrumental in overcoming adversity by industries transitioning towards sustainable operations. Through policy reform and education as a priority, the global community will be in a better position to further cement its commitment to embracing green chemistry principles, setting the stage for a cleaner, healthier world (Singh et al., 2023).

REFERENCES

1. Abbasi, R., Shineh, G., Mobaraki, M., Doughty, S., & Tayebi, L. (2023). Structural parameters of nanoparticles affecting their toxicity for biomedical applications: A review. *Journal of Nanoparticle Research, 25*. https://doi.org/10.1007/s11051-023-05690-w (accessed on 24th October 2025).
2. Achieng, N. M. (2025). Green chemistry in pharmaceuticals: Reducing environmental impact. *Newport International Journal of Public Health and Pharmacy*. https://www.semanticscholar.org/paper/fcf7c9c0751460e055a99f3481f52db214e8d25d (accessed on 24th October 2025).
3. Ankley, G. T., Cureton, P. M., Hoke, R. A., Houde, M., Kumar, A., Kurias, J., Lanno, R. P., et al. (2020). Assessing the ecological risks of per- and polyfluoroalkyl substances: Current state-of-the science and a proposed path forward. *Environmental Toxicology and Chemistry, 40*, 564–605. https://doi.org/10.1002/etc.4869 (accessed on 24th October 2025).
4. Arnold, B. S., Gross, T. S., McDonald, K., Spulveda, M. S., et al. (2001). *Relations of Environmental Contaminants, Algal Toxins, and Diet with the Reproductive Success of American Alligators on Florida Lakes*. https://core.ac.uk/download/11016994.pdf (accessed on 24th October 2025).
5. Azim, E., Boyd, K. A., Doi, L., & Little, D. C., et al. (2012). *Scoping a Public Health Impact Assessment of Aquaculture with Particular Reference to Tilapia in the UK*. https://core.ac.uk/download/1438606.pdf (accessed on 24th October 2025).
6. Babaahmadifooladi, M., Du Laing, G., Jacxsens, L., Van de Wiele, T., et al. (2020). *Gap Analysis of Nickel Bio Accessibility and Bioavailability in Different Food Matrices and its Impact on the Nickel Exposure Assessment*. https://core.ac.uk/download/288817544.pdf (accessed on 24th October 2025).
7. Banerjee, P., Kemmler, E., Dunkel, M., & Preißner, R. (2024). ProTox 3.0: A webserver for the prediction of toxicity of chemicals. *Nucleic Acids Research, 52*, W513–W520. https://doi.org/10.1093/nar/gkae303 (accessed on 24th October 2025).
8. Beasley, G., & Kneale, P. (2001). *Macroinvertebrates, Heavy Metals and PAHs in Urban Watercourses*. https://core.ac.uk/download/51015.pdf (accessed on 24th October 2025).
9. Blake, A., Bell, D. V. J., Corbett, C. J., Batteate, C. M., Olivetti, E., Giove, S., et al. (2016). *Advancing Alternative Analysis: Integration of Decision

Science. https://core.ac.uk/download/323079787.pdf (accessed on 24th October 2025).

10. Butler, G., Cooper, J. M., Leifert, C., Lueck, L., et al. (2007). *Effects of Organic and 'Low Input' Production Methods on Food Quality and Safety*. https://core.ac.uk/download/10925455.pdf (accessed on 24th October 2025).
11. Cabernard, L., Pfister, S., Oberschelp, C., & Hellweg, S. (2021). Growing environmental footprint of plastics driven by coal combustion. *Nature Sustainability, 5*, 139–148. https://doi.org/10.1038/s41893-021-00807-2 (accessed on 24th October 2025).
12. Christie, D. A., & Tansey, E. M. (2004). *Environmental Toxicology: The Legacy of Silent Spring*. https://core.ac.uk/download/30696194.pdf (accessed on 24th October 2025).
13. Cuhra, M., Bøhn, T., & Cuhra, P. (2016). Glyphosate: Too much of a good thing? *Frontiers in Environmental Science, 4*. https://www.semanticscholar.org/paper/78bf17947fef78de9df3d144bcb1f84aa23300a3 (accessed on 24th October 2025).
14. de Jong, E., Visser, H. A., Sousa Dias, A., Harvey, C., & Gruter, G. J. M. (2022). The road to bring FDCA and PEF to the market. *Polymers, 14*, 943. https://doi.org/10.3390/polym14050943 (accessed on 24th October 2025).
15. Erythropel, H. C., Zimmerman, J. B., de Winter, T. M., Petitjean, L., Melnikov, F., Lam, J. C. H., Lounsbury, A. W., et al. (2018). The Green ChemisTREE: 20 years after taking root with the 12 principles. *Green Chemistry, 20*, 1929–1961. https://doi.org/10.1039/c8gc00482j (accessed on 24th October 2025).
16. Esanakula, J. R., Rathod, C., Babu, N. B. K., Srikanth, D. V., & Bhaskar, S. (2025). Applications of nanotechnology in electric vehicles: Enhancing energy efficiency and reducing environmental impact. *Journal of Environmental Nanotechnology*. https://www.semanticscholar.org/paper/7d33adfb1160ad86d3632535a721c71236d9ccad (accessed on 24th October 2025).
17. Greenhaw, K., Weatherford, K., & Rabinowitz, R. (2013). *Small Businesses, Public Health, and Scientific Integrity*. https://core.ac.uk/download/71361208.pdf (accessed on 24th October 2025).
18. He, X., Deng, H., & Hwang, H. M. (2018). The current application of nanotechnology in food and agriculture. *Journal of Food and Drug Analysis, 27*, 1–21. https://doi.org/10.1016/j.jfda.2018.12.002 (accessed on 24th October 2025).

19. Heinemann, N., Alcalde, J., Miocic, J., Hangx, S., Kallmeyer, J., Ostertag-Henning, C., Hassanpouryouzband, A., et al. (2021). Enabling large-scale hydrogen storage in porous media–The scientific challenges. *Energy & Environmental Science, 14*, 853–864. https://doi.org/10.1039/d0ee03536j (accessed on 24th October 2025).
20. Jarvis, A. P., Kelly, M. G., Lofts, S., Merrix, L., et al. (2009). *Ecological Indicators for Abandoned Mines, Phase 1: Review of the Literature*. https://core.ac.uk/download/58316.pdf (accessed on 24th October 2025).
21. Jeevanandam, J., Barhoum, A., Chan, Y. S., Dufresne, A., & Danquah, M. K. (2018). Review on nanoparticles and nanostructured materials: History, sources, toxicity and regulations. *Beilstein Journal of Nanotechnology, 9*, 1050–1074. https://doi.org/10.3762/bjnano.9.98 (accessed on 24th October 2025).
22. Landrigan, P. J., Stegeman, J. J., Fleming, L. E., Allemand, D., Anderson, D. M., Backer, L. C., Brücker-Davis, F., et al. (2020). Human health and ocean pollution. *Annals of Global Health, 86*, 151. https://doi.org/10.5334/aogh.2831 (accessed on 24th October 2025).
23. Maertens, A., Luechtefeld, T., Knight, J., & Hartung, T. (2024). Alternative methods go green! Green toxicology as a sustainable approach for assessing chemical safety and designing safer chemicals. *ALTEX, 41*(1), 3–19. https://www.semanticscholar.org/paper/3aaaeae752abe1c96eff22e02dec332b5c6f8328 (accessed on 24th October 2025).
24. Mahlapuu, M., Håkansson, J., Ringstad, L., & Björn, C. (2016). Antimicrobial peptides: An emerging category of therapeutic agents. *Frontiers in Cellular and Infection Microbiology, 6*. https://doi.org/10.3389/fcimb.2016.00194 (accessed on 24th October 2025).
25. McPherson, A., Nimpuno, N., & Sadique, T. (2009). *Greening Consumer Electronics: Moving Away from Bromine and Chlorine*. https://core.ac.uk/download/71351620.pdf (accessed on 24th October 2025).
26. Mittal, D., Kaur, G., Singh, P., Yadav, K., & Ali, S. A. (2020). Nanoparticle-based sustainable agriculture and food science: Recent advances and future outlook. *Frontiers in Nanotechnology, 2*. https://doi.org/10.3389/fnano.2020.579954 (accessed on 24th October 2025).
27. Mrozik, W., Rajaeifar, M. A., Heidrich, O., & Christensen, P. A. (2021). Environmental impacts, pollution sources and pathways of spent lithium-ion batteries. *Energy & Environmental Science, 14*, 6099–6121. https://doi.org/10.1039/d1ee00691f (accessed on 24th October 2025).
28. Muratov, E., Bajorath, J., Sheridan, R. P., Tetko, I. V., Filimonov, D., Poroikov, V., Oprea, T. I., et al. (2020). QSAR without borders. *Chemical*

Society Reviews, 49, 3525–3564. https://doi.org/10.1039/d0cs00098a (accessed on 24th October 2025).

29. Naidu, R., Biswas, B., Willett, I. R., Cribb, J., Singh, B. K., Nathanail, C. P., Coulon, F., et al. (2021). Chemical pollution: A growing peril and potential catastrophic risk to humanity. *Environment International, 156*, 106616. https://doi.org/10.1016/j.envint.2021.106616 (accessed on 24th October 2025).
30. Niazi, S. K., & Mariam, Z. (2023). Computer-aided drug design and drug discovery: A prospective analysis. *Pharmaceuticals, 17*, 22. https://doi.org/10.3390/ph17010022 (accessed on 24th October 2025).
31. Radošević, K., Cvjetko Bubalo, M., Slivac, I., Srček, V., & Redovniković, I. R. (2016). Green technology meets ecotoxicology. *Croatian Journal of Food Science and Technology, 8*, 120–128. https://www.semanticscholar.org/paper/bd59947d3ef05a9d59fcd6419bb4c5448eb51ace (accessed on 24th October 2025).
32. Rosentrater, K. A., & Suleiman, R. A. (2015). *Current Maize Production, Postharvest Losses and the Risk of Mycotoxins Contamination in Tanzania*. https://core.ac.uk/download/38931868.pdf (accessed on 24th October 2025).
33. Samorì, C. (2010). *Use of Solvents and Environmental Friendly Materials for Applications in Green Chemistry*. https://www.semanticscholar.org/paper/53fae3f8e3e6435b18a4cd01a5db5575cd38546d (accessed on 24th October 2025).
34. Sheldon, R. A. (2017). Metrics of green chemistry and sustainability: Past, present, and future. *ACS Sustainable Chemistry & Engineering, 6*, 32–48. https://doi.org/10.1021/acssuschemeng.7b03505 (accessed on 24th October 2025).
35. Singh, B. J., Chakraborty, A., & Sehgal, R. (2023). A systematic review of industrial wastewater management: Evaluating challenges and enablers. *Journal of Environmental Management, 348*, 119230. https://doi.org/10.1016/j.jenvman.2023.119230 (accessed on 24th October 2025).
36. Singh, G., & Khullar, T. (2023). *Green Chemistry Strategies for Waste Minimization: From Waste to Wealth*. https://core.ac.uk/download/660318997.pdf (accessed on 24th October 2025).
37. Solecki, W. D., & Tiefenbacher, J. P. (1999). *Pollution Prevention: Factors Behind Toxic Release Reduction in the U.S. Paper Industry*. https://core.ac.uk/download/72056478.pdf (accessed on 24th October 2025).
38. Stoica, C., Banciu, A., Idriss, H., Lian, J. Z., Pătraşcu, A. M., Cucurachi, S., Richeter, S., et al. (2025). Green design and life cycle assessment of

novel thiophene-based surfactants to balance their synthesis performance and environmental impact. *Materials, 18*. https://www.semanticscholar.org/paper/fd74ae2d3176fc270cf0719dc9e869427381b00b (accessed on 24th October 2025).

39. Tannert, R., Barth, S., Hildebrandt, J., Taubert, A., & Weber, J. (2025). Sustainability of enzymatic monomer synthesis: Evaluation via comparison of petrochemical and enzymatic alkene epoxidation by life cycle assessment. *Chemsuschem, 18*. https://www.semanticscholar.org/paper/46a88a8a07110cddf6ffcad30f95370b4cb925f5 (accessed on 24th October 2025).

40. Zuin, V. G., Eilks, I., Elschami, M., & Kümmerer, K. (2021). Education in green chemistry and in sustainable chemistry: Perspectives towards sustainability. *Green Chemistry, 23*, 1594–1608. https://doi.org/10.1039/d0gc03313h (accessed on 24th October 2025).

CHAPTER 10

EMERGING TRENDS AND FUTURE DIRECTIONS

Contents

INTRODUCTION

The field of green technology has also drawn increased focus in the recent past as societies struggle to address the pressing concerns of climate change, resource depletion, and environmental degradation. Growing awareness is driving the world towards more sustainable behavior on a large scale in areas like energy, transportation, and city planning. The innovation of new technologies designed to reduce carbon footprints and enhance efficiency signals this new trend. As green technology continues to evolve daily, not only does it provide solutions for existing environmental issues but also builds the framework for future sustainable growth. Policymakers, business entities, and individuals at large require an analysis of key emerging trends and implications because these technologies can potentially change societal paradigms and economic structures fundamentally. Moreover, the contribution of urbanization and increased population cannot be downplayed in green technology. With cities serving as the hubs of economic growth, they are faced with the task of maintaining growth and remaining sustainable. Cities are turning more sophisticated in the implementation of smart technologies that integrate sustainability into their structures, thereby developing greener and more liveable spaces for the residents and tourists. The global shift towards intelligent urbanization is compounded by initiatives from institutions of higher learning that facilitate research and development in green practices. These institutions are agents for change, driving the right knowledge and practices that lead cities to transition to greener solutions. Cities are borrowing from one another, implementing best practices, and developing strategies to incorporate leading technology into city planning and infrastructure in this instance. In addition to the urban perspective, the role of up-and-coming economies cannot be forgotten when green technology directions are being debated in the future. BRICS (Brazil, Russia, India, China, and South Africa) nations--Brazil, Russia, India, China, and South Africa--are leveraging their unique positions to drive sustainable initiatives, often skipping old systems to adopt new alternatives. These advances show how these countries are tackling domestic sustainability issues while adding to international discussion about green technology. By examining smart city plans and collaborative university projects in these countries, it is possible to learn about how developing countries are not only embracing but also leading green technological innovation. Understanding these dynamics will illuminate the directions that are being established by the new economies, ultimately adding to the comprehensive body of knowledge on worldwide sustainability trends and future green technology innovation.

Overview of Emerging Trends

With green technology unfolding in the future, utilization of the Internet of Things (IoT) is one of the high-profile new tendencies that impose a paradigm shift towards sustainability. This groundbreaking use employs advanced sensors and networked equipment to boost efficiency in many areas, including energy management and urban development. Of particular interest is the role of IoT in facilitating smart grids, a cornerstone for optimizing energy usage and reducing waste. Advances on these cyber-physical systems are matched with an emerging need to offset cybersecurity vulnerabilities that accompany increased connectivity. Research has only just begun to analyze the spectrum of cyberattacks being launched against such systems, with an overwhelming imperative for robust security in safeguarding new smart energy infrastructure. Thus, this suggests a research area that aligns with a lot to discover, in this case, emphasizing the competition between technological development and the necessity for effective cybersecurity processes within smart grids (Alomari et al., 2025). Concurrent with IoT, optimization of sustainable maritime transport is another paramount trend in green technology. This area is under significant pressure to continue being efficient as it faces the rising level of global trade demand and environmental compliance. Glimpses of emerging research indicate a collective push towards the integration of alternate fuel and efficient means of operations in shipping and at ports. The emphasis lies in utilizing technological innovation that is compatible with decarbonization and achieving maximum economic efficiency. Most prominently, marine transport stakeholders are coming to adopt multi-objective optimization models, balancing environmental responsibility with profitability. In addition to seeking to reduce emissions, these models seek to advance the sustainability of operations consistent with global ecological goals.

Figure 10.1. In tandem with IoT advancements, sustainable maritime transport optimization has emerged as another cornerstone trend in green technology.

Source: https://i0.wp.com/d1meba.org/wp-content/uploads/2024/01/76 -125-1144-71.png.

Future research will likely concentrate on application of the latest technologies like blockchain and artificial intelligence for further automation of processes in this sector, pointing to how traditional industries can embrace green technology (Xu et al., 2025). Lastly, the convergence of various disciplines in the field of green technology shows a broader trend towards collaborative models of governance of environmental concerns. While governments and institutions witness the mutually dependent nature of regulatory facilitation and technological innovation, there is a shift toward synergistic strategies encompassing market-based interventions. The multi-disciplinary approach not only spurs innovation but also aligns policy interventions with technological capabilities, facilitating comprehensive solutions to the major environmental issues. Stakeholder participation across industries will be crucial to fostering sustainable behaviors and reducing the industry-wide carbon footprint. Therefore, the reciprocal interplay between policy design, management policies, and technological innovation is the most important area of future scholarly inquiry. Joint thinking is pivotal in formulating a green technology-driven sustainable future that can be flexible and resilient in light of ongoing environmental challenges (Xu et al., 2025) (Figure 10.1).

10.1. RENEWABLE ENERGY SOURCES

Increased momentum to combat climate change has driven worldwide shifts toward renewable energy. Renewable sources of energy, including solar, wind, hydro, and biomass, are increasingly becoming crucial not only for greenhouse gas emission reduction but also for energy self-sufficiency. In line with the goal of pursuing carbon neutrality, emerging solutions such as green hydrogen generation are gaining traction. This emergent interest stems from the realization that renewable energy can prove to be a game-changer in decarbonizing hard-to-abate sectors, including heavy industry and transport. The bibliometric analysis conducted indicates exponential growth in renewable energy research following worldwide policy interventions implemented since 2021, marking the imperative for using different sources of energy in terms of building energy security and sustainability in the future (César et al., 2025). Notably, green hydrogen has become the darling of renewable energy trends, especially because it provides a flexible means for storage of energy and grid balancing. Hydrogen generation by processes such as electrolysis is becoming more efficient with the help of artificial intelligence breakthroughs that streamline processes for generating hydrogen. The systematic review practiced emphasizes how AI-optimized algorithms promote the transcendence of technical and financial challenges in hydrogen production, making it an increasingly promising alternative for mass use. Through the use of machine learning, deep learning, and other AI methods, researchers can optimize the efficiency of various production pathways, solidifying hydrogens position as a fundamental element in the transition to a low-carbon energy age (Bhuiyan et al., 2025). Besides, the use of renewable energy sources necessitates reorganization of existing energy systems and policies. Governments are promoting the adoption of such technologies through various interventions, including renewable energy subsidies and carbon tax enactments with the aim to phase out fossil fuels. Such strategic policies provide a sustainable environment to innovate, particularly in the domains of energy storage and hybrid energy systems. Use of multi-objective optimization techniques also adds credibility to the decision-making process by making energy and economic feasibility more reliable. In the future, research must work hard on overcoming the regulatory hurdles and technical setbacks that are slowing down large-scale deployment of renewable energy sources, ultimately building a robust and sustainable energy system (César et al., 2025).

10.1.1. Solar Energy Innovations

Technological innovation in solar energy is a core driving force in the innovation of green technology. Driven by mounting pressure to counteract climate change and deplete fossil fuels, solar energy innovation has intensified over the past decade. Researchers and engineers are increasingly exploring how nanotechnologies can be utilized to advance the efficiency and affordability of solar panels. For instance, perovskite solar cells hold much potential with high efficiency rates of energy conversion but lower production costs. Using nanomaterials not only ensures better performance of solar panels but also allows for more flexibility of their application, e.g., in building-integrated photovoltaics. With the invention and implementation of such advanced technologies, the solar energy sector does much to satisfy world energy demands while leading the way in sustainability, toward the call for energy innovation that addresses future needs for energy supply and the management of the environment (Knol et al. 2025). In addition to improving efficiency and responsiveness, solar energy innovations are transforming the face of energy storage solutions. As intermittent power sources, solar energy requires sophisticated storage facilities to ensure constant power provision under low light levels. Improvements in battery technology, particularly lithium-sulfur and solid-state batteries, provide longer life cycles and higher energy densities compared to their traditional lithium-ion counterparts. These advances could enhance the competitiveness of solar energy with traditional energy sources, enabling the shift towards a cleaner energy system. Apart from this, technological innovation in management of grids and decentralized power systems is also making it possible to integrate solar power into today's energy system in a better manner. Overcoming the problems of storage and grids, the solar power industry not only makes itself more feasible but also provides opportunities for masses to adopt renewable technology on a large scale and hence contributes towards a sustainable future (Byrne et al., 2013). Finally, the future of solar energy innovations lies in tandem with joint efforts among academia, the industry, and the government. Sustainable energy solution development requires concerted research and investment programs meant to increase the pace at which clean technologies are developed. Initiatives that will encourage public-private partnerships and accelerate the funding of research will create an atmosphere conducive to innovation. Also, the use of incentives for the deployment of solar energy and technological research into emerging technologies will be crucial in driving mass deployment. The role of small-scale technology, such as that based on nanotechnology, is significant in making crucial contributions to achieving breakthrough innovations in power generation. In the future, advancing on these innovations and partnerships will be essential in constructing a greener and more sustainable energy future and thus addressing the pressing challenge of climate change and scarcity of resources (Knol et al. 2025) (Figure 10.2).

Figure 10.2. With the growing urgency to address climate change and deplete fossil fuel resources, innovation in solar energy has gained traction in recent years.

Source: https://static01.nyt.com/images/2021/10/13/climate/13cli-energy1/13cli-energy1-mediumSquareAt3X.jpg.

10.1.2. Wind Energy Developments

Of all the renewable energy technologies, wind power has been a leading candidate in the world's transition to clean energy systems. The development of wind energy technologies has followed an exponential path in recent decades on the shoulders of technological advances as well as policy efforts. Improvements in turbine technology, such as increasing rotor diameters and taller heights, have significantly elevated energy capture and efficiency. Apart from that, off-shore wind farms have seen growing popularity, utilizing stronger and more uniform winds out at sea. As countries are putting work into reaching their carbon neutrality, the use of wind power in countries' grids has been an issue of extreme priority. The Intergovernmental Panel on Climate Change points out the urgency of transitions in the near term and shows that wind energy is going to play a key role in limiting greenhouse gas emissions globally (Zhang et al., 2023). However promising the horizon of wind energy is, its application at a large scale faces some hurdles to reach its full potential. One such challenge is the variability of wind, rendering grid stability and energy reliability unpredictable. In order to mitigate this, there is a necessity to enhance technologies of storing energy, such as batteries and pumped hydro storage, to complement electricity generated from wind. In addition, good planning and location of wind farms are essential in minimizing environmental impacts and local resistance. Participation of stakeholders and good environmental assessment can help in soothing issues regarding noise, visual impact, and disturbance to wildlife. The encouragement of co-ownership strategies at the local level can also enhance public acceptability and participation. These issues underscore the necessity of strategic policy structures and technological support to enable wind energy integration into mass-scale energy systems (Zhang et al., 2023). In the future, the destiny of wind energy schemes is on a rising trend, particularly with the increasing emphasis on climate resiliency and energy security. Future technological innovations will likely focus on enhancing turbine efficiency and grid integration to deal effectively with the variability nature of wind energy production. New materials and digital technologies, such as artificial intelligence and machine learning, can also transform practices for maintaining wind farms and predictive analytics. Establishing multinational partnerships to share best practices and research can accelerate wind technology development and deployment. In a diversified energy mix, wind power finds its place in global climate agreements and sustainable development goals, not only to decrease dependence on fossil fuels but also to promote socioeconomic capacity building for local communities. Embracing these innovations will be essential to scale up wind power, thus generating a secure and sustainable future (Calvin et al., 2023).

10.1.3. Biomass and Biofuels Advances

The development of biomass and biofuels is an imperative area in the drive for renewable energy solutions. With industries moving towards more non-petrochemical dependence, there is considerable emphasis on developing biobased substitutes that respond to the imperatives of environmental sustainability. Agricultural residues in the form of lignocellulosic biomass have become a substantial source towards this purpose. Recent developments in integrated biorefineries have maximized the use of constituents of lignocellulose, which facilitates the production of biofuel, bioplastic, and other high-quality bioproducts. This shift also signifies a proactive approach towards environmental challenges and, at the same time, opens economic opportunities in the agricultural sector, which also promotes circular economy in the sense that waste is turned into wealth as resources are utilized that would otherwise result in environmental loss (Mujtaba et al., 2023). With the impact of fossil fuel dependency increasing day by day, carbon neutrality is more urgent a necessity. Current international efforts lag far behind gaining carbon neutrality, indicating the necessity for effective means to cut greenhouse gas emissions. The transition towards biomass and biofuels can contribute significantly towards achieving this, as these clean fuels yield smaller carbon footprints compared to their fossil fuel-based counterparts. Development of methods to convert farm waste into biofuels not only promotes energy independence but also national and greater socio-economic goals by bringing livelihoods and food security through improved waste management practices. Therefore, the discovery of biomass utilization is at the core of not only energy production, but national and international environmental policies as countries proceed towards ambitious carbon-neutral targets (Chen et al., 2022). The future of biomass and biofuels hinges on long-term technological innovation and a concerted effort across industries. There needs to be a collective effort in research that seeks to simplify extraction and conversion processes, optimize feedstock crops, and otherwise enhance the sustainability of biofuel production. This needs to go hand in hand with policy environments that encourage investment in clean technologies and R&D in biomass production. The overlap of energy generation, agriculture, and environmental protection holds potential for innovation that can shift energy paradigms. Drives toward increasing the efficiency and yield of lignocellulosic biomass conversion will not only increase current dependency on fossil fuels but also contribute significantly to global emissions reduction targets. Hence, biomass industry growth will be one of the main pillars in shifting towards a sustainable and responsive energy future (Mujtaba et al., 2023) (Figure 10.3).

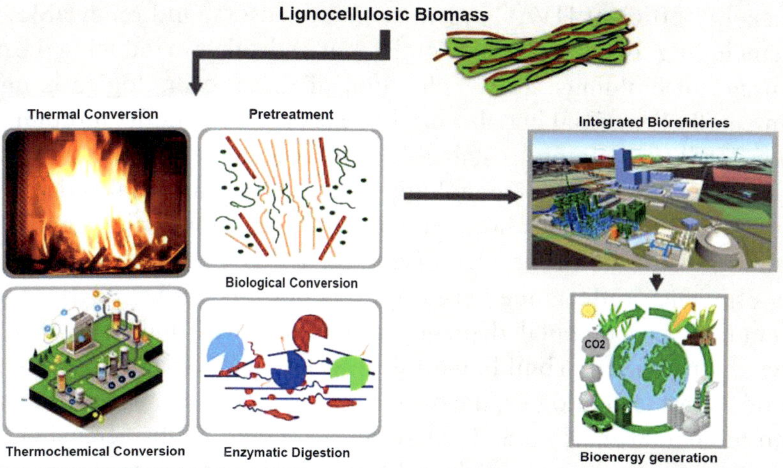

Figure 10.3. Recent innovations in integrated biorefineries have maximized the utility of lignocellulose components, facilitating the production of biofuels, bioplastics, and other valuable bioproducts.

Source: https://pub.mdpi-res.com/catalysts/catalysts-11-00309/article_deploy/html/images/catalysts-11-00309-ag.png?1615446386.

10.2. ENERGY EFFICIENCY TECHNOLOGIES

The search for energy saving technologies has emerged as topmost priority in the backdrop of global warming and the need to mitigate greenhouse gas emissions. Technologies in various sectors are increasingly focusing on such technologies with a view to curbing energy consumption and promoting sustainability. Specifically, advancements in artificial intelligence (AI) and machine learning are emerging as focal technologies that assist in optimizing energy use in complex environments. Through analysis of real-time data and projected energy needs, AI can optimize resource allocation in commercial buildings and industrial operations, creating enormous cost savings with reduced environmental impact. These smart technologies offer a bright future prospect of improvements in energy efficiency, displaying the capability of smart systems to assist in more efficient and greener approaches to industries heavily dependent on energy use, thereby towards the general goals of green technology initiatives (Poka et al., 2025). Moreover, the construction sector is particularly emphasized as one for the implementation of energy efficiency technologies. Since buildings generally constitute a vast percentage of global energy consumption, the integration of green technology is inevitable for sustainability in the future. Technologies such as energy-efficient HVAC systems, smart sensors, and renewable energy devices, including solar panels, all complement each other in reducing the overall energy usage of buildings. The exploration of these technologies is not only environmentally beneficial but also has huge economic benefits through energy efficiency, leading to a more sustainable built environment. Case studies show effective application of these technologies in chosen commercial and residential buildings, which emphasizes how effective they are in reducing energy usage and enabling resilience in the face of climate risks. This holistic approach to building efficiency reflects the necessity to adopt innovative solutions so that contemporary environmental demands can be fulfilled and a greener world constructed. In addition to building efficiency optimization and implementation of state-of-the-art technologies, the overall energy efficiency implications bleed over into municipal services and smart city planning. With increasing energy demands in urban areas, smart grids and energy-efficient cities are paramount to developing sustainable cities. Through the implementation of technologies that adaptively optimize energy delivery and promote the consumption of renewable energy sources, cities can greatly advance urban energy efficiency. In addition, retrofitting existing infrastructure with energy-saving technology is a necessary move toward lessening the green footprint of urbanization. Urban planning in the future focuses on the interconnectedness of energy efficiency technologies within cities in such a way that unified systems support sustainable living and resource management to facilitate a greener, energy-efficient future in urban development (Poka et al., 2025).

10.2.1. Smart Grids and Smart Meters

In the context of new green technology, smart grids and smart meters are principal innovations that are revolutionizing the consumption and use of energy. Smart grids integrate cutting-edge technologies, including the Internet of Things (IoT), to produce dynamic, adaptive power systems capable of optimizing the distribution and utilization of energy. With the improvement of real-time communication between consumers and utilities, smart grids enhance efficiency and enable integration of renewable energy sources. Such a shift is crucial in the quest for sustainability goals and reduced levels of greenhouse gas emissions as traditional power grids commonly cannot manage the variability of renewable energy sources. Smart metering technologies facilitate the filling of this gap by allowing the consumer to efficiently monitor and regulate his or her energy usage, thus promoting energy conservation and load management as outlined in the green technology general framework. The smart meter introduces this technology model further, a stark contrast from conventional forms of energy consumption. They provide consumers with real-time detailed data about their energy usage pattern, so that they may make informed decisions about the way they consume energy. Not only does the data obtained by smart meters assist consumers in learning where and how they could conserve energy, but also provide utilities with the analysis needed to assist in optimizing grid management. As noted in recent research, the incorporation of IoT solutions into energy systems has the capacity to strengthen operating efficiencies and propel the move towards sustainability (Motlagh et al., 2020). Moreover, smart meters manage demand response programs, allowing utilities to have improved management of peak loads, thus helping stabilize the grid and reduce the consumption of fossil fuels. Even though their potential is huge, smart meter and grid installations have some associated challenges. Interoperability, data security, and privacy are some of the significant issues affecting large-scale deployment. The confidentiality and integrity of consumer data must be safeguarded since networked devices introduce vulnerabilities into the energy system. Additionally, the adoption of these technologies depends on the creation of enabler infrastructures, for example, those powered by next-generation 5G networks (Shafique et al., 2020). Such connectivity not only supports trustworthy communication among devices but also supports rising demands for high rates of data and low latency operations essential in resilient grid operations. Solving these issues will be vital for realizing the full potential of smart grids and meters to encourage sustainable energy habits and make a greener world (Figure 10.4).

Figure 10.4. Smart grids integrate advanced technologies, including the internet of things (IoT), to create dynamic, responsive electricity networks capable of optimizing energy distribution and consumption.

Source: https://www.cyient.com/hubfs/Whitepaper/Energy%20Grid%20Optimization/Banner_Image_Resized.jpg.

10.2.2. Energy-Efficient Appliances

As sustainability is on the increase in the current era, energy-efficient appliances make the top trend in green technology. Aimed at reducing energy consumption without compromising performance, these appliances are critical to the maintenance of personal energy bills at a low level while reducing overall environmental deterioration. Their significance is underscored by Intergovernmental Panel on Climate Change reports, which place a high focus on curbing the pace of greenhouse gas emissions and adopting green technologies as a response to climate change mitigation (Calvin et al., 2023). The use of energy-efficient appliances in households not only promotes ecologically friendly habits but also fosters consumer demand for modern conveniences that pay off in the long term. As material progress and intelligent technologies are making giant strides, the avenues for energy efficiency in appliances are expanding by leaps and bounds, setting the stage for a greener and more sustainable tomorrow. Another significant aspect of energy-saving appliances is how they are making energy conservation across all categories, including residential, commercial, and industrial sectors.

For instance, intelligent appliances such as Energy Star-rated refrigerators, washing machines, and dryers are designed to consume less power and water than their conventional counterparts. Such reduced consumption takes pressure off energy grids, particularly during peak loads. Moreover, the use of smart appliances—since they are offered with internet connectivity as well as machine learning algorithms—enables efficient power use through remote monitoring and automatic adjustment based on user patterns and price conditions. These technologies are all part of a larger movement towards energy efficiency, now widely recognized as a requirement for the attainment of sustainable development and resource management in a world of finite resources. Ultimately, broad introduction of these technologies can lower energy-related emissions significantly. In the future, the future of energy-efficient appliances will be shaped by ongoing technological advancements and policy regulations that will continue to optimize energy performance. With technologies like machine learning and artificial intelligence being integrated into appliance design, the potential for instant feedback and better user experiences is certain to promote more acceptance and installation. Moreover, government incentives and policies on adopting energy-efficient technology will promote the transition towards more environmentally friendly choices among consumers. These developments not only hold the promise of reducing energy consumption but also of driving broader system changes in the manner in which societies interact with energy resources. With increasing regulatory pressures and changing expectations among consumers, energy-efficient appliances are bound to take center stage in the future of green technology and sustainable way of life practices (Wang et al., 2022) (Figure 10.5).

Figure 10.5. Appliances such as energy star-rated refrigerators, washers, and dryers are designed to consume less electricity and water than their conventional counterparts.

Source: https://miro.medium.com/1*Q_YdXNnErc1fAxDBNcNPtw.jpeg.

10.2.3. Building Automation Systems

Building Automation Systems (BAS) represent a critical convergence of technology and sustainability, offering tremendous potential for energy savings and reduced environmental impact in commercial and residential buildings. With the application of advanced sensors, automated control, and real-time analysis of data, BAS facilitate the most efficient operation of heating, ventilation, air conditioning, and lighting systems. Such systems not only enhance user comfort and operation efficiency but also optimize energy consumption, making the carbon footprint lower. As green technology becomes increasingly popular in demand, the integration of BAS into building structures becomes an imperative trend that rings true to current sustainability objectives. This is in line with the further transition towards smart cities, where energy management is increasingly dependent on systems that interlink and are focused on stewardship of the environment and technological solutions. The evolution of BAS is closely aligned with the advancement in digital technologies and the promotion of greater sustainability efforts within city environments. Technologies such as the Internet of Things (IoT) and machine learning are enabling BAS to become very adaptive platforms capable of automatically responding to changing energy requirements. For instance, through predictive analytics, the systems are able to predict and adjust accordingly based on occupancy and external weather conditions, optimizing energy usage.

Figure 10.6. Building automation systems (BAS) represent a critical convergence of technology and sustainability, offering considerable opportunities for energy efficiency and reduced environmental impact within residential and commercial buildings.

Source: https://www.factech-expo.com/content/dam/sitebuilder/rxth/factech/images/blogs/FTC_Blog5.jpg/_jcr_content/renditions/original.image_file.2400.1413.file/FTC_Blog5.jpg.

Additionally, the efficiency of BAS is also enhanced by their capability to include renewable power sources such as solar panels and energy storage devices in the building design. Such a synergistic synergy not only supports the transition towards renewable energy usage but also comes together with broader climate action planning, putting focus on the contributory role BAS will have in achieving net-zero emissions goals as highlighted in discussion over climate commitments like UN COP26 (Yogesh et al., 2021). With the paradigm of construction progressing further in the direction of smart interfaces and human-centric solutions, the future directions for BAS also encompass growing security and privacy issues. When various technologies are combined, there is the additional challenge of how data collection and management can be geared to maintaining respect for privacy of individuals and operation security. For instance, the vulnerability in interconnectivity can undermine BAS to cyber-attacks in the event of inadequate protection. Hence, as the emerging dialogue on digital security practices records, better models for safeguarding user data are critical (Wang et al., 2022). This is an area that the need for an equilibrium approach to technology deployment comes to the forefront where the benefit of automation and real-time monitoring alongside proper treatment of data converge. Thus, to overcome these obstacles is essential for the building of public trust and ensuring the long-term utilization of BAS in an era increasingly characterized by technological integration and environmentally friendly measures (Figure 10.6).

10.3. SUSTAINABLE TRANSPORTATION

The vision of sustainable transport becomes ever more pivotal in the face of the short-term imperatives of climate change and urban traffic congestion. With ever-rising global transport emissions, a transition to sustainable modes of transport becomes ever more critical towards improving environmental sustainability and public health. This shift promotes not only the utilization of electric vehicles and alternative fuels but also the integration of new technologies such as artificial intelligence (AI) and the Internet of Things (IoT). These technologies enable smart transportation systems that optimize traffic, enhance route efficiency, and reduce fuel consumption. In this, AI-based solutions are increasingly taking center stage in addressing logistics, predicting maintenance needs, and facilitating the use of renewable energy in transportation, thus initiating a cleaner, more efficient mobility picture (Bhuiyan et al., 2025). Integrating renewable energy sources into transportation is a significant advancement within the overall sustainable transport matrix. Electric and hydrogen vehicles are at the forefront with low-emission alternatives to the traditional gasoline and diesel engines. But widespread use of the technologies is conditional on the development of robust charging and refueling infrastructures. This will require expenditure in physical network infrastructure and strategic planning and the rollout of new energy-efficient equipment. The role of AI in optimizing charging stations' functions and putting energy immediately available during peak hours of demand is a great example of technology facilitating smoother integration of renews into transport systems. It has been proven in research that with better smart systems, hydrogen production and usage can be optimized considerably, paving the way for utilization of these clean sources on a large scale (Bhuiyan et al., 2025). Additionally, the application of IoT technologies in green transport encourages significant innovation in supply chain management, resulting in overall system efficiency. The application of a network of intelligent devices, such as sensors and RFID tags, makes it possible for transport systems to monitor and optimize the movement of goods in real time. Visibility in such manner allows enhanced control over inventory, timely maintenance, and efficient route planning, reducing environmental footprint. As IoT technology grows, its application to smart transportation systems becomes increasingly critical. In addition to enabling painless human-computer interaction, these technologies also enable businesses to use greener methodologies up and down the supply chain, in line with overall transportation sustainability goals.

10.3.1. Electric Vehicles and Charging Infrastructure

The development of Electric Vehicles (EVs) in the transport sector is an important aspect of emerging green technology, particularly with urban centers more and more concerned with being sustainable and reducing emissions. Since conventional cars are such a significant cause of air pollution as well as greenhouse gases, transitioning to EVs is a dramatic shift towards cleaner modes of transportation. The integration of EVs not only addresses ecological concerns but also creates opportunities for technological advances in energy management systems. This encompasses efficiency in charging infrastructure, which must evolve to accommodate more electric vehicles on the streets. Improving the accessibility of charging infrastructure and factoring in the impact of charging efficiency and speed will enhance EVs' usage rates and contribute better to the overall strategy of the sustainability of urban areas. Offering strong charging facilities is mandatory in the scenario of the Internet of Vehicles (IoV), which is widely recognized as a key technology for intelligent transportation systems for smart cities. This convergence of IoV and EV technology spells a sharp increase in the coordination and efficiency of transport networks. The industry experts are of the view that the future advancements in 6G communications will provide the support required to enable this shift, allowing data transmission in real-time for optimizing charging habits and energy usage in EVs. Thus, the investors are prioritizing investments that enable the deployment of dense network infrastructures with the capability of controlling energy consumption between various nodes while examining the energy efficiency of communication protocols and EV running. An integrated strategy for energy optimization will become the key one as cities adopt these emerging technologies to enhance mobility in cities and maintain their low carbon footprint (Wang et al., 2021). Furthermore, enhanced battery technology is a fundamental factor in propelling the mass use of electric vehicles and the viability of charging infrastructure. Today's lithium-ion batteries, while being successful, are limited in energy density, life, and expense. As manufacturers aim to design next-generation batteries in order to meet burgeoning customer needs, innovation is focused on new chemistries and visionary designs that promise to increase performance metrics, which in turn will drive the electric car economy. As Zeng et al. (2019) observes, the ongoing advancement of battery technologies, combined with the push for infrastructure enhancement, will determine the pace at which EVs flood the market. To be able to benefit from these developments, sustained research investment and industry cooperation will be needed to overcome existing challenges and make electric cars, with efficient batteries and supported by robust charging networks, lead the future of sustainable transport (Figure 10.7).

Figure 10.7. The expansion of electric vehicles (EVs) in the transportation sector serves as a critical component of emerging green technologies, particularly as cities increasingly prioritize sustainability and reduced emissions.

Source:

10.3.2. Public Transportation Innovations

In recent times, innovations in public transport have become a cornerstone of moving forward with green technology, and they have had a big impact on urban planning and energy consumption patterns. Among innovations, electric trains and buses are a game-changer, where cities such as Amsterdam and Los Angeles have introduced fleets that run solely on renewable fuels. This transition not only addresses the necessity to reduce greenhouse gas emissions but also significantly enhances the efficiency of public transport systems as a whole. The impact of these technologies extends beyond pollution reduction; they also invest in the health of the wider population by encouraging improved air quality, thereby reducing respiratory disease linked to urban smog. Hence, the implementation of electric public transport systems is a commendable response in urban transport and environment sustainability, a marked trend among future green technology initiatives towards creating a sustainable city environment. In addition, the utilization of smart transport systems has revolutionized the way public transport is managed and utilized. Technologies such as real-time tracking and smartphone apps enable more user engagement and effectiveness, enabling passengers to better plan their trips and utilize transit assets more efficiently. These technologies align with objectives stated in international summits, most recently the COP26 climate summit, where leaders pledged action towards immediate and significant action against climate change. By leveraging digital infrastructure in monitoring and streamlining transit flow, cities are able to reduce congestion and emissions associated with it—steering towards more sustainable urban space. This digitalization of public transport brings into perspective the key role technology plays in revolutionizing societal habits, citing that despite obstacles, public transport innovation guided by responsibility is important for steering ecologically sound practices into urban spaces (Yogesh et al., 2021). Besides, multimodal transport integration tells the evolving tale of public transportation innovations. By merging diverse modes of transportation—such as cycling, carpooling, and conventional public transportation—cities create synergistic relationships that become less burdensome and more accessible to the populace. The trend allows for the door to more eco-friendly commuter behavior to be opened, since citizens get to employ a range of transport units that decrease their carbon footprint. These integrated strategies are essential in achieving climate targets set by international agreements like the Paris Accord, uncovering the key role that optimized public transport plays in achieving net-zero emission goals. Through these innovations, the economic benefits of local cities can be stimulated while the environmental stewardship obligations are also fulfilled. Lastly, innovations in public transit not only revolutionize how people engage with transit but are also key movers towards building sustainable communities (Calvin et al., 2023).

10.3.3. Alternative Fuels and Technologies

The shift to new technologies and energy sources is increasingly becoming the norm in addressing the global energy crisis and mitigating climate change. Hydrogen, for instance, has become of interest as a potential candidate for the decarbonization of transport, heat, and power generation. Historically, hydrogen has previously been prematurely labeled as a game-changer in the offing; however, innovations in the last few years in electrolysis and fuel cell technologies are propelling its use as a clean energy source. Pairing of renewable energy sources such as wind and solar to produce green hydrogen contributes to its appeal as a likely game-changer in the transition towards a carbon-free world. As countries are investing in hydrogen infrastructure, the possibility of its large-scale deployment rises proportionally, thus being a significant factor for reducing greenhouse gas emissions and encouraging cleaner energy policies worldwide (Staffell et al., 2018). Biofuels are becoming more in vogue as well as a clean source of energy that is capable of reducing the dependence on fossil fuels. Biofuels derived from organic waste such as farm residues, waste fuels, and even algae are a renewable resource for powering vehicles and generating electricity. The emergence of advanced biofuels, which are formulated to provide more energy efficiency and reduced carbon emission, is critical in maintaining strict environmental requirements.

Figure 10.8. Hydrogen, for instance, has emerged as a promising candidate for decarbonizing not only the transport sector but also heating and power generation.

Source: https://www.activesustainability.com/media/816080/co2-hidrogeno-verde.jpg.

However, the environmental viability of producing biofuels is debatable because land conversion and competition over resources might erase their perceived environmental benefits. Sustainable sourcing must become the focus of responsible management and investment in technology to assure optimal biofuels potential without aggravating food security conditions. Thus, as one of a multifaceted array of alternative fuels, biofuels offer a feasible, though complex, solution for existing energy issues (Riahi et al., 2016). Research into hybrid and electric vehicle technologies also points to the diverse world of alternative fuels. These technologies increasingly employ lithium-ion batteries for electric vehicles (EVs) and sophisticated fuel cell technology for hydrogen-based vehicles. Constant improvement in battery capacity and charging systems is instrumental in making EVs more practical and convenient. Above all, investments in intelligent grid technology are capable of facilitating the integration of renewable sources of energy, making it easier to provide the increasing demand for electricity generated by sustainable means. Such advancements also have a regulatory sidekick, as it picks up speed, for phasing out traditional internal combustion engines. Complementarity with policy support mechanisms and alternative fuel technologies is crucial since they together form a gradient for sustainable transport in the future through integrating different sources of energy and reducing fossil fuel dependence (Staffell et al., 2018) (Figure 10.8).

10.4. WASTE MANAGEMENT SOLUTIONS

Increased consciousness regarding waste as a valuable resource is changing waste management practices across industries, particularly in aquaculture and fisheries. The traditional definition of waste has been shifted to a waste-to-wealth strategy, whereby byproducts such as fish waste and aquaculture effluent are turned into useful products. This new approach not only minimizes environmental pollution but also increases sustainable economic development. For instance, valorization processes of fish waste allow the recovery of proteins and oils to be utilized in animal feed or bioplastics, thus making resource utilization more efficient in these industries. Integrating environmental biology lies at the core of optimizing such processes and planning monitoring and assessment strategies for waste management programs, according to recent research. These advancements constitute a paradigm shift in how industries can employ waste to enable sustainable business (Meena et al., 2024). At the broader levels of production and industrial procedures, emerging trends in energy management are becoming increasingly relevant in waste management solutions. The integration of artificial intelligence (AI) within production systems is enhancing energy efficiency, a key concern for reducing waste. With the application of AI-powered demand forecasting, industries can schedule energy consumption in alignment with production schedules, thereby restricting energy use in non-essential processes and subsequent waste. Moreover, industrial cyber-physical system development helps maximize the monitoring of resources to enable manufacturing companies to optimize processes dynamically. These developments are all consistent with sustainable development guidelines, in that they restrict the total carbon input of manufacturing while maximizing economic effectiveness. Thus, the recent technological progress, particularly in the aspect of Industry 4.0 and the Green Deal, is elaborating on intensive waste minimization and management strategies in factory environments (Rojek et al.,

2024). The concept of circular economy also depicts future waste management trends with an emphasis on recycling and reusing material to establish a closed-loop framework. This model not only addresses the weaknesses of traditional linear manufacturing but also challenges businesses to reimagine the product life cycle and the waste they generate. With the design of products as being more disassemblable, recyclable, or upcyclable, businesses can significantly reduce the volume of waste that ends up in landfills and the environment. Additionally, consumers can be encouraged to participate in recycling programs through education and incentives to foster a culture of sustainability. As markets increasingly accommodate such values, waste management systems will automatically change to embrace more holistic strategies. That change is essential for addressing both environmental concerns and increasing calls for environmentally conscious practices across industries, which will result in a healthier world (Meena et al., 2024).

10.4.1. Recycling and Upcycling Technologies

The incorporation of upcycling and recycling technologies into contemporary waste management is a paradigm shift towards resource sustainability under the scope of green technology. Recycling transforms used materials to new ones, hence conserving natural resources and reducing pollution that otherwise accompanies production processes. The practice has evolved, employing advanced technologies to enhance efficiency and the quality of recovered products. Examples range from the development of automated sorting machinery and chemical recycling that have a greater recovery rate for complex waste streams. As the world's need for sustainable methodology increases, the emergence of literature on such technologies offers insight into new methodology that not only saves environmental footprint but also encourages economic growth by creating employment within the recycling sectors. Upcycling, which is a term similar to yet distinct from recycling, is concerned with the innovative remaking of waste materials into products that are greater in value and quality. This practice is now more in the fore as consumers and businesses alike attempt to address the staggering problems of waste overload and environmental degradation. A number of studies and practitioners advise upcycling as an effective action to reduce material and energy consumption, therefore fostering sustainable consumption and production practices. The increased emphasis on upcycling activities has been mirrored in the increasing body of academic literature dedicated to studying definitions, trends, and upcycling barriers. Despite limited upcycling literature, the potential to further clarify its scope, particularly in creating sustainable business models and consumerism that drive the circular economy (Sung et al., 2015) forward, is clearly evident. The intersection of recycling and upcycling technologies emphasizes the importance of consumer attitudes and knowledge in achieving sustainability goals. Studies have shown that little appreciation exists among consumers regarding the sustainability impacts of their purchases, particularly in industries such as the fashion sector. Uncertainty surrounding the definition of sustainability and heterogeneity among consumer drivers are critical barriers to promoting effective recycling and upcycling behaviors. In an effort to create a sustainability culture, policies to educate consumers and incorporate habituated behavior such as recycling and upcycling into the social norm are necessary. These range from making sustainability options more transparent to altering social norms surrounding clothing consumption and disposal. Through conquering these issues, stakeholders can facilitate a more sustainable future that engages both suppliers and consumers in sound environmental stewardship (Axelrod et al., 2016).

10.4.2. Waste-to-Energy Processes

The worldwide waste management problem has generated interest in novel technologies, with Waste-to-Energy (WtE) processes being regarded as a promising area of focus in green technology. Not only do these processes decrease the quantity of waste that goes to landfills but also generate renewable energy, thereby addressing two related environmental problems at the same time. The technologies utilized in WtE typically include incineration, anaerobic digestion, and gasification, all of which transform waste into electricity, heat, or biofuels. Adoption of such technologies equates to strategic transformation towards circular economy principles under which waste items are reused as worthful resources. Also, the advancement in emission control technologies makes the processes more sustainable by largely preventing the emission of harmful pollutants, which reflects the ability of WtE to be core to waste management in the future. This alignment with circular economy principles is critical in the scenario where landfills are dwindling and energy demands are increasing. The efficiency of Waste-to-Energy processes transcends the production of energy; they also foster bigger sustainability impacts. Waste-to-Energy processes can cut greenhouse gas emissions markedly relative to other waste disposal methods. By utilizing energy otherwise lost to decomposition in landfills, WtE lowers methane emissions, an extremely potent greenhouse gas. Besides this, the generated heat can also be utilized to provide district heating networks to the population from a clean and sure source of heat. Studies have identified that the integration of WtE in municipal waste management infrastructure can lead to a significant reduction of the carbon footprint of urban areas. Therefore, policymakers now view WtE as not only a tool for waste management but also as an actual solution to achieve climate targets and promote sustainable urban development. This is aligned with trends towards more adaptable urban systems. Although there are seemingly benefits of Waste-to-Energy processes, challenges still exist and have to be addressed to implement such technologies fully within the mainstream energy market. Public acceptance and understanding are often significant barriers, precipitated by historical concerns over emissions and purported health risks. Such public education campaigns are therefore required on a grand scale to inform communities about the safety and efficacy of modern WtE facilities. Furthermore, ensuring regulatory frameworks promote innovation without neglecting rigorous environmental standards is critical. New technologies, including carbon capture and storage (CCS) innovations and enhanced recycling processes, can be followed by WtE technologies and further increase the environmental effectiveness of such technologies. By the facilitation of collaboration among stakeholders like government agencies, private sector individuals who are innovators, and communities, the potential of the Waste-to-Energy processes in the future can be optimized and contribute meaningfully to sustainability against the backdrop of the greater war against climate change, as pointed out by Calvin et al. (2023) (Figure 10.9).

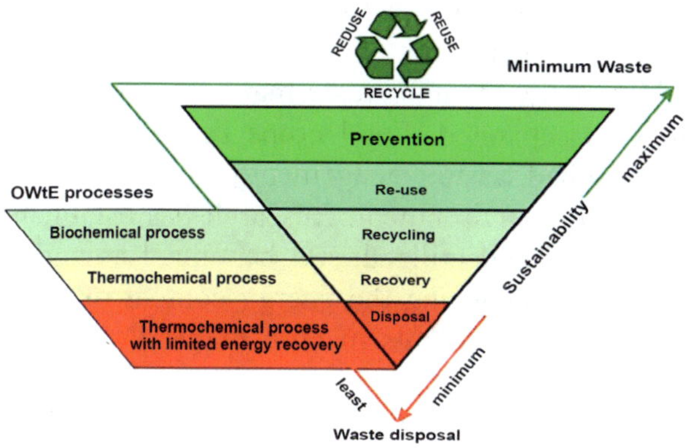

Figure 10.9. The efficacy of waste-to-energy processes is not limited to energy production alone; it also contributes to broader sustainability goals.

Source: https://www.mdpi.com/energies/energies-17-03797/article_deploy/html/images/energies-17-03797-g001.png.

10.4.3. Composting and Organic Waste Management

Increased levels of organic waste in municipal solid waste streams have generated interest in composting as a useful waste management option.

Composting not only avoids sending waste to the landfill but also transforms organic waste into useful resources that are beneficial for the soil. Microbial decomposition of organic matter is involved in composting, which leads to the production of compost, which makes the soil fertile, holds water, and captures carbon. Most recent advances in composting technology, such as anaerobic digestion and vermicomposting, offer new technologies to enhance efficiency and scalability. With the use of these technologies, municipalities are better equipped to manage organic waste in sustainable ways, reducing environmental impacts of traditional waste disposal processes while supporting local gardens and agriculture use. Despite inherent advantages of composting, its use and acceptance by citizens are hindered by issues. Among the key challenges is a belief that composting is time-consuming or odorous, dissuading individuals and communities from embracing it. Education and outreach programs play a significant role in informing the public about why they should compost and how to initiate composting at the household level or on an organizational scale. Moreover, municipalities with curbside composting initiatives have achieved higher participation rates if they are supported by outreach efforts. The better they understand the benefits of organic waste management, the greater the studies emphasize composting as an integral part of a larger strategy for sustainable urban planning and waste reduction, reflecting what Ferronato et al. (2019) discussed in terms of the environmental effects of ineffective waste management. Additionally, organic waste management and food security are issues that cannot be separated. Food crops being contaminated by heavy metals, typically from bad waste management practices, pose a critical danger to human health and the environment. Research has established that organic waste can be composted and utilized as a beneficial amendment to mitigate such adverse effects by enhancing the energy content of soil and promoting the utilization of essential nutrients in crop plants. As Kumar et al. (2019) discuss, organic waste management using sustainable methods can have a direct influence on the quality of food that is being produced, thereby promoting both food security and public health. Thus, integrating composting operations within larger environmental policies and agricultural programs is crucial towards the development of a more strengthened ecosystem and promoting sustainable development, ultimately backing a secure and safer food supply chain.

10.5. WATER CONSERVATION AND MANAGEMENT

With rising water scarcity worldwide, the adoption of advanced technologies in water saving and water management practices is increasingly becoming an imperative necessity. Among the most significant developments emerges from the Internet of Things (IoT) technology that has revolutionized water resource monitoring and management. Through the utilization of sensor networks and smart devices, organizations are able to gather unheard-of amounts of information regarding their water systems. Increased awareness makes intelligent decisions possible, efficient water allocation, and active conservation. For instance, smart irrigation systems not only conserve water but also optimize use based on real-time data on weather and soil moisture content. Such innovations offer a sustainable approach to water resource management, balancing short-term priorities of consumption with long-term environmental considerations and also improving efficiency in urban water supply systems, eventually attaining greater resilience to future water shortage possibilities (Tyagi et al., 2024). Moreover, produced water (PW) handling from the oil and gas sector is a growing area of interest given the adoption of sustainable water practices. The huge quantities of PW generated present challenges as well as opportunities for innovative applications outside traditional disposal. From a holistic Energy-Water-Food Nexus, the potential in PW can be tapped for irrigation of farms, livestock farming, and recharging of groundwater, effectively recycling this resource to address global environmental and food security challenges. In addition, emerging technologies such as constructed wetlands and nanotechnology are ushering in more effective treatment options for PW. In addition to enhancing the water quality, these technologies enable us to achieve a better circular water resource management, which is in support of sustainability goals and acting positively towards the overall SDGs (Achkar et al., 2024). The consequences of integrating such advanced technologies into water conservation and management extend far beyond short-term optimization of resources. Smart communities through water management solutions encourage hands-on engagement in sustainable habits, sustaining a culture of conservation. With greater awareness of their water consumption habits, changes in behavior can continue to reduce usage, complementing conservation efforts even more. Besides, collaboration among NGOs, governments, and tech firms can promote the transfer of best practices and innovation across sectors, creating a stronger strategy for water scarcity mitigation. In adopting a collective, technology-driven strategy, communities are not only able to meet their current water needs but also safeguard these important resources for future generations with mounting environmental pressures (Tyagi et al., 2024; Achkar et al., 2024).

10.5.1. Water Purification Technologies

The immediate need for effective water purification technologies comes to the forefront of the unprecedented scale of water scarcity reaching millions of individuals globally. Conventional technologies such as filtration, coagulation, and chlorination are required; however, they are typically not equipped to handle sophisticated contaminants, such as emerging ones like pharmaceuticals and personal care products. In recent times, photocatalysis has been seen as an emerging technology in this area, leveraging photocatalysts such as cobalt oxide (Co_3O_4) for their ability to degrade various organic contaminants upon light irradiation. Not only can these technologies potentially enhance degradation efficiencies, but they also hold the potential for sustainable water treatment by reducing reliance on chemical disinfectants. As the requirement for purer water grows with each passing day, improvements in photocatalytic processes are of vital significance to introduce new routes to optimize the efficiency and sustainability of water purification technology in accordance with green technology trends (Dang et al., 2025). The integration of nanotechnology into water purification processes is another vital trend driving innovations in the field. Nanomaterials, such as carbon nanotubes and graphene oxide, have revolutionized water treatment processes with their provision of high surface areas and tunable properties for enhanced contaminant removal. Nanofiltration membranes are particularly remarkable because of their ability to selectively remove contaminants at low energy expenses.

The application of nano-adsorbents also enables the removal of various types of pollutants, including heavy metals and microplastics, which induce severe environmental hazards. As continued research and development in nanomaterial-based technologies are made, their integration into existing infrastructures holds prospects for cost-effectiveness as well as improved purification output. Utilization of nanotechnology in water treatment is thus not merely innovation but a revolution towards achieving sustainable water security as part of emerging trends in green technology (Krupanidhi et al., 2019). Future research areas in water purification technology are also directed more towards overall solutions where many advanced technologies are combined to provide superior performance. For instance, hybriding nanofiltration with photocatalytic processes can lead to hybrid systems with optimal pollutant removal efficiency with minimum energy requirements. This area also avoids the limitations related to single technologies, such as membrane fouling or high chemical consumption requirement. Additionally, the creation of more efficient biosynthesis processes can offer cleaner production pathways for efficient nano-adsorbents and photocatalysts as well, enhancing the application of additional sustainability practices. Along with advancing technology, scientists will have to collaborate with industry players and government agencies to bypass regulatory issues and scale the technologies. The continued effort to research new trends in this sector holds the promise of making a significant contribution towards a more sustainable and resilient solution to global water purification challenges (Dang et al., 2025; Krupanidhi et al., 2019) (Figure 10.10).

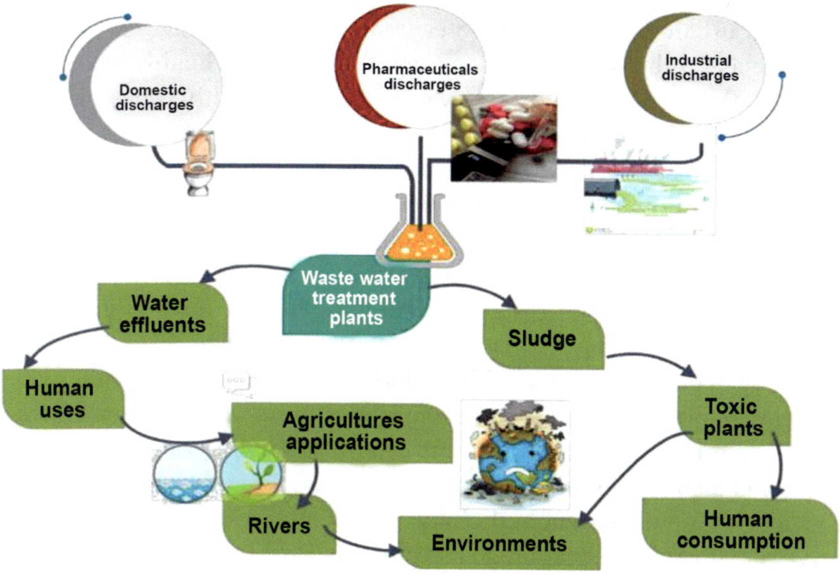

Figure 10.10. Conventional methods such as filtration, coagulation, and chlorination are essential; however, they often struggle to address complex pollutants, including emerging contaminants like pharmaceuticals and personal care products.

Source: https://www.mdpi.com/applsci/applsci-11-06659/article_deploy/html/images/applsci-11-06659-g001.png.

10.5.2. Rainwater Harvesting Systems

Rainwater harvesting systems represent a key step ahead in green technology, a pragmatic answer to growing concern with water shortages and environmental degradation. By capturing and storing rainwater for diverse applications, such systems not only conserve water resources but also reduce pressure on municipal water supplies and sewage treatment facilities. Rainwater harvesting within the context of urban planning is increasingly being viewed as a foremost element of sustainable building design. As the frequency of droughts and the variability of climatic conditions increase, the importance of effective management of water needs comes to the forefront. This is consistent with trends in green technology where new solutions are needed to ensure resource sustainability and environmental longevity, contributing to the creation of self-sustaining cities. The global application of rainwater harvesting systems is highlighting a revolutionary step towards sustainable construction and environmental balance.

The systems are capable of significantly reducing stormwater runoff issues, which cause urban flooding and water contamination. They also reverse the adverse effects on the local water streams, thus creating healthier aquatic life. In addition, the integration of rainwater harvesting with renewable energy sources has gained momentum as a latest trend. For example, the merging of this technology with solar panels can lead to a systemic resource management strategy where buildings generate and collect their own energy and water. This is especially relevant in green building construction approaches that culminate in the implementation of circular economy principles, where waste is minimized to the barest essentials, and resources are continuously reused, yielding both ecological and economic benefits (Chen et al., 2023). The future of rainwater harvesting systems is better technology and greater awareness of their benefits to the general public. The more people and communities become aware of the necessity for water conservation, the more they will adopt these systems. Governments also beginning to appreciate the benefits of rainwater harvesting as a sustainable development and climate-resilient option, with many offering incentives or subsidies to encourage its adoption. Besides, ongoing research into the materials and designs of such systems that will enhance their effectiveness and trustworthiness will pave the way for broader application and adoption in different climates and regions of the world. With focus on innovative practice and stakeholder collaboration, rainwater harvesting systems can potentially be part of an integral solution to not only alleviate water shortage but also provide a sustainable future to the building industry (Vickram et al., 2023).

10.5.3. Smart Irrigation Techniques

With increasing world water deficits, smart irrigation technologies are a breakthrough in agricultural practices, encouraging better water efficiency without sacrificing crop yields. They utilize technology and data manipulation to optimize the process of irrigation, thus saving water. Methods such as soil moisture monitoring, weather forecasting, and irrigation controlled by computers allow farmers to manipulate the watering schedule of each crop, rather than relying on traditional methodologies. The relevance of these procedures becomes more pronounced since they can adapt to various climatic patterns and changes in water availability, thereby becoming valuable resources in the fight against drought and resource depletion. Smart irrigation is not only a response to scarcity; it is a proactive response to developing sustainable agricultural practices that are economically viable and ecological (Aznar et al., 2019). In addition, the integration of digital technologies in smart irrigation practices is transforming traditional agricultural systems, leading to a more integrated farming system. Virtual platforms offer farmers real-time information utilized in irrigation decision-making and in making effective use of resources. For instance, mobile applications can offer farmers localized weather data and soil moisture measurements that enable them to make informed irrigation decisions based on the environmental condition. In addition to facilitating efficient water management, digitalization also promotes the resilience of farm practices to climatic changes (Addom et al., 2019). Moreover, such innovations also promote inclusivity because they involve different farming communities, such as youth and women, in developing agriculture.

Figure 10.11. Innovations such as soil moisture sensors, weather forecasting systems, and automated irrigation controllers allow farmers to tailor their watering schedules to specific crop needs, rather than relying on traditional methods.

Source: https://forwardfooding.com/wp-content/uploads/2024/11/Smart-Irrigation-Systems-scaled-1-scaled-e1731315114263.jpeg.

As smallholder farmers are implementing such technologies at an increasing pace, their livelihoods will improve and productivity will increase, which will ultimately lead to food security. In the future, the future of smart irrigation practices is in the research and development that is being carried out persistently in green technology. There is an imperative call for concerted action towards the development of knowledge regarding the efficiency of various irrigation systems in particular agricultural contexts. Research should be directed to determining the cost-effectiveness of implementing these approaches, taking into account the expenses and benefits associated with their implementation. Also, the ways intelligent irrigation can be utilized to counteract climate change through water reduction and increased efficiency need to be evaluated (Aznar et al., 2019). By forming an overall perception of the factors, the stakeholders will be able to ensure large-scale upscaling of intelligent irrigation. Not only this, but this will also promote sustainable agriculture and achieve broader environmental goals, ensuring a safer and food-safe future (Figure 10.11).

10.6. GREEN BUILDING PRACTICES

Green building practices are an evolutionary revolution in the construction business, taking into account environmental sustainability and economic viability. Green building practices entail a range of measures including energy-efficient design, utilization of sustainable materials, waste minimization, and use of renewable energy sources. The evolution of green building falls into three stages: the preliminary exploration of external ideas, the development of the associated technologies, and the application of the technologies in the mature stage. As discussed, the evolution of China's green building development is divided into three phases: the beginning, the growth phase, and the maturity (Chen et al., 2025). This stepwise evolution depicts the rise in sensitivity and sophistication within the industry, where players are becoming increasingly aware of reducing the environmental footprints of construction projects. In a world with global initiatives towards sustainability, reducing the carbon footprint of buildings has become a pressing issue. Buildings account for over 40% of the world's CO_2 emissions and are therefore one of the chief causes of climate change. This problem is most pronounced in developing nations, such as Bangladesh, where economic development and urbanization challenge the enforcement of sustainable construction techniques. Despite developments in environmentally friendly technologies, there is resistance to their application due to challenges in high financial costs, challenges in accessing funds, and insufficient government incentives (Sani et al., 2025). Overcoming these challenges needs a concerted effort, where policymakers, industry players, and financial institutions come together to create supportive conditions for green building adoption. The future of green building activity is based on innovation and adaptive measures to bridge the unique challenges of diverse socio-economic environments. With evolving trends of green technologies, it is important for players to advocate policies that promote environmentally sound construction. Projects such as supplemental investments for green infrastructure, easier access to green finance, and more powerful public-private partnerships can provide a healthy climate for development. In addition, raising awareness of the long-term economic and environmental benefits of green buildings can drive demand. By building on successful global models and addressing homegrown challenges, the green building sector can contribute significantly to securing larger-scale sustainability goals in conjunction with meeting mounting urban housing demand. By such concerted action, the advances of future green technology can come with the urgency of environmental stewardship and climate resilience.

10.6.1. Sustainable Materials and Resources

The growing conversation about green resources and materials has become a foundation of green technology, particularly because the world is grappling with comprehending the magnitude of resource depletion and climate change. This shift to sustainability not only targets resource extraction but also lifecycle, thus affecting the manner in which products are produced, used, and ultimately disposed of. Rising trends point towards the use of renewable materials like bioplastics and organic composites providing environmentally friendly substitutions to traditional materials extracted in most cases unsustainably. Prioritizing the use of renewable sources reduces carbon emissions, keeps industry aligned with the regulatory climate by reducing environmental footprint, and fulfills consumer needs for environmental sustainability. Hence, the application of sustainable materials is poised to transform manufacturing processes and propel a wider move towards a circular economy. When exploring the transformation of sustainable resources, one must explore innovations that enhance resource efficiency along with supporting environmental balance. For instance, innovations such as upcycling and valorization of waste have gained momentum, converting waste into assets, hence minimizing waste generation. Such a strategy not only addresses the issues of landfill saturation but also improves economic resilience by creating new markets and jobs on the basis of green activities. Furthermore, recycling technology development allows for recovery of raw material from end-of-life products, making supply chains more sustainable. Sustainable material interdependence with progressively developing process technologies ushers in a paradigm shift offering vast prospects for resource conservation and environmental sustainability and encourages industrial practice conformity with global sustainability goals. Evolving trends in sustainable materials and resources call for an interdisciplinary approach with social, economic, and environmental domains. The call to reimagine product design with a sustainability lens operates to underscore the need for stakeholder collaboration across policymakers, consumers, and manufacturers. As grounded in existing research, enhancing systemic issues throughout the developing and developed worlds is instrumental to driving innovation and resilience against climate problems. An integrated vision for the interconnectedness of ecosystems, health, and urban environments will drive the advancement of sustainable practices across industries. Ultimately, the shift towards sustainable material and resources is not a single-minded technological focus but a many-sided movement towards enhancing human health and creating supportive environments for disadvantaged populations (Buckley et al., 2014; Long et al., 2013) (Figure 10.12).

Figure 10.12. By prioritizing renewable sources, industries can reduce their carbon footprint, align with regulatory frameworks aimed at reducing environmental impact, and respond to consumer demand for eco-friendly products.

Source: https://cdn.azeusconvene.com/wp-content/uploads/Featured-slider_How-to-Reduce-Carbon-Footprint-at-Work-A-Practical-Guide-for-Businesses.jpg.

10.6.2. LEED Certification and Standards

In green technology, Leadership in Energy and Environmental Design (LEED) certification is a fundamental requirement for promoting sustainable building design. As a global standard, LEED is a system of guidelines that serves as a model for designers and building owners striving to lower their carbon footprint through effective construction practices. The standards address all manner of factors such as energy efficiency, water usage, and indoor environmental quality. Thus, besides encouraging more sustainability, the application of LEED-certified methods also encourages a sense of responsibility within the construction industry. However, though publicized and embraced far and wide, LEED-certified space continues to represent only a very small fraction of the overall commercial real estate market today, averaging below 1% across various different cities all over the United States (Fuerst et al., 2011). The policy environment also influences the urban development context to favor greater utilization of LEED certification. Policies most often involve incentives such as tax relief or expedited permitting procedures and are designed to induce developers to pursue more sustainable building projects.

Although some initial evidence indicates that green building incentives offered at the city level are correlated positively with the percentage of LEED-certified areas, the advantage does not extend to all metropolitan regions. In most cases, one may get a distorted view of the effect of such policies due to the non-random assignment-induced biases. With an understanding of such dynamics, it becomes imperative to analyze the market size and economic conditions at the national level as they greatly affect the success of LEED certification programs (Kontokosta et al. 2025). With the ever-growing demand for green building practices, LEED certification will have to change in line with emerging trends in the green technology sector. As more awareness regarding climate change and environmental degradation grows, pressure for more stringent sustainability standards is mounting. LEED's future trends can be the incorporation of advanced technology such as building automation systems, which optimize energy usage in real time, and the introduction of more recyclable materials. Apart from this, sustained collaboration between private and public domains will also be vital in furthering more effective incentives that will be capable of inspiring the building industry to greater conformity with LEED ratings. As a result, sustained advancements in LEED certification will not only reflect current technological advancements but also have implications on subsequent practices in green building worldwide.

10.6.3. Passive Design Strategies

Passive design strategies play vital roles in green technology, becoming essential steps towards designing energy-efficient structures whose elegant fusion of beauty and sustainability is something to behold. These methods take advantage of natural climatic factors and leverage such elements as sunlight, wind, and natural insulation towards enhancing occupants' comfort with diminished reliance on mechanical systems. Through the placement of buildings to maximize the use of natural light and strategically locating windows and insulation, architects can reduce heating and cooling needs. It is a sign that designers are becoming increasingly aware that sustainability isn't just about environmental issues; it's also about the beauty of good design. As urban environments have come under increasing criticism for their environmental footprints, the need for integrating passive design principles that ensure functionality along with aesthetic value has never been greater. The urgency lies in how to balance efficiency and beauty and how to answer growing demands for more sustainable building practices in urban development (Zhong et al., 2025). Moreover, the application of passive design principles not only addresses environmental requirements but also responds to evolving user expectations for comfort and proximity to nature. Application of biophilic design principles—facilitating contact with the natural world—promotes mental well-being and overall higher productivity. By incorporating features such as natural ventilation systems, green roofs, and living walls, the space is healthy for users and harmful to the environment to a smaller degree. By reducing the energy footprint of buildings without reducing livability, architects and designers are then in a position to deliver creative solutions that portray increasing public interest in sustainability considerations. It is away from the traditional design that prioritizes functionality at the expense of environmental conservation.

Figure 10.13. Many buildings require retrofitting to meet modern sustainability standards, demanding innovative adaptations that respect the original architecture while enhancing energy performance.

Source: https://powerline.net.in/wp-content/uploads/2023/05/PL-80-678x381.jpg.

Therefore, the emergence of passive design strategies is a future development direction for architecture practice in favor of both health and environmental consciousness (Liu, 2015). In addition, the implementation of passive design strategies is not without numerous challenges, particularly in terms of existing building retrofitting and urban regeneration. A great deal of existing building stock has to be rescaled to meet prevailing sustainability requirements through innovative alterations that remain compatible with the existing fabric but optimize efficiency. The economic implications of implementing these measures in combination can also dissuade stakeholders from performing significant refurbishment, even when the long-term cost benefits in terms of increased energy efficiency are present. As the technology of green technology itself continues to evolve, a strong need for comprehensive learning frameworks exists that equip professionals to employ passive design measures as optimally as possible. By placing research and interdisciplinarity at the forefront, the engineering and architecture professions can design buildings that not only address present needs but also adapt to future challenges in urbanization and sustainability of the environment. This integrated design will pave the way toward more resilient and sustainable architectural practice as society treads through the complexities of climate change (Zhong et al., 2025) (Figure 10.13).

10.7. CONCLUSION

Finally, the trajectory of green technology is also guided by a series of upcoming trends that are indicative of increased emphasis on sustainability and environmental stewardship. Most notable among these trends is the incorporation of smart digitalization algorithms into engineering applications. The algorithms provide more accuracy and efficiency to environmental analyses, leading to more competent decision-making processes. As big data and machine learning solutions gain more acceptance throughout sectors, environmental predictive modeling becomes increasingly significant in terms of opportunities for proactive risk identification and optimization of resource usage. Their incorporation not only advances the field but also meets global sustainability efforts, aligning with the argument of technology in addressing climate change challenges, as highlighted in geotechnical engineering innovation research (Hasan et al., 2025). In addition, nanotechnology transformation plays a pivotal role in future green technology, particularly in materials science and energy applications. The development of environmentally friendly nanomaterials has opened doors for innovation, which has led to the development of efficient energy solutions and high-performance materials with lower environmental impact. Research lays stress on utilizing green methods in synthesizing nanoparticles that not only contribute towards sustainable activities but also enhance the efficiency of applications ranging from water purification to plant pests in agriculture (Zhao et al., 2015). Through a consideration of sustainable approaches in nanotechnology, researchers can ensure that innovation does not come at the expense of environmental integrity, contributing to an integrated mode of doing technology. Lastly, as green technology improves, policymakers, entrepreneurs, and institutions of learning must collaborate harmoniously. The synergy is a prerequisite for countering the all-around challenges posed by climate change and resource planning. Cross-disciplinary projects can have the possibility of initiating innovation through the melding of knowledge from various disciplines and thus bringing about an integrative strategy towards sustainable development. Green technology research and development policies will serve as central drivers for getting these innovations onto the ground level. Emphasizing the importance of sustainable practices ensures that future developments not just enhance efficiency and resilience in infrastructure but also pay major dividends towards a more equitable and sustainable global world for generations to come (Hasan et al., 2025).

10.7.1. Summary of Key Trends

The green technology ecosystem is evolving extremely rapidly because of the increasing demand for increased connectivity and higher-level capabilities in data handling. The most striking among these is the move from fifth-generation (5G) to the touted sixth-generation (6G) wireless communication technologies. The development of 6G, scheduled for implementation between 2027 and 2030, is aimed at enhancements in the nature of higher data rates, reduced latency, and better quality of service than its predecessor. Not only does the process facilitate existing use of green technology, but it also opens the door for novel paradigms of wireless communications that embrace artificial intelligence and other next-generation technologies. The synergy of these advanced capabilities has the power to upgrade energy consumption and distribution efficiency, and this will play a pivotal role in the realization of a sustainable future in the green technology frontier (Chowdhury et al., 2020). Augmented by the advancement of communication technology, blockchain systems are increasingly becoming synonymous with their disruptive potential in the energy sector. Such technology offers a safe and open platform for managing energy transactions, driving innovation through peer-to-peer (P2P) energy trading, and enhancing decentralized marketplaces integration. Systematic review of blockchain initiatives reveals an overabundance of chances and practical applications across all domains of energy-related, from electric car charging to Internet of Things (IoT) integration. Through making tamper-proof and automatic processes possible through smart contracts, blockchain diminishes trust problems and transaction efficiency, thus promoting the utilization of renewable energy technologies. The integration of blockchain with mainstream uses in energy, however, will be influenced by overcoming existing technical limitations and marketplace hurdles (Andoni et al., 2018). Another key trend shaping the future of green technology is the marriage of machine learning and big data analytics with sustainability efforts. As environmental challenges rise and withering pressure of compliance, industries are now increasingly applying high-end analytics to optimize the use of resources and mitigate waste. Predictive modelling facilitated by machine learning helps companies predict demand, become more energy efficient, and ensure overall operational sustainability. The use of such technologies not only leads to cost savings but also enables the use of cleaner energy sources and processes. Furthermore, the complementarity between green technology and data analytics is in tandem with the broader imperative to transition towards a circular economy, wherein waste is reduced and resources are reclaimed efficiently, thus ensuring a sustainable future (Chowdhury et al., 2020; Andoni et al., 2018).

10.7.2. Future Implications for Society

With the development of society, the issues of green technology extend well beyond its environmental dimensions to reach the economic, social, and political domains. Green technologies such as renewable energy technologies, energy efficiency innovation, and waste reduction strategies are adopted, which is a paradigm shift when considering sustainable development. This paradigm shift matters particularly in developing economies where adoption of green technology can foster long-term economic growth and lower poverty rates. By transferring the benefits of these technologies to local levels, firms are able to create employment opportunities and foster community resilience, hence establishing a foundation for sustainable prosperity. Green business model conversion of the conventional business models not only reflects a response to the imperative need for environmental sustainability but also provides competitive advantages in a rising global market trending towards sustainability practice (Md. Islam, 2023). Moreover, the trajectory of society's future will be influenced in a substantial way by how policy and technology interconnect. Governments and institutions must adapt by creating systems that allow for the implementation of green technologies such that innovation and sustainability become mutually compatible. Such systems can involve motivating businesses to adopt greener strategies, providing research and development funding, and having policies in place that allow for environmentally sound strategies. They will help create a resilient business operations system together with addressing issues such as climate change and resource depletion. With the emergence of green technologies, they will automatically impact organizational strategy and societal values to put sustainability at the center of both public and private agendas. The role played by policymakers in this regard cannot be overstated, as they will become responsible for leading society towards a green future (Lim, 2022). Finally, the consequences of society upon embracing green technology are long-term shifts in consumerism, education, and public awareness. As individuals become more conscious of the impact their shopping has on the environment, there will be a definitive shift towards supporting companies that are sustainable. This demand from consumers for green goods will spur still more innovation and investment in sustainable technologies, fueling a virtuous circle that benefits from constantly re-affirming the commitment to sustainability. Further, schools and teachers will provide a crucial role in instilling in future generations a values-based appreciation of protecting the environment. Incorporating sustainability-focused curricula in schools and universities encourages students to begin considering the critical thinking around their environmental impact and how they can contribute towards leading societys transformation towards a green economy. This fundamental shift in mindset is important to enable a collective commitment towards sustainable practices much further down the line (Md. Islam, 2023).

10.7.3. Call to Action for Continued Innovation

The imperative for continuous innovation in green technology is occasioned by the more observable impacts of environmental degradation on human life and health. The intersection of economic development and ecological sustainability demands immediate action against climate change, loss of ecosystems, and depletion of natural resources. With increasing global population and urbanization, there is a need for sustainable practices in order to promote ecological balance in pursuit of prosperity. Investment in new technologies—everything from alternative energy solutions to sustainable transport systems—will be at the forefront of transforming the way that society interacts with the environment. Interdisciplinary collaboration between policymakers, engineers, and scientists will need to be incentivized to ensure collective effort is translated into practical outcomes. Through the nurturing of culture for innovation, society can create avenues for environmentally friendly progress that balance economic goals with protection of the environment, thus ensuring a sustainable future. Secondly, the call to action is beyond that of any technology in isolation; it calls for a governance and values regime shift to sustainable ways. To address the pressing environmental challenges, a good structure that ensures policy coherence is critical. This includes integrating health, economic, and environmental policies by such organizations as the United Nations (UN) and other international organizations. Ensuring policies consider long-term effects of using the environment for short-term financial gains is fundamental in curving future risks. As recommended, "considerable potential exists to associate action for the reduction of environmental damage with enhanced health outcomes" for nations of all types of economics. Governments, enterprise, and civil society need to work together to invent new pathways for sustainable development, remapping the value and utilization of resources in order to maximize planetary health and human wellbeing, thereby solving the concomitant problems of resource utilization and environmental sustainability in the process. Lastly, continuous innovation in green technology can only be achieved through a strong commitment to learning and informed public involvement. The call for green solutions must be answered with a human population endowed with the wisdom to participate in environmental stewardship in an active manner. This entails the integration of sustainability literacy into educational systems in order that forthcoming generations understand the consequences of their actions. Public education on the merits and practices of green technologies can initiate individual actions that collectively result in significant systemic changes. Besides, as reiterated through the prism of health impacts occasioned by environmental degradation, there is an urgent need for an integrated strategy that harmonizes technological advance with education programs for a sustainable world. A synergy among different actors, from health professionals to ecologists, is vital to the achievement of robust communities that are capable of confronting the future environmental issues, driving sustainable transformations for generations to come.

REFERENCES

1. Achkar, J. E., Malhas, R., & Alsaba, M. (2024). Innovative produced water management: A nexus approach for sustainable oil and gas industry–A critical review. *Day 2 Wed, March 06, 2024*. https://www.semanticscholar.org/paper/03e6a5b0de87cce64b566d1da40258c19b325373 (accessed on 24th October 2025).
2. Addom, B. K., Hailu, M., Totapally, S., Tsan, M., et al. (2019). *The Digitalization of African Agriculture Report 2018–2019*. https://core.ac.uk/download/220111121.pdf (accessed on 24th October 2025).
3. Alomari, M. A., Al-Andoli, M. N., Ghaleb, M., Thabit, R., Alkawsi, G., Alsayaydeh, J., & Gaid, A. (2025). Security of smart grid: Cybersecurity issues, potential cyberattacks, major incidents, and future directions. *Energies*. https://www.semanticscholar.org/paper/d82ee847bcd7f54684e2c9a1844382994057017f (accessed on 24th October 2025).
4. Andoni, M., Robu, V., Flynn, D., Abram, S., Geach, D., Jenkins, D., McCallum, P., et al. (2018). Blockchain technology in the energy sector: A systematic review of challenges and opportunities. *Renewable and Sustainable Energy Reviews, 100*, 143–174. https://doi.org/10.1016/j.rser.2018.10.014 (accessed on 24th October 2025).
5. Atadoga, A., Umoga, U. J., Lottu, O. A., & Sodiya, E. O. (2024). Tools, techniques, and trends in sustainable software engineering: A critical review of current practices and future directions. *World Journal of Advanced Engineering Technology and Sciences, 11*, 231–239. https://doi.org/10.30574/wjaets.2024.11.1.0051 (accessed on 24th October 2025).
6. Axelrod, R., Beard, N., Birtwistle, G., Birtwistle, A., Blake, J., Bly, S., Byun, S., et al. (2016). *Sustainable Clothing: Challenges, Barriers and Interventions for Encouraging More Sustainable Consumer Behavior*. https://core.ac.uk/download/82981701.pdf (accessed on 24th October 2025).
7. Aznar Sánchez, J. Á., Batlles delaFuente, A., Fidelibus, M. D., Velasco Muñoz, J. F., et al. (2019). *Rainwater Harvesting for Agricultural Irrigation: An Analysis of Global Research*. https://core.ac.uk/download/286590637.pdf (accessed on 24th October 2025).
8. Bender, E. M., Gebru, T., McMillan-Major, A., & Shmitchell, S. (2021). *On the Dangers of Stochastic Parrots*. https://doi.org/10.1145/3442188.3445922 (accessed on 24th October 2025).
9. Bentsion, V. (2016). *The Impact of the BRICS' Universities on Internationalization of Smart Cities Technologies*. https://elar.urfu.

ru/bitstream/10995/48242/1/rrfp_2016_1_005.pdf (accessed on 24th October 2025).

10. Bhuiyan, S. M. Y., Chowdhury, A., Hossain, M. S., Mobin, S. M., & Parvez, I. (2025). AI-driven optimization in renewable hydrogen production: A review. *American Journal of Interdisciplinary Studies.* https://www.semanticscholar.org/paper/dcc4cc6d47fd5cb583bc3ce7f36c2b6ce1610866 (accessed on 24th October 2025).

11. Buckley, C., Cloete, E., Godfrey, L., Hildebrandt, M., et al. (2014). *The State of Green Technologies in South Africa.* https://core.ac.uk/download/30610831.pdf (accessed on 24th October 2025).

12. Byrne, R., & Ockwell, D. (2013). *Building Pathways of Pro-Poor Energy Access: PV-Powered Electricity Services in Kenya.* https://core.ac.uk/download/30610851.pdf (accessed on 24th October 2025).

13. Calvin, K., Dasgupta, D., Krinner, G., Mukherji, A., Thorne, P., Trisos, C. H., Romero, J., et al. (2023). *IPCC, 2023: Climate Change 2023: Synthesis Report. Contribution of Working Groups I, II and III to the Sixth Assessment Report of the Intergovernmental Panel on Climate Change [Core Writing Team, H. Lee and J. Romero (eds.)].* IPCC, Geneva, Switzerland. https://doi.org/10.59327/ipcc/ar6-9789291691647 (accessed on 24th October 2025).

14. Chen, H., Jia, Z., & Pan, Z. (2025). Analysis of the development history of green building design in China based on CiteSpace. *Highlights in Science, Engineering and Technology.* https://www.semanticscholar.org/paper/8894e463d822e5ccd53c4b330431fee3be314cc3 (accessed on 24th October 2025).

15. Chen, L., Hu, Y., Wang, R., Li, X., Chen, Z., Hua, J., Osman, A. I., et al. (2023). Green building practices to integrate renewable energy in the construction sector: A review. *Environmental Chemistry Letters, 22,* 751–784. https://doi.org/10.1007/s10311-023-01675-2 (accessed on 24th October 2025).

16. Chen, L., Msigwa, G., Yang, M., Osman, A. I., Fawzy, S., Rooney, D. W., & Yap, P. S. (2022). Strategies to achieve a carbon neutral society: A review. *Environmental Chemistry Letters, 20,* 2277–2310. https://doi.org/10.1007/s10311-022-01435-8 (accessed on 24th October 2025).

17. Chowdhury, M. Z., Shahjalal, M., Ahmed, S., & Jang, Y. M. (2020). 6G wireless communication systems: Applications, requirements, technologies, challenges, and research directions. *IEEE Open Journal of the Communications Society, 1,* 957–975. https://doi.org/10.1109/

ojcoms.2020.3010270 (accessed on 24th October 2025).

18. Dang, V. D., Nhung, N. T. H., Rabani, I., Tran, N. T., Thuy, B. T. P., & Truong, H. B. (2025). Advances in Co_3O_4 nanomaterial-based photocatalysts for water purification: Mechanisms, green synthesis, activation of oxidants, waste-derived sources, and computational insights. *RSC Advances, 15*, 19088–19103. https://www.semanticscholar.org/paper/a0e94a5bbbe107f696b494d3c27b4e6fca7f0eec (accessed on 24th October 2025).
19. Dwivedi, Y. K., Hughes, L., Kar, A. K., Baabdullah, A. M., Grover, P., Abbas, R., Andreini, D., et al. (2021). Climate change and COP26: Are digital technologies and information management part of the problem or the solution? An editorial reflection and call to action. *International Journal of Information Management, 63*, 102456. https://doi.org/10.1016/j.ijinfomgt.2021.102456 (accessed on 24th October 2025).
20. Ferronato, N., & Torretta, V. (2019). Waste mismanagement in developing countries: A review of global issues. *International Journal of Environmental Research and Public Health, 16*, 1060. https://doi.org/10.3390/ijerph16061060 (accessed on 24th October 2025).
21. Fuerst, F., Kontokosta, C., & McAllister, P. (2011). *Taking the LEED? Analyzing Spatial Variations in Market Penetration Rates of Eco-Labeled Properties*. https://core.ac.uk/download/107868.pdf (accessed on 24th October 2025).
22. Gatta, P. P. (2022). *The State of World Fisheries and Aquaculture 2022*. FAO eBooks. https://doi.org/10.4060/cc0461en (accessed on 24th October 2025).
23. Gopal, S., & Najam, A. (2012). *Connecting the Dots: Information Visualization and Text Analysis of the Searchlight Project Newsletters*. https://open.bu.edu/bitstream/2144/22895/1/Searchlight-Pardee-Report.pdf (accessed on 24th October 2025).
24. Hasan, M. M., Hasan, M. R., Jiang, Y., Liu, J., Feng, H., He, B., Rahman, M. M., et al. (2025). Innovation in geotechnical engineering equipment, intelligent digitalization, and green zero-carbon technologies: Current developments and future directions at TYUT. *EJSMT*. https://www.semanticscholar.org/paper/14c4d17f53c09082f4820de4b16018704f2f0f9f (accessed on 24th October 2025).
25. Islam, M. T. (2023). Newly developed green technology innovations in business: Paving the way toward sustainability. *Technological Sustainability, 2*, 295–319. https://doi.org/10.1108/techs-02-2023-0008 (accessed on 24th October 2025).

26. Knol, W. H. C. (2025). *Small-Scale Technologies for Energy Innovations: Role and Implication Directions*. https://core.ac.uk/download/pdf/6832363.pdf (accessed on 24th October 2025).

27. Kontokosta, C., Fuerst, F., & McAllister, P. (2025). *Taking the LEED? Analyzing Spatial Variations in Market Penetration Rates of Eco-Labeled Properties*. https://core.ac.uk/download/pdf/6575572.pdf (accessed on 24th October 2025).

28. Krupanidhi, Y. G., & Ramya, Y. A. (2019). Nanotechnology in water purification: Innovations in filtration and desalination. *World Journal of Advanced Research and Reviews*. https://www.semanticscholar.org/paper/21f2e7e6f85daa7eb27841b03a7adf310eaf85d0 (accessed on 24th October 2025).

29. Kumar, P., Lee, S. S., Zhang, M., Tsang, Y. F., & Kim, K. H. (2019). Heavy metals in food crops: Health risks, fate, mechanisms, and management. *Environment International, 125*, 365–385. https://doi.org/10.1016/j.envint.2019.01.067 (accessed on 24th October 2025).

30. Lee, I. (2016). A conceptual framework of the Internet of Things (IoT) for smart supply chain management. https://www.semanticscholar.org/paper/abd92f1855d108c7069ac75a833787057d03efb6 (accessed on 24th October 2025).

31. Lim, W. M. (2022). Ushering a new era of *Global Business and Organizational Excellence*: Taking a leaf out of recent trends in the new normal. *Global Business and Organizational Excellence, 41*, 5–13. https://doi.org/10.1002/joe.22163 (accessed on 24th October 2025).

32. Liu, Y. (2015). *Application of Research Study in the Teaching of Civil Engineering*. https://www.semanticscholar.org/paper/e60e8583bb326e0d27fd4b3b6e84864580e02096 (accessed on 24th October 2025).

33. Lloyd-Jones, D. M., Allen, N. B., Anderson, C. A. M., Black, T., Brewer, L. C., Foraker, R. E., Grandner, M. A., et al. (2022). Life's essential 8: Updating and enhancing the American Heart Association's construct of cardiovascular health: A presidential advisory from the American Heart Association. *Circulation, 146*. https://doi.org/10.1161/cir.0000000000001078 (accessed on 24th October 2025).

34. Long, J., & Vages, N. (2013). *A Critical Scan of Four Key Topics for the Philanthropic Sector: A Study by the Rockefeller Foundation and Accenture Development Partnerships*. https://core.ac.uk/download/71362647.pdf (accessed on 24th October 2025).

35. Meena, D. K., Das, B. K., Muzaddadi, A. U., & Sahoo, A. (2024). Waste

to wealth–A transformative perspective in greening the environment with a focus on fisheries and aquaculture: Emerging cutting-edge trends. *Proceedings of the Zoological Society of India*. https://www.semanticscholar.org/paper/fc00a5495c46d8043b0b5124d20c38e64b108fcc (accessed on 24th October 2025).

36. Motlagh, N. H., Mohammadrezaei, M., Hunt, J. D., & Zakeri, B. (2020). Internet of Things (IoT) and the energy sector. *Energies, 13*, 494. https://doi.org/10.3390/en13020494 (accessed on 24th October 2025).

37. Mujtaba, M., Fraceto, L. F., Fazeli, M., Mukherjee, S., Savassa, S. M., de Medeiros, G. A., do Espírito Santo Pereira, A., et al. (2023). Lignocellulosic biomass from agricultural waste to the circular economy: A review with focus on biofuels, bio composites and bioplastics. *Journal of Cleaner Production, 402*, 136815. https://doi.org/10.1016/j.jclepro.2023.136815 (accessed on 24th October 2025).

38. Poka, V. (2025). Energy-efficient cloud networking: Strategies for sustainable distributed system. *International Research Journal of Modernization in Engineering Technology and Science*. https://www.semanticscholar.org/paper/9b656a5f2e234044a70981fe742260bfe82d9b96 (accessed on 24th October 2025).

39. Riahi, K., van Vuuren, D. P., Kriegler, E., Edmonds, J., O'Neill, B. C., Fujimori, S., Bauer, N., et al. (2016). The Shared Socioeconomic Pathways and their energy, land use, and greenhouse gas emissions implications: An overview. *Global Environmental Change, 42*, 153–168. https://doi.org/10.1016/j.gloenvcha.2016.05.009 (accessed on 24th October 2025).

40. Rodríguez-Aburto, C., Poma-García, J., Montaño-Pisfil, J., Morcillo-Valdivia, P., Solís-Farfán, R., Curay-Tribeño, J., Pilco-Nuñez, A., et al. (2025). Applications of renewable energies in low-temperature regions: A scient metric analysis of recent advancements and future research directions. *Energies*. https://www.semanticscholar.org/paper/24f968c6b9eae11e49bcb0e1218138c836946e59 (accessed on 24th October 2025).

41. Rojek, I., Mikołajewski, D., Mroziński, A., & Macko, M. (2024). Green energy management in manufacturing based on demand prediction by artificial intelligence—A review. *Electronics*. https://www.semanticscholar.org/paper/81a4e98ddc1892f447c956ab33a296f6dd7f9f47 (accessed on 24th October 2025).

42. Sani, J., Mallick, K. K., Sadhukhan, P. K., Banu, M. A., Shikder, R. R., & Biswas, T. K. (2025). Achieving carbon-neutral construction: Global trends and Bangladesh's sustainable future. *American Journal of Environmental Economics*. https://www.semanticscholar.org/paper/4d3db6b6d8d3a9c1d

c64e380b364aa8cdf37584b (accessed on 24th October 2025).

43. Shafique, K., Khawaja, B. A., Sabir, F., Qazi, S., & Mustaqim, M. (2020). Internet of Things (IoT) for next-generation smart systems: A review of current challenges, future trends and prospects for emerging 5G-IoT scenarios. *IEEE Access, 8*, 23022–23040. https://doi.org/10.1109/access.2020.2970118 (accessed on 24th October 2025).
44. Staffell, I., Scamman, D., Velazquez Abad, A., Balcombe, P., Dodds, P. E., Ekins, P., Shah, N., et al. (2018). The role of hydrogen and fuel cells in the global energy system. *Energy & Environmental Science, 12*, 463–491. https://doi.org/10.1039/c8ee01157e (accessed on 24th October 2025).
45. Sung, K. (2015). *A Review on Upcycling: Current Body of Literature, Knowledge Gaps and a Way Forward.* https://core.ac.uk/download/30636736.pdf (accessed on 24th October 2025).
46. Tyagi, R., Kathuria, K., Sharma, P., & Sharma, P. (2024). Optimizing water management and conservation through IoT. *2024 2nd International Conference on Disruptive Technologies (ICDT).* https://www.semanticscholar.org/paper/6c7e8cde19c7f48cd5082daed43dcbcda30e04a4 (accessed on 24th October 2025).
47. Vickram, A. S., & Lakshmi, V. S. (2023). Advancements in environmental management strategies and sustainable practices for construction industry: A comprehensive review. *International Journal of Civil Engineering, 10*, 10–22. https://doi.org/10.14445/23488352/ijce-v10i8p102 (accessed on 24th October 2025).
48. Wang, J., Zhu, K., & Hossain, E. (2021). Green Internet of Vehicles (IoV) in the 6G era: Toward sustainable vehicular communications and networking. *IEEE Transactions on Green Communications and Networking, 6*, 391–423. https://doi.org/10.1109/tgcn.2021.3127923 (accessed on 24th October 2025).
49. Wang, Y., Su, Z., Zhang, N., Xing, R., Liu, D., Luan, T. H., & Shen, X. (2022). A survey on Metaverse: Fundamentals, security, and privacy. *IEEE Communications Surveys & Tutorials, 25*, 319–352. https://doi.org/10.1109/comst.2022.3202047 (accessed on 24th October 2025).
50. Whitmee, S., Haines, A., Beyrer, C., Boltz, F., Capon, A., de Souza Dias, B. F., Ezeh, A., et al. (2015). Safeguarding human health in the Anthropocene epoch: Report of The Rockefeller Foundation–Lancet Commission on planetary health. *The Lancet, 386*, 1973–2028. https://doi.org/10.1016/s0140-6736(15)60901-1 (accessed on 24th October 2025).
51. Xu, L., & Chen, Y. (2025). Overview of sustainable maritime transport optimization and operations. *Sustainability*. https://www.semanticscholar.

org/paper/d704a9cc47720aceabeabaa23611d1ab488d7c14 (accessed on 24th October 2025).

52. Zeng, X., Li, M., Abd El-Hady, D., Alshitari, W., Al-Bogami, A. S., Lü, J., & Amine, K. (2019). Commercialization of lithium battery technologies for electric vehicles. *Advanced Energy Materials, 9*. https://doi.org/10.1002/aenm.201900161 (accessed on 24th October 2025).

53. Zhang, H., Luo, M., Pei, T., Liu, X., Wang, L., Zhang, W., Lin, L., et al. (2023). Unequal urban heat burdens impede climate justice and equity goals. *The Innovation, 4*, 100488. https://doi.org/10.1016/j.xinn.2023.100488 (accessed on 24th October 2025).

54. Zhao, X., Wang, M. C., Huang, J. C., Qiang, Y., & Yoon, I. (2015). Nanomaterials and nanodevices. *The Scientific World Journal, 2015*. https://www.semanticscholar.org/paper/974cab1ac9251ba04f709a03a5a0915581f1c4a0 (accessed on 24th October 2025).

55. Zhong, S., & Pillai, M. D. (2025). The pursuit of sustainability and sophisticated in both inside and outside styles, including strategies to produce environmentally conscious and visually stunning surroundings. *Journal of Neonatal Surgery*. https://www.semanticscholar.org/paper/3ac149971f676177b8c4985bf6e7d08878560775 (accessed on 24th October 2025).

INDEX

D

E